Glow Discharge Processes

Glow Discharge Processes

SPUTTERING AND PLASMA ETCHING

Brian Chapman

A WILEY-INTERSCIENCE PUBLICATION

JOHN WILEY & SONS

New York • Chichester • Brisbane • Toronto • Singapore

Library of Congress Cataloging in Publication Data

Chapman, Brian N
Glow discharge processes.

"A Wiley-Interscience publication."
Includes bibliographical references and index.
1. Sputtering (Physics) 2. Glow discharges.
I. Title. II. Title: Plasma etching.

QC702.7.P6C48 537.5′2 80-17047
ISBN 0-471-07828-X

Printed in the United States of America

20 19 18 17 16 15 14 13

Preface

This book is based on a series of seminars held in 1978 and 1979. The seminars were intended to give some more insight into several practical glow discharge processes that are being increasingly used, particularly in the semiconductor industry. I hope that the text will serve as a useful general introduction to some of the scientific principles involved in these processes.

Glow discharges, like so many topics in science, are incompletely understood. Results are often misinterpreted, contradictory, or irrelevant. Glow discharge science has its own sub-language of special terms, with names that are often misleading, and with meanings which cannot be assumed to be constant from author to author! One can easily understand the need for the precision of scientific writing and sympathise with the conditions, provisos and double negatives of the author who is taking care not to make any definite statement which might be wrong. This is probably as scientific literature must be when one is close to the borders of knowledge and ignorance, but it is rather daunting to a newcomer to that particular branch of science.

Many of you will have had the experience of wanting to learn something about a particular area of science you're not familiar with, and so you go along for advice to the chap in your company or university who is considered the local expert. More often than not you come away with a list of references, in just about all of which it is assumed that you know the subject pretty well! And this is a particular problem in multi-disciplinary subjects such as sputtering where you are as likely to meet some electrical engineering phase angles as you are some organic chemistry.

This book is trying to be an introductory book. It attempts to thread a path through all the basic material you need before you can read the much more erudite reviews on the subject. In an effort to spare readers from attacks of mental indigestion, I have selected those aspects which appear to be more useful for first-time acquaintance, and even these are dealt with too briefly.

This text was written for readers with a wide range of backgrounds, and the emphasis is on concepts rather than on rigorous detail. I have usually restricted discussions to the general application of an idea, and have often ignored exceptions. So don't take my arguments and results too literally or too seriously; they are meant to give a qualitative 'feel' for the technology, not to comprise a scholarly tome. It has been my experience, particularly whilst I was teaching at Imperial College, that the qualifications and detail are best left until later when the basic concept has been assimilated. For similar reasons, fine distinctions such as between torr and mmHg, have been ignored. After reading this book, I hope that readers will go on to learn more about the specific aspects of glow discharges that they are interested in, and then will be able to return and pick out all the faults in this book. Apart from the mechanical errors of text preparation, I'm sure there are also errors in my understanding. I would be obliged if readers would write to me care of the publishers, and advise me of such errors; we have a duty not to promote misunderstanding.

In a readable technical book of finite thickness, it's necessary to assume some knowledge of the reader. I've tried to assume only that which should be common knowledge to science and engineering graduates. Further background and other items are discussed in the appendices.

A few words of confession about units. I have tried to use the cgs, imperial and hybrid units commonly encountered in the technology. Unfortunately, in moments of panic, I revert to the MKS system with which I learned. This only adds further confusion to an already tortured situation and I can only apologize and join the guilty. But be reasonable – see it my way.

The author of a book is just one of a small army of people who contribute to the innumerable details of its production. In gratitude, first thoughts go to the many people who have established an understanding of plasma physics and glow discharge processes, and without whom this book would not exist. Thanks are due to authors and publishers for permission to use copyrighted material; formal acknowledgment follows.

Harold Winters, John Vossen, John Thornton and John Coburn (Harold is nearly always last, so this time I've inverted the usual order) kindly let me force them into reviewing my work. I thank them for their time and their many helpful comments, and for their endurance in reading my first manuscript. Others of my colleagues have also been very patient with me.

Many folks, unfortunately too numerous to mention, worked very hard and co-operatively in the mechanical preparation of the book. I thank them all. Sadly, I would like to acknowledge particularly the contribution of the late Don Brown, who prepared most of the artwork.

Don Marchese, Frank Bresnock, Walt Koste, and several of my other friends helped me in more ways than they realise, and were always ready to lend an ear to my complaints and frustrations. I look forward to reciprocating their kindnesses.

Finally, and most of all, to Carol and Toby. I needed to say nothing to them, nor they to me. Say no more! Now we can get back to the plans, the 505 and the radio 'trolled robot.

Brian Chapman

New Fairfield, Connecticut

Reproduction and Copyright Acknowledgments

My thanks are due to the many authors and publishers who have permitted me to reproduce the figures and tables herein. Copyright of the authors or publishers, as relevant, is acknowledged. A list of publishers and related publications follows:

Academic Press (including Adv. Electronics & Electron Phys.)
Airco Temescal
American Chemical Society
American Elsevier Publishing Co.
American Institute of Mechanical Engineers (Trans. Met. Soc. AIME)
American Institute of Physics (J. Chem. Phys., Physics of Fluids, J. Appl. Phys., Appl. Phys. Lett., Rev. Sci. Instrum., J. Vac. Sci. Tech.)
American Physical Society (Phys. Rev.)
Dover Publications
Electrochemical Society (J. Electrochem. Soc.)
Heinemann Educational Books
IBM Corp. (IBM J. Res. Develop.)
Institute of Electrical & Electronics Engineers (Proc. IEEE)
Japanese J. Appl. Phys.
Joint Institute for Laboratory Astrophysics (JILA)
Litton Industries (General Mills Report)
McGraw Hill Book Company
M.I.T. Press
North Holland Publishing Co. (Nucl. Instrum. & Methods)
Optical Society of America (Optics & Spectroscopy)
Oxford University Press
Plenum Publishing Corp.
RCA Corp. (RCA Review)
Royal Society (Proc. Roy. Soc.)
Societé Francaise du Vide (Le Vide, Les Couches Minces)
Society for Applied Spectroscopy (Appl. Spectroscopy)
Solid State Technology
Standard Telecommunication Laboratories
Taylor & Francis (Wykeham Publications)
Tokuda Seisakusho
John Wiley & Sons

Contents

Glow Discharge Processes

Chapter 1: Gases

In all the plasma processes discussed in this book, namely deposition and etching by physical sputtering and by plasma enhanced chemistry, the gaseous environment plays a major role. It's clear just by looking at a plasma that it's a very busy place, with emission from excited atoms, with bombardment from ionized species, etc. However, the overwhelming majority of the gas in those processes is neutral and in the electronic ground state.

In this chapter we shall be looking at this gaseous environment, both on collective and individual atomistic bases. Many of the concepts covered here will be applied, in later chapters, to the electrons and to the ionized and dissociated atoms and molecules in the plasma. The following is a brief resumé of the relevant points in kinetic theory.

MASSES AND NUMBERS OF ATOMS

The simplest gas atom is the hydrogen atom – a proton surrounded by an electron. There aren't many hydrogen atoms around because they prefer to form pairs and become hydrogen molecules. A hydrogen atom is very small and very light: it has a mass of $1.66\ 10^{-24}$ g. Therefore there are $6.02\ 10^{23}$ atoms in 1 gram of atomic hydrogen, i.e. in 1 gram molecule. Since a gram molecule is a quantity of material corresponding to the molecular weight in grams, and since the molecular weight of an element is essentially the ratio of the mass of one of its atoms to that of a hydrogen atom, then it follows that one gram molecule of any material contains $6.02\ 10^{23}$ molecules–which is Avogadro's number.

KINETIC ENERGY AND TEMPERATURE

When a set of atoms is put into an enclosure, the atoms will absorb energy from the environment, for example by radiation from the walls. The energy of the gas is stored in the form of translational kinetic energy (and in the case of molecules, also in vibrational and rotational states). Because the gas molecules are in motion, frequent collisions occur both amongst the molecules and between the molecules and the wall, and these collisions cause a continual interchange of energy. This energy has some steady state distribution, which can be derived by statistical techniques. One would expect that the overall kinetic energy of the gas, and the average kinetic energy of each constituent molecule, would depend on the absolute temperature T (Figure 1-1):

$$\frac{1}{2} m\overline{c^2} = \frac{3}{2} kT$$

where m is the mass of a molecule, $\overline{c^2}$ is the mean square speed, and k is the relevant constant of proportionality, known as Boltzmann's constant and having the value of 1.38 10^{-16} ergs/deg K.

Figure 1-1. $\frac{1}{2}m\overline{c^2} = \frac{3}{2}kT$

MEAN SPEED $\overline{c}$

It follows that the mean square speed is given by 3 kT/m. However, a more useful parameter is the mean speed $\overline{c}$ which is not equal to the square root of $\overline{c^2}$ (since 'mean' and 'root mean square' are differently defined) but can be shown to have a similar value:

$$\overline{c} = \left(\frac{8kT}{\pi m}\right)^{1/2}$$

For argon atoms at 20°C, $\overline{c}$ has a value of 3.94 10^4 cm/sec, which is a little more than 880 mph. So whilst the atoms are moving very rapidly, they're not travelling so much faster than a jumbo jet, and are considerably slower than Concorde at 1450 mph. (Hopefully the planes have a larger mean free path!) Superman – "faster than a speeding bullet" – is neck and neck with Concorde (Figure 1-2).

MAXWELL-BOLTZMANN DISTRIBUTION

It is implicit in our usage of a mean speed $\overline{c}$ that atoms travel both slower and faster than $\overline{c}$, which is as one would expect in any multiple collision process. By considering the random nature of these collision processes, Maxwell and

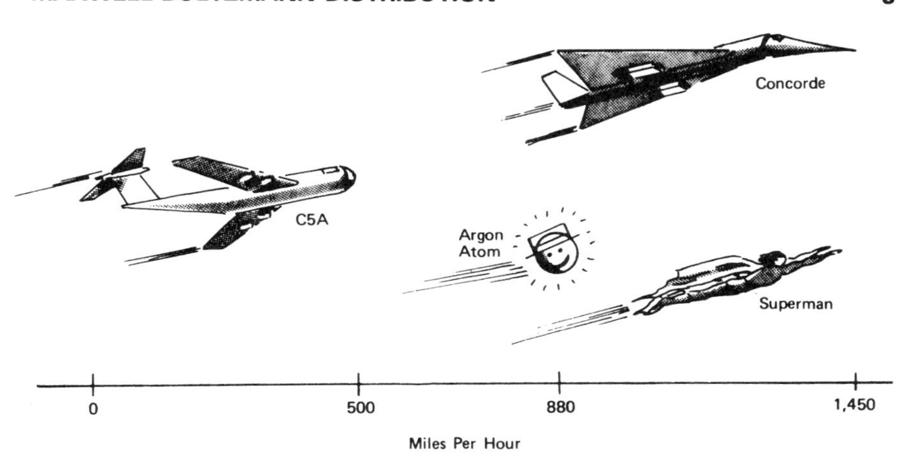

Figure 1-2. Relative speeds of argon atoms and other familiar bodies

Boltzmann (independently) were able to show that the number dn of molecules out of a total of n that have speeds between c and c + dc, is given by:

$$\frac{dn}{dc} = \frac{4n}{\pi^{1/2}} \left(\frac{m}{2kT}\right)^{3/2} c^2 \exp\left(-\frac{mc^2}{2kT}\right)$$

This Maxwell-Boltzmann distribution function is shown in Figure 1-3. Although the detail of the distribution is not important for our present purposes, we should point out that while some atoms travel very much faster and some very

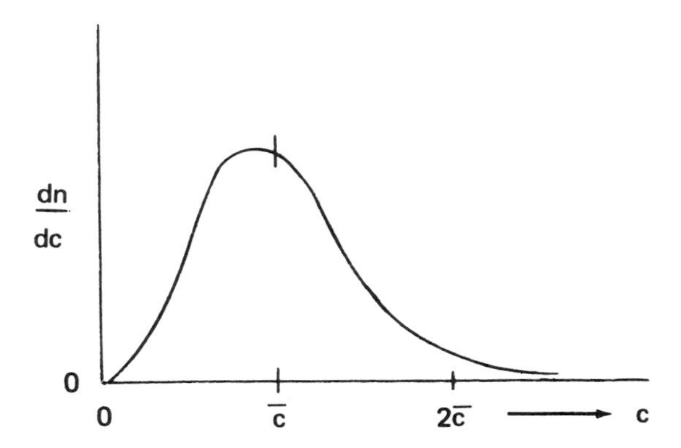

Figure 1-3. Maxwell-Boltzmann distribution

much slower than the mean, nearly 90% of them at any time have speeds between half and twice $\overline{c}$. Again, without wanting to stress the detailed form of the Maxwell-Boltzmann function, it is worthwhile to note that the mean speed ($\overline{c}$) which we discussed earlier (and indeed the mean of any other parameter) can be obtained from the distribution function since:

$$\overline{c} = \frac{1}{n}\int c \, dn$$

PRESSURE

If we place a molecule in an enclosure, then because it has some energy, it will continually be in motion, bouncing off the walls of the enclosure and off other molecules. Each time a molecule hits a wall, it will exert a force and the total force per unit wall area due to all the atoms is known as *pressure* (Figure 1-4).

Consider a section A of the surface of a wall (Figure 1-5). Imagine an x-axis perpendicular to the wall, and let A have an area of 1 cm^2. Consider the elastic impact (an assumption of kinetic theory) of an atom of velocity $\dot{x}$ on the wall. Its momentum is $m\dot{x}$, and since it bounces back at the same speed, the change of momentum is $2\,m\dot{x}$. If the volume density of atoms with velocities between $\dot{x}$ and $\dot{x} + d\dot{x}$ is $n(\dot{x})d\dot{x}$, then the total number that strike the wall per unit time are contained in a cylinder of base A and length numerically equal to $\dot{x}$, and this is just $n(\dot{x})\dot{x}d\dot{x}$. The rate of change of momentum due to these is $2\,mn(\dot{x})\dot{x}^2\,d\dot{x}$. But force is equal to the rate of change of momentum (force = mf = mdv/dt), and therefore the total force per unit area on the wall (otherwise known as pressure) is the integral of this over all velocities $\dot{x}$:

$$p = \int_0^\infty 2mn(\dot{x})\dot{x}^2\,d\dot{x}$$

The function $n(\dot{x})$ is just the one-dimensional version of the Maxwell-Boltzmann distribution, and substituting this and integrating yields:

$$p = nm\overline{\dot{x}^2}$$

But

$$\overline{\dot{x}^2} + \overline{\dot{y}^2} + \overline{\dot{z}^2} = \overline{c^2}$$

Since $\dot{x}$, $\dot{y}$, and $\dot{z}$ are symmetrical, then $\overline{\dot{x}^2} = \overline{c^2}/3$, so the pressure on the wall is given by:

$$p = \frac{nm\overline{c^2}}{3}$$

This expression tells us directly how the pressure in a plasma processing chamber depends directly on the density of gas atoms n, their mass m, and their mean square speed. It is also clear why a temperature increase, for example due to one of the plasma processes, leads to a pressure increase: higher temperature means higher kinetic energy which means higher $\overline{c^2}$.

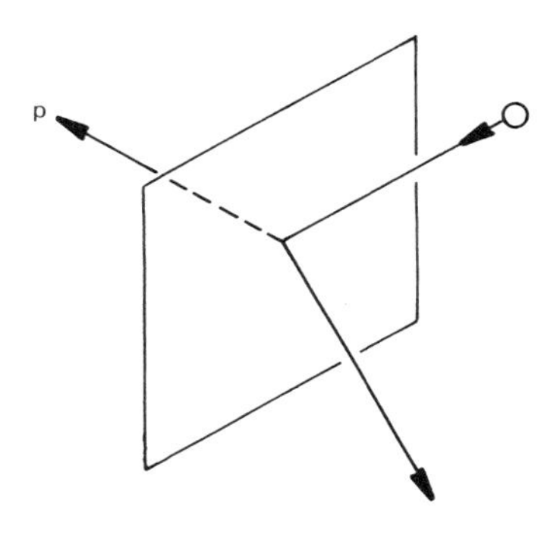

Figure 1-4. Wall collisions cause pressure

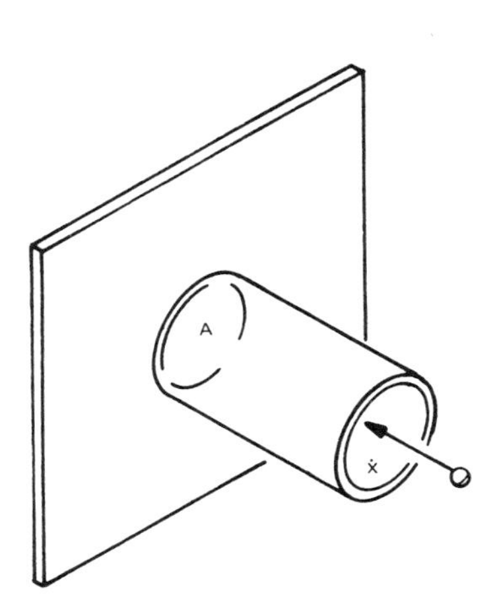

Figure 1-5. $p = \frac{nm\overline{c^2}}{3}$

PARTIAL PRESSURES

With more than one gas in a system, the total pressure is the sum of the partial pressures that each of the gases alone would have exerted in the system. This follows from the last section since, if two gases are described by subscripts 1 and 2 respectively, then:

$$p = \frac{1}{3}m_1 n_1 \overline{c_1{}^2} + \frac{1}{3}m_2 n_2 \overline{c_2{}^2}$$

$$= p_1 + p_2$$

PRESSURE UNITS

The pressure units commonly used in plasma processing at present are the torr and the millitorr, and the equivalent units of mm Hg and μ (microns), which is just an abbreviation of μm Hg. These are defined by:

$$1 \text{ standard atmosphere} = 760 \text{ mm Hg} = 760 \text{ torr}$$
$$1 \text{ mtorr} = 1\ 10^{-3} \text{ mm Hg} = 1\ \mu\text{m Hg} = 1\mu$$

We shall see later that most plasma processes take place between 1 mtorr and 1 torr.

Much of the kinetic theory discussed above has been in cgs units, where the pressure unit is 1 dyne/cm^2. Using the density of mercury (13.6 g/cc), we find that 1 standard atmosphere = 760 torr = 1.0 10^6 dynes/cm^2. Actually, the S.I. (Systēme Internationale) pressure unit which we're now supposed to use is the pascal (Pa), which is 1 newton/metre^2. This has an obvious relationship to our cgs unit:

$$1 \text{ Pa} = 1 \text{ N/m}^2 = 10 \text{ dynes/cm}^2 = 7.5 \text{ mtorr}$$

Presumably we shall all be converting to pascals over the next decade or so.

AVOGADRO'S LAWS

The pressure expression derived above can also be written as

$$p = \frac{1}{3}\frac{N}{V}\overline{mc^2}$$

where N is the total number of molecules in a volume V. Earlier we saw that a gram molecule of any gas contains the same number (Avogadro's number) of molecules. Consider two gases at the same pressure p and at the same temperature, so that $\overline{mc^2}$ has the same value for each. If there is 1 gram molecule of each gas, N has the same value in each case. Hence it follows from the pressure relation that they occupy the same volume V; it seems that equal volumes of all gases at the same temperature and pressure contain the same number of molecules. In particular, a gram molecule of any gas at standard temperature (0°C) and pressure (1 atmosphere) contains 6.02 10^{23} molecules and occupies a fixed volume, which happens to be 22.4 litres – a bit less than 1 cubic foot.

NUMBER DENSITY OF GASES

It follows from Avogadro that any gas at standard temperature and pressure (STP) contains 2.7 10^{19} molecules/cc. A more useful number to remember is the equivalent figure at 1 torr, which is 3.5 10^{16}. So although plasma processes take place at reduced pressures, they are still very crowded environments! The lowest plasma process operating pressure usually encountered is about 1 mtorr, and even at this low pressure there are 3.5 10^{13} molecules/cc.

Those involved in the semiconductor industry and other industries which utilize fine dimension geometries may be interested to consider the number of gas atoms contained in an etched cubic hole, 1 μm wide and 1 μm deep, in a substrate. At the same operating pressure of 1 mtorr, there would be about 35 gas molecules (Figure 1-6), a much easier number to come to terms with!

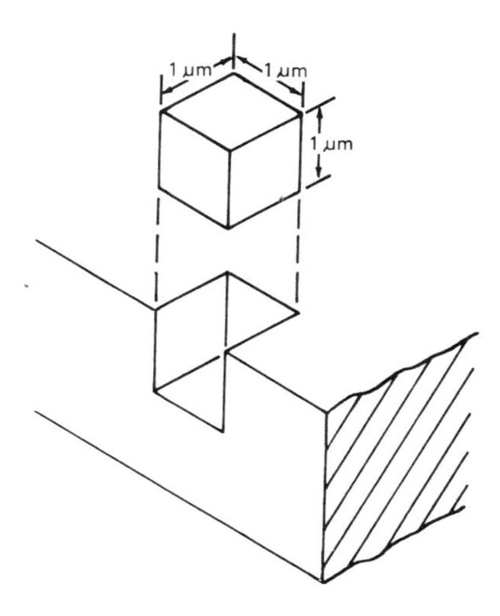

Figure 1-6. A 1μm cube contains 35 gas molecules at 1 mtorr

IMPINGEMENT FLUX

We saw earlier how the flux (i.e. number per unit area per unit time) of atoms bombarding the walls of a chamber exerts a pressure. If we consider Figure 1-5 and the argument used earlier to derive the pressure relationship, then we can show that:

$$\text{Flux/unit area} = \int_0^\infty n(\dot{x})\dot{x}d\dot{x}$$

and on integration, this yields an impingement flux of $n\bar{c}/4$ per unit area (Figure 1-7).

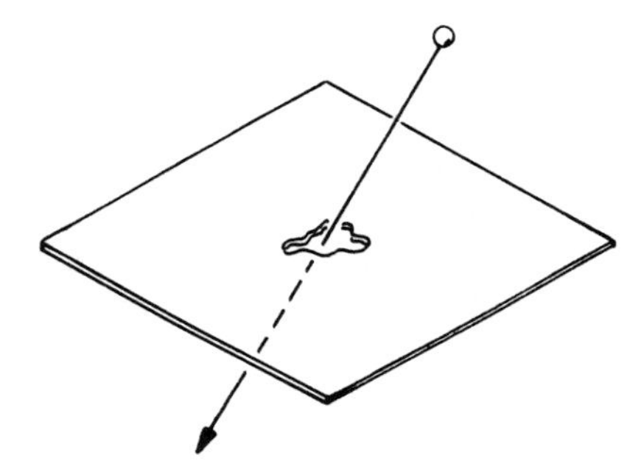

Figure 1-7. Flux = $\frac{n\bar{c}}{4}$ per unit area

For argon at 500 millitorr and 20°C, the impingement flux is $1.8\ 10^{20}$ atoms per cm^2 per second. Even the base of the μm size hole we considered earlier will be subjected to $1.8\ 10^{12}$ argon atoms per second at 500 mtorr, dropping only to $3.6\ 10^9$ at 1 mtorr. Generally several gas species are present in the vacuum chamber during a plasma process, causing every surface in the chamber, *including the substrate*, to be constantly bombarded by a large flux of many gas species; this may be particularly significant during thin film deposition.

MONOLAYER FORMATION TIME

The preceding surface impingement rate leads us to the concept of *monolayer formation time*. The substrate is being bombarded by many gas species other than the depositing thin film or the reactive etching species. The time to form a monolayer of any species on the surface is based on the assumption that every atom striking the surface remains there, which is not really true except at low temperatures, but does give us an accurate idea of the fluxes involved. The formation time is obviously inversely proportional to the impingement flux. A typical sputter deposition rate is one monolayer per second ($\sim 10^{15}$ atoms per cm^2 per second for a typical atom of 3 Å diameter). This rate corresponds to a gas pressure of not much more than 10^{-6} torr. Hence, if a contaminant gas has a partial pressure of only 10^{-6} torr, every welcome sputter depositing atom arriving at the substrate is accompanied by a contaminant atom! Additionally, the argon used as the sputtering gas may have a partial pressure typically four orders of magnitude higher, so that each single condensing atom of the growing thin film may be accompanied by 10 000 argon atoms. It is therefore hardly surprising that some of this argon is trapped in the growing film, though as we'll see later, we can control the amount of this trapped argon. By comparison, in an industrial vacuum evaporation process whereby aluminium is deposited at about 1 μm/min in a pressure of 10^{-4} torr – which would conventionally be regarded as poor vacuum – the fluxes of depositing aluminium and background gases at the substrate are about equal. It's all a question of relative rates.

MEAN FREE PATH

The mean free path is the average distance travelled by a gas atom between collisions with other gas atoms (Figure 1-8) and obviously decreases at higher pressures (greater densities of gas molecules). At 1 mtorr and room temperature,

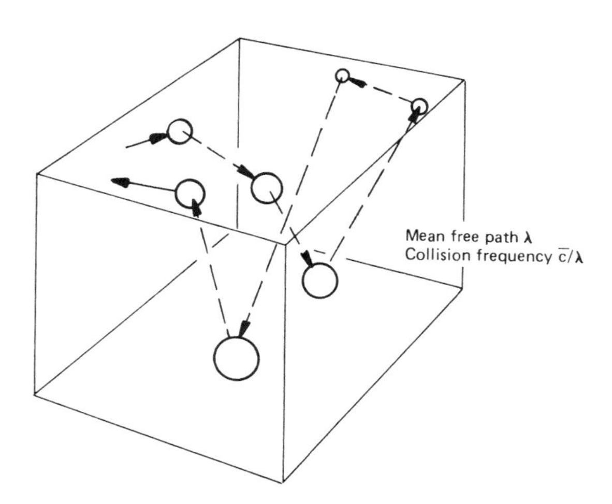

Figure 1-8. Random motion and collisions of an atom

the mfp of argon is about 8 cm and most other gases are within a factor of three of this. (These figures are relevant only for atoms of thermal energy).

Kinetic theory assumes that there are no interactions between gas atoms other than during collisions. The probability of collision, and hence mean free path, depends on atom size, and so will vary from gas to gas.

It's easy to derive an approximate value for a mean free path. If an ideal gas atom has a diameter of 3 Å, it will collide if the centre comes within 3 Å of the centre of another atom. The collision area therefore has a radius of 3 Å or an area of 2.8 10^{-15} cm^2, and this will sweep out a volume of gas as the atom moves. At 1 mtorr, equivalent to 3.54 10^{13} atoms/cc, there is one atom in each 2.8 10^{-14} cm^3. So the average distance that the atom of 2.8 10^{-15} cm^2 collision area must move to sweep out a volume of 2.84 10^{-14} cm^3 is 10 cm, which is close to the actual value for argon quoted above.

PROBABILITY OF COLLISION

Just as there is a distribution of atom speeds, so there is a distribution of free path lengths between collisions. We might like to know, for example, the probability P(x) of an atom travelling a distance x without colliding. But, considering events over the next small increment of length Δx,

$$\begin{aligned} P(x + \Delta x) &= P(x) \, x \, (\text{probability of travelling } \Delta x \text{ without colliding}) \\ &= P(x)\,(1 - \Delta x/\lambda) \end{aligned}$$

where 1/λ is the probability per unit length of making a collision. By expanding the above expression and integrating, we find that:

$$P(x) = \exp\left(-\frac{x}{\lambda}\right)$$

since P(0), the probability of travelling zero distance without colliding, is clearly unity.

By a similar argument, and using the result derived above, the number dn of atoms colliding between x and x + dx is

$$dn = n_O \, x \exp\left(-\frac{x}{\lambda}\right)\frac{dx}{\lambda}$$

where n_O is the total number of atoms. Let's now calculate the mean free path $\overline{x}$ between collisions. By the usual averaging process,

$$\overline{x} = \frac{\int x \, dn}{\int dn}$$

$$= \frac{\int_0^\infty \frac{n_o}{\lambda} x \exp\left(-\frac{x}{\lambda}\right) dx}{\int_0^\infty \frac{n_o}{\lambda} \exp\left(-\frac{x}{\lambda}\right) dx}$$

$$= \lambda \text{ (integrating by parts)}$$

So we have the useful results that if the mean free path between collisions is λ, then the probability of collision per unit length is $1/\lambda$, and the probability of travelling x without collision is $\exp(-x/\lambda)$.

Let's apply this finding to the sputter deposition of copper in a diode system. Suppose that the substrate is 5 cm away from the target electrode, and let's guess that the mean free path of copper in argon is also 5 cm at 1 mtorr. Under these circumstances, the probability of a sputtered atom reaching the substrate without colliding with an argon atom is

$$\exp -\frac{x}{\lambda} = \exp -\frac{5}{5} = 0.37$$

which is at least a reasonable chance. However, we need raise the pressure to only 14 mtorr before the chance is less than the proverbial one in a million. Under these circumstances, transport is by diffusion.

COLLISION FREQUENCY

We've already met the idea of a collision frequency in the discussion of mean free path. It is the average number of collisions which a gas atom makes per unit time and so is equal to $\bar{c}/\lambda$ (Figure 1-8). In argon at 50 mtorr pressure, the collision frequency is $4\ 10^5$ per second.

ENERGY TRANSFER IN BINARY COLLISIONS

We shall consider here the collision between two particles of masses m_i and m_t. Assume that m_t is initially stationary and that m_i collides with velocity v_i at an angle θ to the line joining the centres of m_i and m_t at the moment of collision (Figure 1-9).

By conservation of linear momentum

$$m_i v_i \cos\theta = m_i u_i + m_t u_t \tag{1}$$

By conservation of energy

$$\tfrac{1}{2} m_i v_i^2 = \tfrac{1}{2} m_i (u_i^2 + v_i^2 \sin^2\theta) + \tfrac{1}{2} m_t u_t^2 \tag{2}$$

Eliminating u_i in (2) from (1) gives:

$$m_i v_i^2 \cos^2\theta = \frac{m_i}{m_t^2}(m_i v_i \cos\theta - m_t u_t)^2 + m_t u_t^2$$

The fractional energy transferred from mass m_i to mass m_t is:

$$\frac{E_t}{E_i} = \frac{\frac{1}{2}m_t u_t^2}{\frac{1}{2}m_i v_i^2} = \frac{m_t}{m_i v_i^2}\left(\frac{2m_i v_i}{m_t+m_i}\cos\theta\right)^2$$

$$= \frac{4m_i m_t}{(m_i+m_t)^2}\cos^2\theta$$

and this has a maximum of $\cos^2\theta$ when $m_i = m_t$.

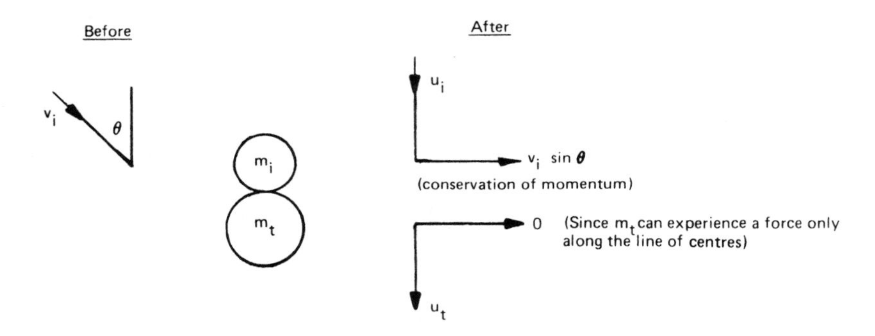

Figure 1-9. Velocity components before and after an elastic collision

Note that we have not needed to know the detailed nature of the interaction between the two masses, so that the result is as true for two atoms as it is for two billiard balls; the laws of conservation of momentum and energy are quite universal. The only implicit assumption is that the potential energies of the two colliding bodies are the same before and after the interaction; i.e. the collision must be elastic.

This is a more important result than might at first appear. It means that there can be very efficient energy transfer in collisions between atoms of the same gas, and explains why, in the example we discussed earlier, the speeds of 90% of gas atoms in steady state conditions are within a factor of two of the mean speed.

However, there are further implications. The term $4\,m_i m_t/(m_i + m_t)^2$ is known as the *energy transfer function* and clearly has the value 1 when $m_i = m_t$. But what happens when the masses are very unequal? For example, how about when an energetic electron strikes a carbon monoxide molecule? If $m_i \ll m_t$, then the transfer function becomes just 4 m_i/m_t. In this case the mass ratio is about 28 *x* 1840, so that the energy transfer function has a value of about 10^{-4}. This means that very little energy can be transferred from the electron to the gas molecule. We shall return to this point when we discuss glow discharges later.

There is another interesting result to be derived. Suppose a very heavy particle hits a very light particle in a head-on collision, i.e. $m_i \gg m_t$ and $\theta = 0$. Then the energy transfer function simplifies to 4 m_t/m_i.

Hence

$$\frac{\tfrac{1}{2} m_t u_t^2}{\tfrac{1}{2} m_i u_i^2} = \frac{4 m_t}{m_i}$$

and

$$u_t = 2 v_i$$

The result is that the light particle speeds away at twice the impact velocity.

GAS FLOW

We've so far been talking about a fixed number of gas atoms in a sealed enclosure. The reality in plasma processing is that we have a steady state population of atoms, being continuously fed by gas flow and continuously pumped by a vacuum system. In several plasma processes, in particular plasma etching and plasma deposition, the inlet gas is consumed and the need for a gas flow to replenish the consumption is apparent. On the other hand, in conventional sputtering there is no consumption of the etching gas. However, one still needs to pump the process chamber to remove contaminants that might, as an example, be desorbed from the chamber walls and fixtures by the various types of fast particle bombardment present in plasma processing. A gas inlet flow is then needed to maintain the chosen operating pressure.

TYPES OF GAS FLOW

The nature of the gas flow in vacuum systems can vary considerably, according to conditions. At high pressures, when the mean free path of a molecule is very short, the behaviour of an individual atom is completely dominated by collisions with other atoms. The energy transfer function discussed earlier enhances this cooperative behaviour. In a pumped system of this type, the atoms will all tend to move along in a stream, dragging each other along by their internal 'friction'. This is known as *viscous flow*, and can be either *laminar*, where the gas streams all move parallel to each other, or *turbulent*.

At very low pressures, when the mfp becomes very large, collisions with other molecules become rare and so take place mainly with the walls of the vessel. Molecules do not usually bounce off a surface in the way that light is reflected, but are instead absorbed, if only for a brief period. After this brief period, the molecule is again liberated from the surface (desorbed); however, it has now lost knowledge of whence it came, so that the subsequent desorption direction is completely random and follows the well-known cosine distribution. The cooperative flow behaviour of the viscous regime is thus lost, and is replaced by *molecular flow*, where there is virtually no interaction.

If d is the characteristic dimension of a vacuum system or of a particular piece of tubing, then viscous flow will obtain if $\lambda \ll d$, and molecular flow if $\lambda \gg d$. Dushman (1962) has suggested that a more pragmatic division is $\lambda < 0.01d$ for viscous flow, and $\lambda > d$ for molecular flow. Each of these regions independently is reasonably well understood, but between these two there is a third region known as *transition* or *Knudsen* flow. The region, characterized by $0.01d < \lambda < d$, is rather complex to analyze and is understood only semi-empirically. Let us see how our plasma processes fit into this pattern. A typical figure for λ would be about 5 10^{-3} cm at 1 torr and room temperature. A typical plasma processing chamber might have a diameter of about 50 cm. Most plasma processes take place between 1 millitorr and 1 torr:

Pressure (torr)	*λ(cm)*	*λ/d*	*Gas Flow*
10^{-3}	5	0.1	transition
10^{-2}	0.5	0.01	transition
10^{-1}	0.05	0.001	viscous
1	0.005	0.0001	viscous

So it seems that most practical processes above about 10 mtorr are in viscous flow, with transition flow taking over below this pressure. Note, however, that the situation might be different in narrow gas feed pipes (Figure 1-10).

PUMPING SPEED AND THROUGHPUT

There are two main ways of describing the amount of gas flowing, for example, through a pumping port in a vacuum system. The *pumping speed* S measures the volume of gas passing per second, but the number of molecules contained therein naturally depends on the pressure of that volume of gas. A frequently more useful term is thus the *throughput* Q, equal to pS, pressure times the pumping speed, and hence proportional to the flux of molecules passing (Figure 1-11). Throughput is alternatively known as *flow rate*, particularly when referring to the inlet rate of a reaction gas.

Based on our earlier choice of units, the appropriate units for throughput and flow rate are of torr litre/second (tl/s). However, another commonly used unit is the standard cc per minute (sccm), standard referring to standard temperature (0°C) and standard pressure (1 atmosphere or 760 torr). It follows from our earlier considerations that:

$$1 \text{ sccm} = \frac{6.023\ 10^{23}}{22414} \text{ molecules per minute}$$

$$= 2.69\ 10^{19} \text{ molecules per minute}$$

It can also easily be seen that

$$1 \text{ torr litre/sec} = 79.05 \text{ sccm}$$

$$= 2.13\ 10^{21} \text{ molecules per minute}$$

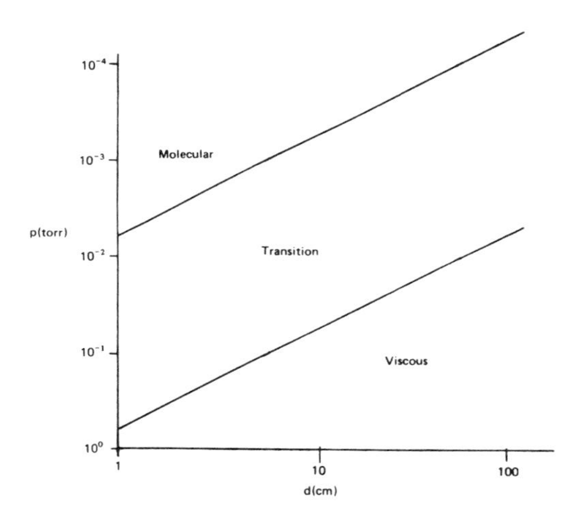

Figure 1-10. The various types of gas flow, based on $\lambda = 5$ cm at 1 mtorr

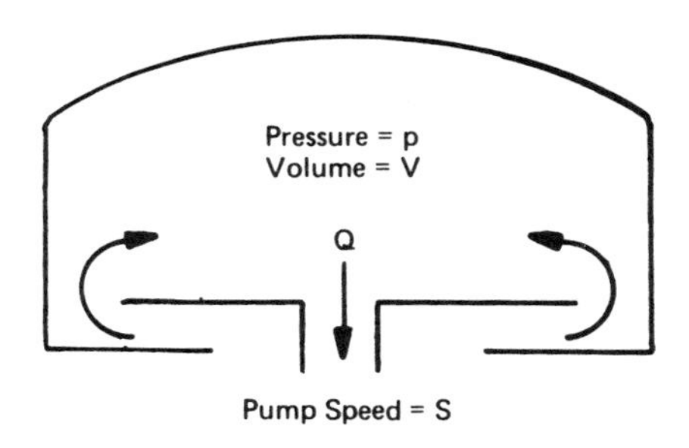

Figure 1-11. $Q = pS = nvA$

MEASUREMENT OF GAS FLOW RATE

It often occurs that we need to measure gas flow rates in a plasma system. This might be because the system is not fitted with a flow meter, or because an installed meter is either not calibrated for the gas of interest or is of dubious reliability. This measure of flow rate can usually be done without formal recalibration, at least to a very good approximation.

The prerequisites are a knowledge of the volume of the chamber and of the pressure within it. The volume of the chamber can be reasonably estimated by physical measurement, but a much more accurate technique is to expand a small known volume of gas at atmospheric pressure into the evacuated chamber and note the resulting pressure. For example, using 10 cc of gas at STP, the final chamber pressure is 392 millitorr. Since the number of gas molecules is conserved, then $10 x 760 = 392\ 10^{-3}$ V. So V = 19.4 litres.

To measure a given flow rate, the chamber is first evacuated and sealed. At time t = 0, the gas is allowed to flow into the chamber and the resulting pressure rise is noted as a function of time. This should result in a linear increase in pressure, at least up to a few torr. Suppose, in the example above, a pressure rise of 3.1 millitorr per second is recorded. Knowing the chamber volume, the flow rate is thus $3.1\ 10^{-3}\ x\ 19.4$ tl/s, which is $6\ 10^{-2}$ tl/s or 4.8 sccm.

RESIDENCE TIME

A parameter which depends directly on flow rate is the *residence time* τ of a gas molecule, i.e. the mean time it remains in the process chamber before being pumped away. If the volume of the chamber is V litres and this is being pumped at the rate of S litres/second, then the mean residence time is just V/S. Since S is not measured directly, we use instead:

$$\tau = \frac{V}{S} = \frac{pV}{pS} = \frac{pV}{Q}$$

To give a practical example, a plasma deposition system might operate at p = 1 torr, V = 100 litres, and Q = 160 sccm (2 tl/s). Then

$$\tau = \frac{1\ x\ 100}{2} = 50 \text{ seconds}$$

So the residence time can be quite substantial.

FLOW VELOCITY

It's sometimes useful to know the velocity of a gas, and by this we mean the velocity of the stream, rather than the much higher instantaneous velocities of the constituent atoms; the flow velocity is analogous to the drift velocity of conduction electrons in a solid or gas. Consider the high pressure radial flow reactor shown in Figure 1-11. If v is the flow velocity of the gas stream down the central pumping port, n is the molecular density of the gas, A is the cross-sectional area of the port, and Q the gas throughput, then

$$Q = nvA$$

For typical values of Q = 160 sccm = 2 tl/s, n (≡1 torr) = 3.5 10^{16}/cc, and A = 10 cm^2, then v = 204 cm/sec, much slower than typical thermal speeds of 10^4 cm/sec.

CONDUCTANCE

The conductance F of an orifice or a length of tubing is a measure of its ability to transmit a gas flow. It is defined by:

$$Q = F\Delta p$$

where Δp is the pressure differential across the element. Gas conductance is thus rather analogous to electrical conductance, and the relationship above is rather like Ohm's Law, with Q being equivalent to the current, and Δp to the applied potential.

A problem in using the concept of gas conductance is that, unlike the electrical analogue, its value changes according to conditions, which is a manifestation of the nature of the flow changing from molecular to transition to viscous. Nevertheless it is a useful concept if we remember the limitations. The same limitations apply to the closely related parameter of pumping speed. And although the conductances of certain geometries can be calculated for molecular flow or for viscous flow, they cannot as yet be calculated for the transition region, which is where, as we have seen, our plasma systems often operate.

The glow discharge processes we're considering take place at relatively high pressures as far as high vacuum pumps are concerned, and so limiting conductances are often used to reduce the operating pressure in the process chamber to a pressure more compatible with the working range of the pump.

Let's consider how this works: initially the process chamber is evacuated to some base pressure limited by real and virtual leaks of magnitude q. If the base pressure is p_o and the pump speed S, then (Figure 1-12 a):

$$q = p_o S$$

A high vacuum system for plasma applications might typically have a pumping speed of 1000 l/s and be able to initially evacuate the system to 10^{-7} torr. In this case the throughput of the pumping system (equal to the leak rate) would be 10^{-4} torr litres per second, or about 2 10^{17} molecules per minute.

Let's now suppose that the plasma process calls for a flow rate of Q, and that by adding this flow of gas the chamber pressure rises to p_1, which must still be in the efficient working range of the pump. Let's also assume that the pumping speed remains constant at S; both of these last conditions would be satisfied up to a pressure of about 1 mtorr in an oil diffusion pump. Then (Figure 1-12 b):

$$q + Q = p_1 S$$

The hypothetical practical process detailed above may call for a gas flow of Q = 80 sccm or 1 torr litre/second. Then

$$p_1 = \frac{q + Q}{S} = \frac{10^{-4} + 1}{1000} = 1 \text{ mtorr}$$

We may have satisfied the flow requirement, but perhaps a higher operating pressure p_2 is required (if a lower operating pressure is required – hard luck!). This higher pressure is realized by adding a limiting conductance F (Figure 1-12 c) at the throat of the pump such that:

$$F(p_2 - p_1) = F(Q + q)$$

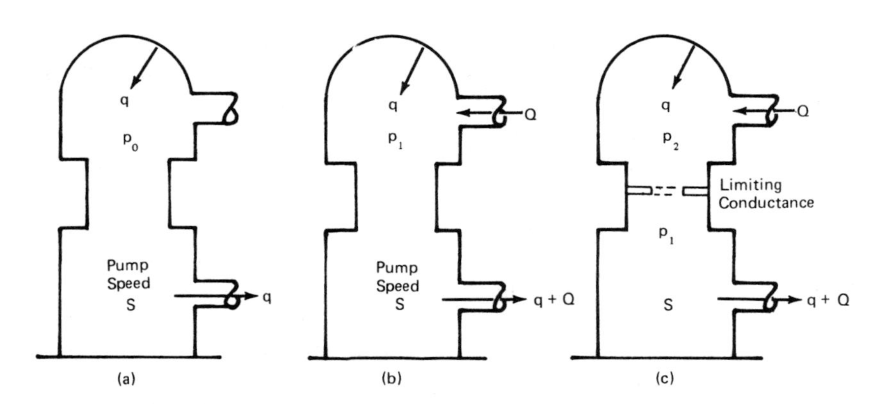

(a) At base pressure p_0, $q = p_0 S$

(b) Now add required gas flow Q, $q + Q = p_1 S$

(c) Then throttle to establish required pressure p_2, $q + Q = F(p_2 - p_1)$

Figure 1-12. Pressure-flow relationships

So, if the required p_2 is 100 mtorr, then F needs to be about 10 torr litres/second, reducing the effective pumping speed by three orders of magnitude. But note that the pressure in the pump and the real pumping speed are unchanged by the throttling operation, remaining at p_1 and S respectively.

GENERAL REFERENCES

J. D. Cobine, *Gaseous Conductors*, Dover, New York (1958)

S. Dushman, *Scientific Foundation of Vacuum Technique*, 2nd edition, ed. J. M. Lafferty, Wiley, New York and London (1962)

J. K. Roberts and A. R. Miller, *Heat and Thermodynamics*, Blackie, London and Glasgow (1960)

Chapter 2. Gas Phase Collision Processes

In the last chapter we considered the behaviour, both microscopically and macroscopically, of gases. Although that consideration was primarily of neutral ground state molecules, we shall see later that some of the concepts are useful in dealing with charged particles also.

Although the glow discharge is an integral part of each of the several processes that we're dealing with, the discharge is performing only a functional role in each case. We are able to group these processes together because the type of discharge is reasonably common: the degree of ionization is typically 10^{-4}, and the current densities are of the order of 1 mA/cm^2, so that essentially we still have a neutral ground state gas that can be described by the gas laws. The prime feature of these discharges is that of ionization, with perhaps as many as 10^{18} electron-ion pairs being produced per second. But the different processes make use of ionization in different ways. For example, in the physical process of sputtering, the main requirement is an adequate supply of ions that can be accelerated onto a target to produce sputtering. In plasma etching, a principal requirement is a process to dissociate relatively stable gas molecules into chemically active species which can then react with the substrate; this dissociation is efficiently carried out by electron impact, and in turn these electrons rely on the ionization process for their sustenance. Inevitably some of our processes require the synergism of both the physical and chemical aspects of the discharge, thus foiling any attempt at neat categorization on our part.

Unfortunately, practical glow discharges are rather complex environments, particularly the chemical discharges used in plasma etching and deposition, and they are far from being well understood. So before we launch into the confusing (and sometimes conflicting) detail of specific discharges, let's look at some more collision phenomena, which are common to all of the glow discharge processes covered in this book. We shall also see the usefulness of the ideas of collision cross-section, and of electron volts as energy units.

COLLISION CROSS-SECTION

Before we look further at collision processes, let's examine an alternative to the mean free path concept that we established in the last chapter, and see why it isn't such a useful parameter after all.

Consider, in Figure 2-1, an electron approaching a volume of gas contained in a slab Δx thick and cross-sectional area A. Except for the very slowest electrons, the atoms of the gas move so much more slowly than the electrons that we can assume the atoms are relatively stationary. If the gas density is n molecules per unit volume and the effective collision area of each molecule is q, then the probability of a collision in that slab in $n \Delta x\, q$ – the fraction of the cross-sectional area occupied by molecular targets. So the collision probability is nq per unit length. This tacitly assumes that $n\Delta xq \ll 1$, so that none of the 'targets' overlap. We can always assure this by making Δx vanishingly small and integrating. But at sufficiently low pressures the inequality is satisfied even for finite Δx.

It might at first appear that λ and q are simply related by $nq = 1/\lambda$. Although this would not be a simple relationship anyway, it also hides the real significance of the concept of *collision cross-section*, as the collision area is normally known. An electron approaching an atom sees the atom not as a solid ball but as an

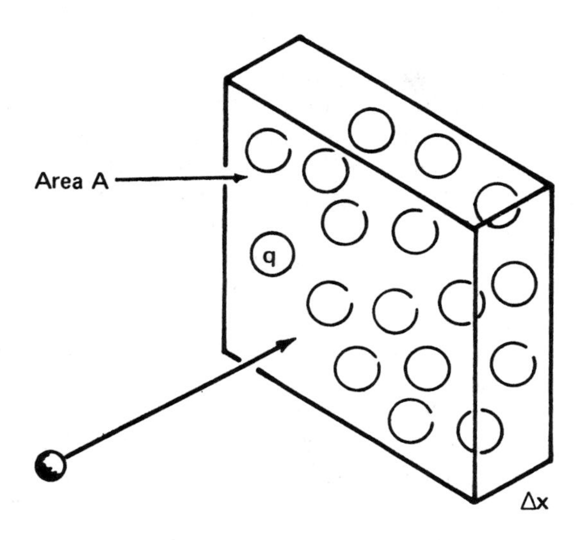

Figure 2-1. Probability of collision $= \frac{n\,A\,q}{A}\,\Delta x = nq$/unit length

assembly of electrons and ions in largely empty space. An analogy is that if the nucleus of the atom were represented in size by a cricket ball, then the electron orbits would be around the perimeter of the field. Probably the situation wouldn't be too different in baseball! The interaction between the primary electron and the electrons and ions of the atom take place via strong electrostatic forces, so the precise outcome depends on the detail of the approach of the electron. This introduces an element of probability or uncertainty into the interaction in contrast to the certainty of collision between two converging billiard balls (or elastically colliding atoms). This probability is implicit in our definition of collision cross-section q. It is also not surprising that q depends on approach velocity, and this is easily dealt with by making q a function of electron velocity or energy. This dependence can be intuitively rationalised in various ways, for example by arguing that the interaction time decreases with increasing velocity.

So it appears that collision cross-section is a more versatile parameter than mean free path λ, which is therefore conventionally reserved for elastic collisions between thermal molecules. And although we introduced the collision cross-section concept via electron-atom interactions, it's clearly of more general application.

ELASTIC AND INELASTIC COLLISIONS

Collision processes can be broadly divided into elastic and inelastic types, according to whether the internal energies of the colliding bodies are maintained. Particles usually have two types of energy: kinetic energy due to their motion and equal to $\frac{1}{2}mv^2$ for translational motion, and internal or potential energy which may be in the form of electronic excitation, ionization, etc.

An *elastic* collision is one in which there is an interchange of kinetic energy only. An *inelastic* collision has no such restriction, and internal energies change also.

In the last chapter, we established an energy transfer function:

$$\frac{4m_i m_t}{(m_i + m_t)^2}$$

which gives the maximum energy transferable in an elastic binary collision. As a consequence, the fraction of energy transferred from an electron to a nitrogen molecule was about 10^{-4}. But now allow the collision to be inelastic, so that the molecule struck gains internal energy of ΔU (Figure 2-2). Then, using almost the same equations as previously:

Momentum conservation

$$m_i v_i \cos\theta = m_i u_i + m_t u_t$$

Energy conservation

$$\tfrac{1}{2}m_i v_i^2 = \tfrac{1}{2}m_i(u_i^2 + v_i^2\sin^2\theta) + \tfrac{1}{2}m_t u_t^2 + \Delta U \tag{1}$$

Eliminating u_i as before gives

$$m_i v_i^2 = \frac{m_i}{m_t^2}(m_i v_i \cos\theta - m_t u_t)^2 + m_t u_t^2 + 2\Delta U \tag{2}$$

which simplifies to

$$2m_t u_t v_i \cos\theta = \frac{m_t}{m_i}(m_t + m_i)\, u_t^2 + 2\Delta U \tag{3}$$

ΔU and u_t are the only variables. To maximize ΔU,

$$2\frac{d}{du_t}(\Delta U) = 2m_t v_i \cos\theta - \frac{m_t}{m_i}(m_t + m_i)2u_t = 0$$

$$\text{i.e. } v_i \cos\theta = \frac{(m_t + m_i)u_t}{m_i}$$

Substituting back into (3), we obtain

$$2\Delta U = \left(\frac{m_t m_i}{m_t + m_i}\right) v_i^2 \cos^2\theta$$

Hence the fraction of the kinetic energy of the first particle that can be transferred to the internal energy of the second, has a maximum value of

$$\frac{\Delta U}{\tfrac{1}{2}m_i v_i^2} = \frac{m_t}{m_t + m_i}\cos^2\theta$$

So whereas the maximum elastic energy transfer from an electron to a nitrogen molecule was only 0.01%, by inelastic means this may rise to more than 99.99%, since when $m_t \gg m_i$, this *inelastic energy transfer function* tends to 1.

From equation (3), it seems that the fractional kinetic energy transfer is

$$\frac{\tfrac{1}{2}m_t u_t^2}{\tfrac{1}{2}m_i v_i^2} = \frac{m_t u_t^2}{m_i}\left(\frac{2m_t u_t}{\frac{m_t}{m_i}(m_t + m_i)u_t^2 + 2\Delta U}\right)^2$$

Excluding the cases when ΔU can be negative, this function has a maximum value when the denominator has a minimum value given by $\Delta U = 0$; then the fractional energy transfer reverts to the $4m_i m_t/(m_i + m_t)^2$ value as expected. So even though a good deal of potential energy can be transferred when $m_i \ll m_t$, it is still impossible to transfer any significant amount of kinetic energy.

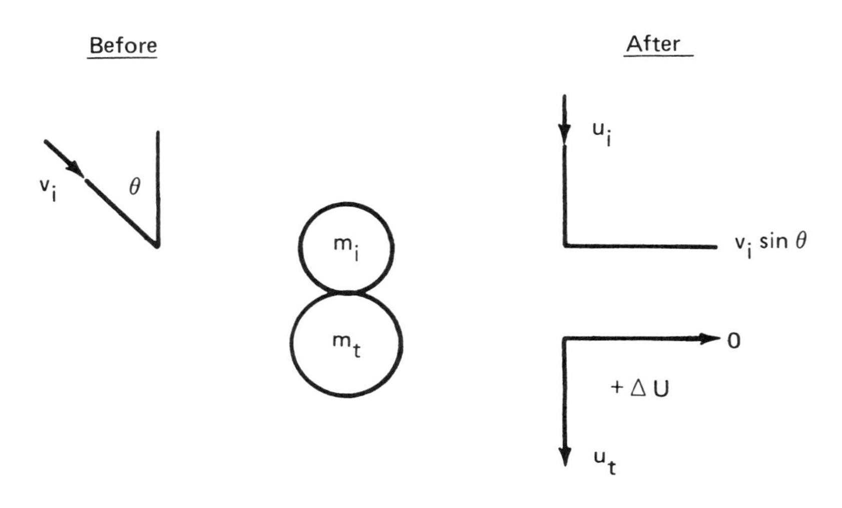

Figure 2-2. Kinetics of an inelastic collision

THE MAIN COLLISION PROCESSES

The gas phase environment of a glow discharge contains electrons, various types of ions, neutral atoms and molecules, and photons. In principle we should consider collisions between all possible pair permutations, but fortunately some collisions are more important than others in the glow discharge environments that we are considering in this book. Collisions involving electrons are dominant in determining the macroscopic behaviour of the glow discharge, and therefore we begin by considering these.

Electron Volts

Electrons are, of course, readily accelerated or decelerated by electric fields. If an electron is accelerated between 2 points of potential difference X volts, then by definition it loses eX joules of potential energy and gains an equivalent amount of kinetic energy; e is the electron charge of $1.6\ 10^{-19}$ coulombs. This eX joules of kinetic energy is also conveniently referred to as X *electron volts* or X eV, and so:

$$1\text{eV} = 1.6\ 10^{-12}\ \text{ergs}$$

This derived energy unit of electron volts turns out to be generally useful and can be applied to neutral as well as other charged particles.

Elastic Collisions

The simplest collisions are elastic, so that kinetic energy is conserved. But since the electron and any atom have such different masses, we know from the energy transfer function $4m_i m_t/(m_i + m_t)^2$ that the transfer of energy is negligible; so the electron just changes direction without significantly changing speed (Figure 2-3). Where the electron is moving in an electric field, elastic collisions generally have the effect of restricting its velocity in the direction of the field, in the same way that drift velocity of a conduction electron in a solid is restricted by lattice collisions. In both cases, the colliding atom is virtually unaffected. We can use the collision cross-section concept to describe the probability of elastic collision, as in Figure 2-4 which shows how the chance of an electron being scattered in argon depends on the energy of the electron. To translate the cross-section into something more easily considered, it appears from Figure 2-4 that the elastic cross-section for an argon atom to a 15 eV electron is about $2.5\ 10^{-15}\ \text{cm}^2$. So, at 10 mtorr when there are $3.54\ 10^{14}$ atoms/cc, the probability of elastic collision for a 15 eV electron is 0.89/cm.

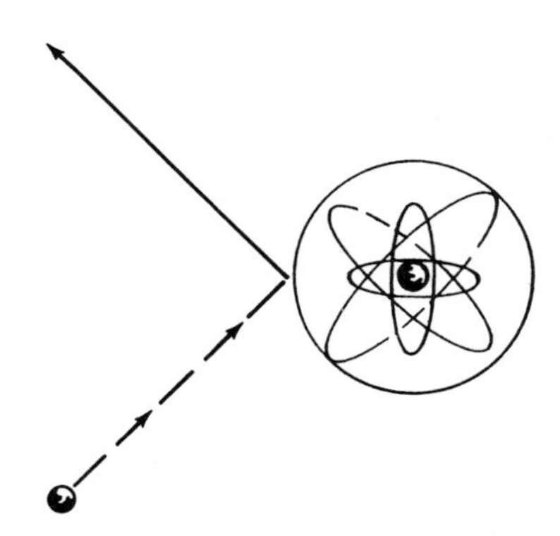

Figure 2-3. Electron – atom elastic collision

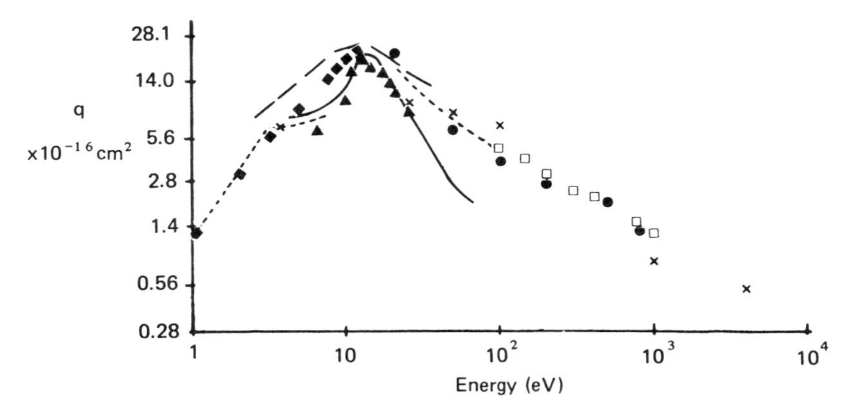

Figure 2-4. Cross-section for elastic scattering of electrons in argon. From DuBois and Rudd 1975; includes data from various authors

Ionization

All other types of electron collisions are inelastic. The most important of these in sustaining the glow discharge is *electron impact ionization* (Figure 2-5) in which the primary electron removes an electron from the atom, producing a positive ion and two electrons, e.g.

$$e + Ar \rightarrow 2e + Ar^+$$

The two electrons produced by the ionizing collision can then be accelerated by an electric field until they, too, can produce ionization. It is by this multiplication process that a glow discharge is maintained.

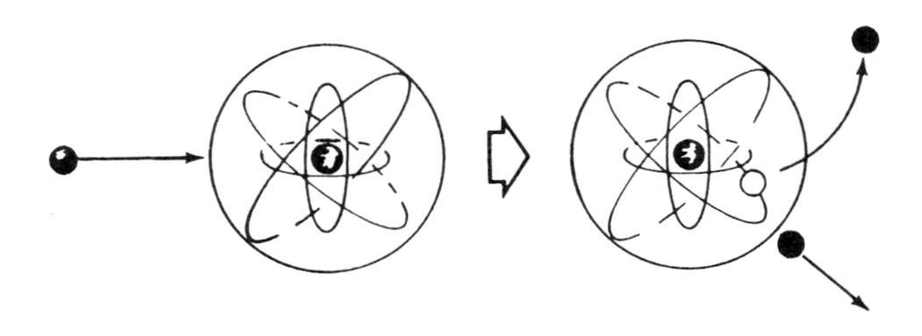

Figure 2-5. Electron impact ionization

There is a minimum energy requirement for this ionization process to occur, equal to the energy to remove the most weakly bound electron from the atom, and this is known as the *ionization potential*; for xenon this has a value of 12.08 eV. Below this threshold energy, the ionization cross-section is clearly zero, but rises as soon as the electron exceeds the ionization potential (Figure 2-6). The energy dependence of the ionization cross-section for xenon over a rather larger electron energy range is shown in Figure 2-7. This is fairly typical for the inert gases, quickly rising above the threshold to a maximum around 100 eV and then falling. Cross-sections for other noble gases are shown in Figure 2-8. Note that this figure gives the cross-sections in units of πa_0^2 rather than cm^2. In this unit, a_0 is the radius ($0.53\ 10^{-8}$ cm) of the first Bohr orbit of hydrogen; πa_0^2 is the area of a hydrogen atom and has the value $8.82\ 10^{-17}\ cm^2$, a unit of useful size.

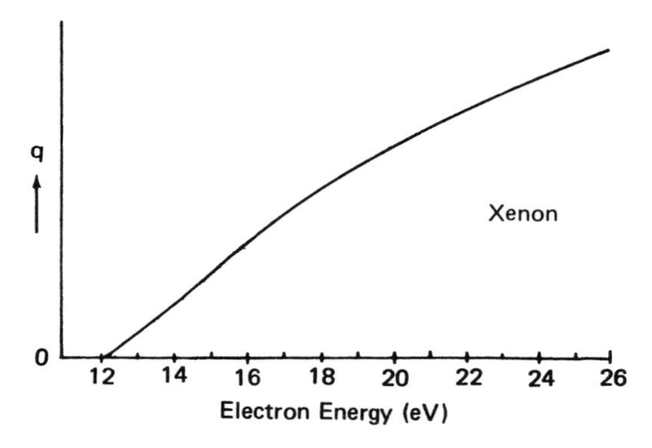

Figure 2-6. Ionization cross-section for xenon near threshold (Rapp and Englander-Golden 1965)

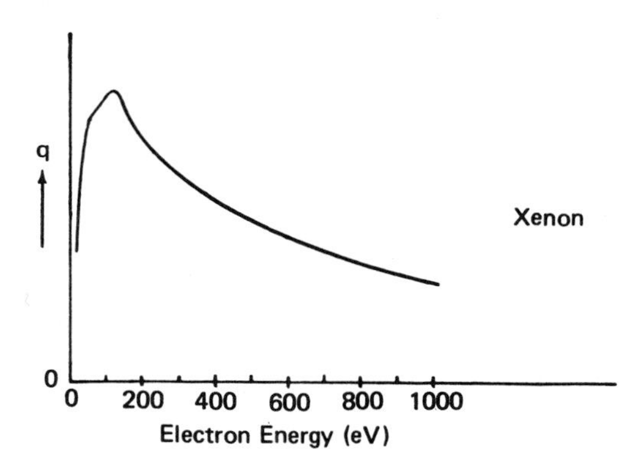

Figure 2-7. Ionization cross-section for xenon, 0–1000 eV (Rapp and Englander-Golden 1965)

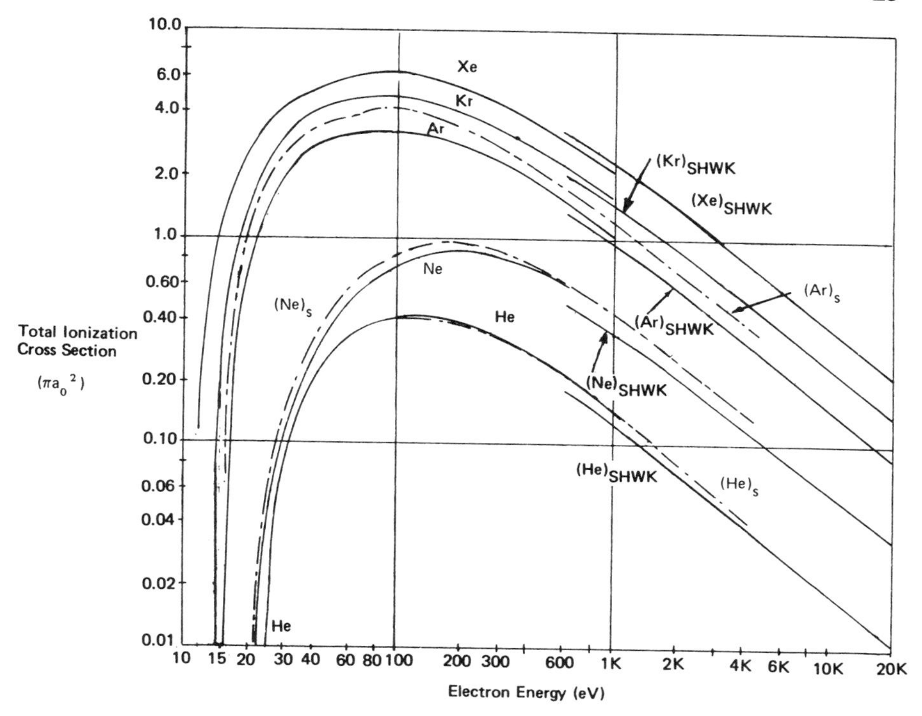

Figure 2-8. Ionization cross-sections of the noble gases (from Rapp and Englander-Golden 1965; includes data from (S) Smith 1930 and (SHWK) Schram et al. 1965. Similar values for Ar, He and Ne have been obtained by Fletcher and Cowling 1973); n.b. $\pi\, a_O{}^2 = 8.82 \times 10^{-17}$ cm^2

It is not only by electron impact that ionization is produced. In principle, the ionization could be due to any suitable energy input, and the possibilities in the discharge must therefore include thermal and photon activation. In the present context, 'thermal activation' means energy received by impact with neutral ground state gas atoms or with the atoms of the walls, and for our 'cold' plasmas, the temperature does not greatly exceed ambient. *Photoionization* can be significant, however. In order to relate the two relevant commonly used photon parameters of wavelength and energy, a useful and reasonably good approximation is:

$$1\text{eV} \equiv 12345\ \text{Å}$$

since

$$E = h\upsilon = \frac{hc}{\lambda}$$

The photoionization cross-section for argon, shown in Figure 2-9, exhibits a threshold at 15.8 eV as expected, rising rapidly to a cross-section of about 3.7 10^{-17} cm^2 before decreasing for higher photon energies, with further discontinuities as fresh ionization thresholds are encountered. By comparison, note (Figures 2-8 and 2-6) that the electron impact cross-section around threshold is virtually zero, but rises to a maximum of about 2.6 10^{-16} cm^2 at around 100 eV.

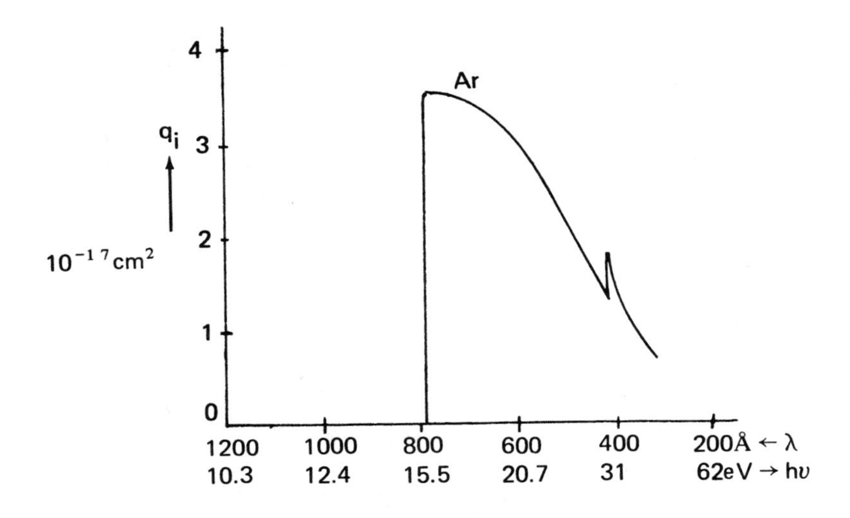

Figure 2-9. Photoionization cross-section of argon (Weissler 1956)

Although the photoionization cross-section seems to decrease to rather small values, this does not mean that there is very little ionization. The reason is two-fold: firstly, the excess energy (between the incident photon and the relevant ionization energy) appears mostly as kinetic energy of the emitted electron, since little kinetic energy can be transferred to the gas atom as we know from the energy transfer function, and this energetic electron can now cause further ionization; secondly, the 'hole' in the atom left by the ionization process will be filled by an electron transition from a higher level and the accompanying photon emission usually causes more ionization – and hence the ejection of more electrons – on its way out of the atom (– the Auger effect is an example). Similar arguments can be used to explain photoelectron emission from chamber walls and all internal surfaces, including electrodes. As a result, one would expect a large proportion of photon energy to lead ultimately to ionization. Although some photons will be lost from the chamber through transparent windows or through quartz chamber walls, most windows do not transmit below about 2500 Å, i.e. above about 5 eV. The maximum photon energies found in plasma processes correspond to about the peak-to-peak driving voltage used, i.e. about 1000 eV, and are due mostly to ion and electron impact on chamber surfaces. These energetic photons will penetrate on hitting another surface, so that electrons produced may not be able to escape; in this case the photon energy is dissipated in heat.

There are two other ionization processes frequently occurring in the glow discharge, and these are described below (ion-neutral collisions, and metastable-neutral collisions).

Excitation

In the ionization process, a bound electron in an atom is ejected from that atom. A less dramatic transfer of energy to the bound electron would enable the electron to jump to a higher energy level within the atom with a corresponding quantum absorption of energy. This process is known as *excitation*, and as with ionization can result from electron impact excitation (Figure 2-10), photo-excitation, or thermal excitation, although the latter is rare in 'our' cold discharges.

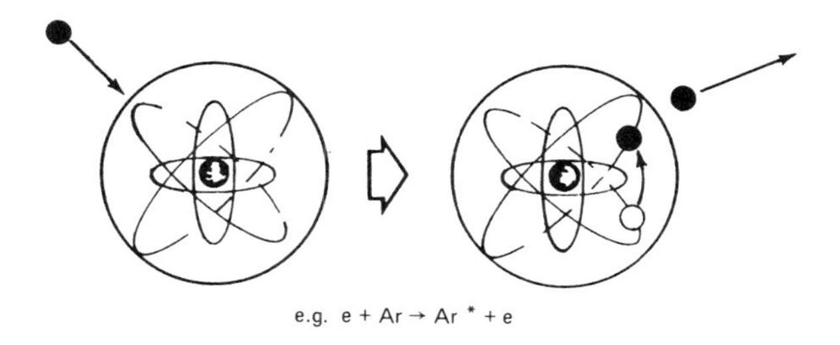

e.g. $e + Ar \rightarrow Ar^* + e$

Figure 2-10. Electron impact excitation

The excited state is conventionally represented by an asterisk superscript, e.g. Ar^*. As with ionization, there is a minimum energy for excitation to occur. The value of the *excitation potential* for argon is 11.56 eV, somewhat less than the ionization potential, as would be expected since excitation raises an electron to a higher (less tightly-bound) shell, and ionization completely removes the electron from the atom. In an exciting collision, the primary electron loses kinetic energy equal to the excitation potential and will also be deflected.

In Figure 2-11, the energy dependencies of the electron excitation cross-sections of the 2p levels of argon are shown. These rise from a threshold at 12.90 eV to a maximum at 21 eV electron energy, of 4 10^{-17} cm^2.

Figure 2-12 shows the excitation cross-section for atomic hydrogen. Note, by comparison, the corresponding cross-section in Figure 2-13 for molecular hydrogen, which has a much lower threshold due to the possibilities of vibrational and rotational excitation.

As with ionization, excitation can also be caused by photons; the cross-sections are of the same order as the ionization cross-sections discussed earlier.

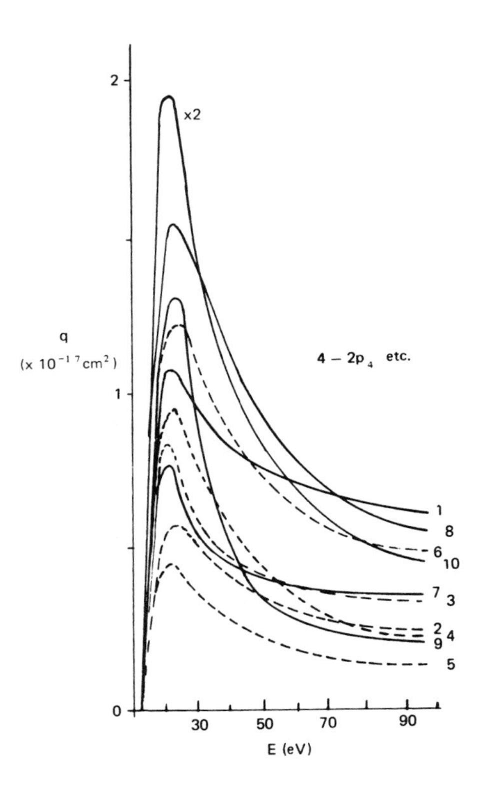

Figure 2-11. Excitation cross-sections of the 2p levels of argon (Zapesochnyi and Feltsan 1966)

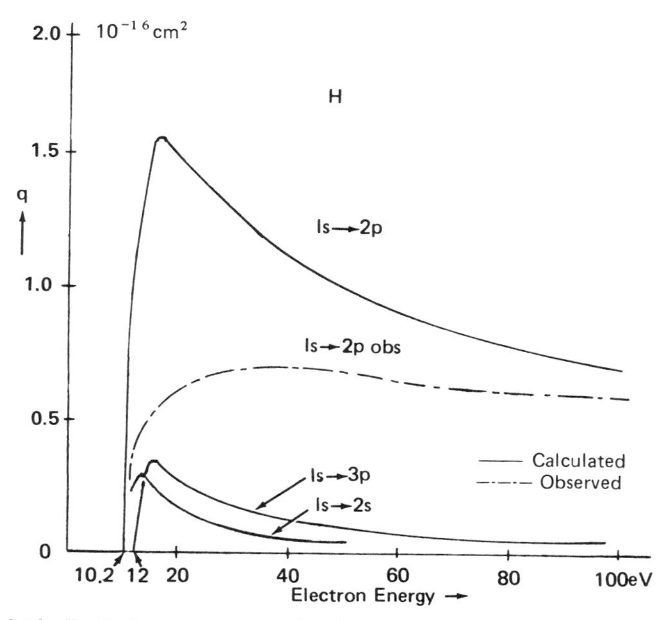

Figure 2-12. Excitation cross-section for electrons in atomic hydrogen (von Engel 1965)

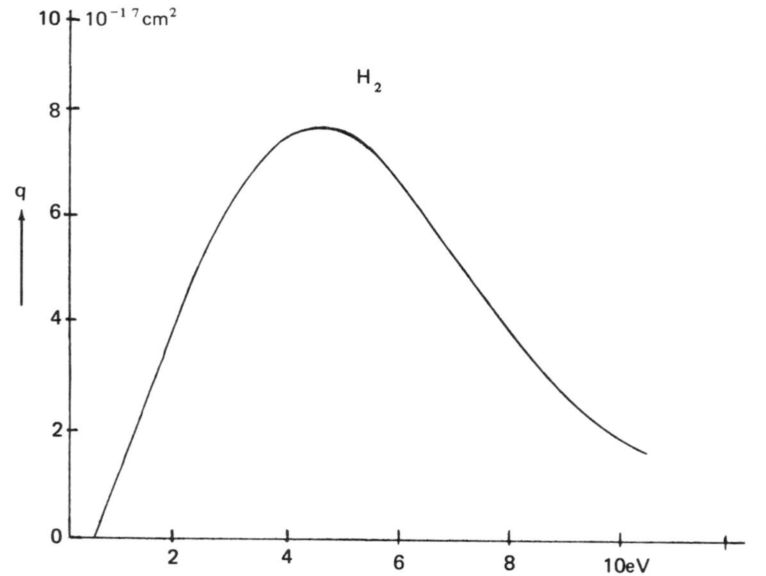

Figure 2-13. Excitation cross-section for electrons in molecular hydrogen (Frost and Phelps 1962)

Relaxation

One of the immediately self-evident features of a glow discharge is that it glows! This glow is due to the *relaxation* or de-excitation of electronically excited atoms and molecules – the inverse of the excitation process just discussed. These excited states are rather unstable and the electron configuration soon returns to its original (ground) state in one or several transitions, with lifetimes varying enormously from nanoseconds to seconds. Each transition is accompanied by the *emission* of a photon of very specific energy, equal to the difference ΔE in energy between the relevant quantum levels (Figure 2-14). Our eyes are sensitive only to wavelengths between about 4100 Å (violet) and 7200 Å (red), corresponding to electron transitions of 3.0 eV and 1.7 eV respectively, but with suitable detection equipment, photons from deep uv (atomic transitions) to far infra red (molecular vibrational and rotational transitions) can be detected. The technique of optical emission spectroscopy is thus very useful for detecting and determining the presence of various atoms in the glow. Figure 2-15 shows the emission spectrum from a glow discharge of $CF_4 + O_2$, a gas mixture commonly used in plasma etching.

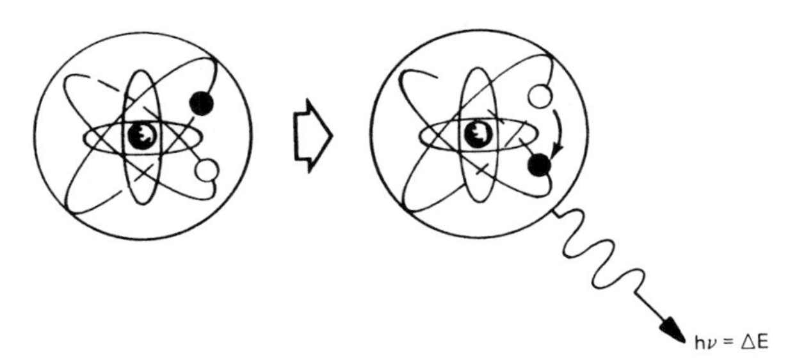

Figure 2-14. Relaxation (or de-excitation)

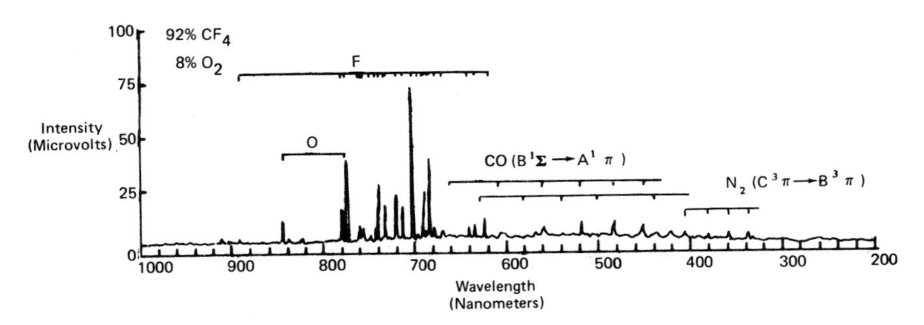

Figure 2-15. Optical emission from a discharge of $CF_4 + O_2$ (Harshbarger et al. 1977)

Recombination

Just as relaxation is the inverse of excitation, so recombination is to ionization – an electron coalesces with a positive ion to form a neutral atom (Figure 2-16). However, there is a problem:

Assume (Figure 2-17) that the electron has a mass m and has a velocity, relative to the ion of mass M, of v before recombination. Let their joint velocity after coalescence be u. The potential energy of the atom has decreased by U_i, the relevant ionization energy. Then, to continue our earlier applied mathematical exercises:

Conservation of momentum:

$$mv = (m + M)u$$

Conservation of energy:

$$\frac{1}{2} mv^2 = \frac{1}{2} (m + M) u^2 - U_i$$

Therefore, eliminating v,

$$\frac{1}{2} m \left(\frac{m + M}{m}\right)^2 u^2 = \frac{1}{2}(m + M)u^2 - U_i$$

which yields

$$u^2 = - \frac{2U_i m}{(m + M)M}$$

but since U_i, m, and M are all positive, this yields only an unreal solution for u. This means that, in general, a 2 body coalescence is just not possible. In practice, allowing for Heisenberg, this type of recombination is very unlikely. But recombination must occur somehow, because otherwise the ion and electron densities in any ionizing environment, such as a glow discharge, would continuously increase and this is contrary to experience. Some more subtle recombination processes therefore may take place:

3 Body Collision: A third body takes part in the collision process, and this third body allows the recombination process to simultaneously satisfy the conservation requirements of energy and momentum (Figure 2-18). The third body is often a wall, ubiquitous in our plasma processes, or it may be another gas atom. The probability of the gas atom taking part in the process will increase with increasing pressure; the wall is always there.

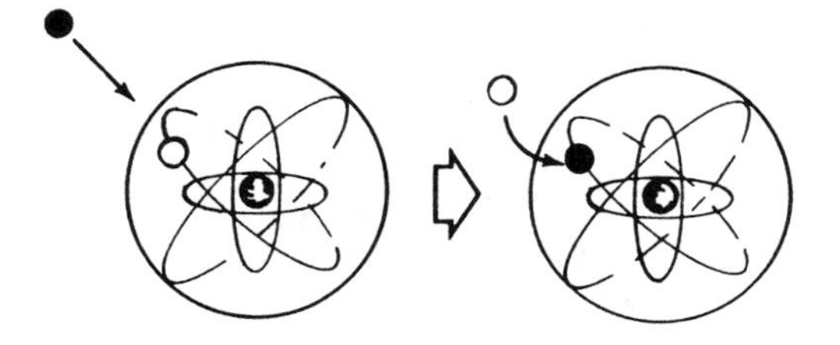

Figure 2-16. Recombination

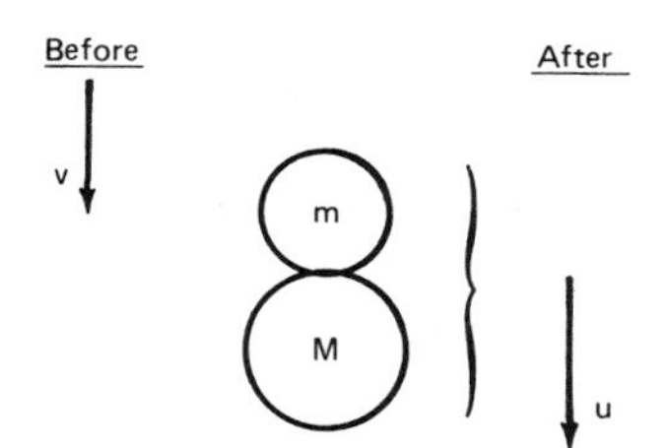

Figure 2-17. Kinetics of recombination

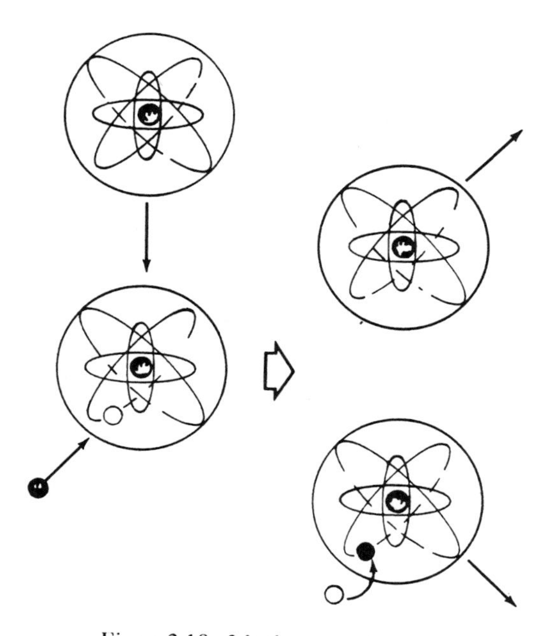

Figure 2-18. 3-body recombination

In a Two Stage Process: The electron attaches to a neutral to form a negative ion. Whilst not very likely, this process has U_i negative and so is possible. The negative ion then collides with a positive ion. The electron transfers and two neutrals are formed (Figure 2-19). The number of collisions between an electron and an atom before forming a negative ion depends on its electronegativity (its affinity for electrons) and ranges from about 10^3 for molecular chlorine to ∞ for the noble gases.

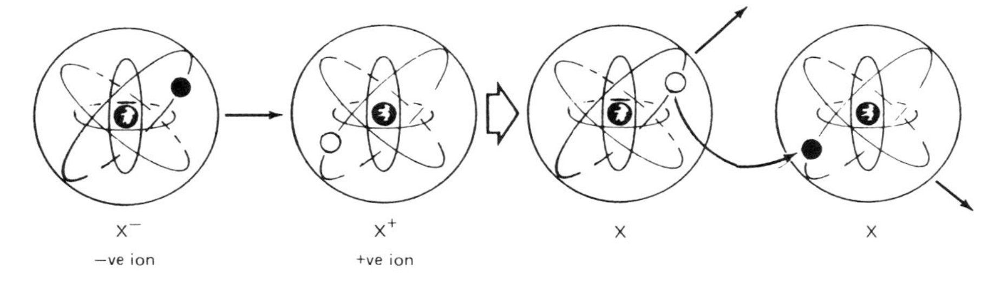

Figure 2-19. Ion-ion recombination

Radiative Recombination: The excess energy in the coalescence process of recombination is carried away by radiation (Figure 2-20). This is really another type of 3 body recombination.

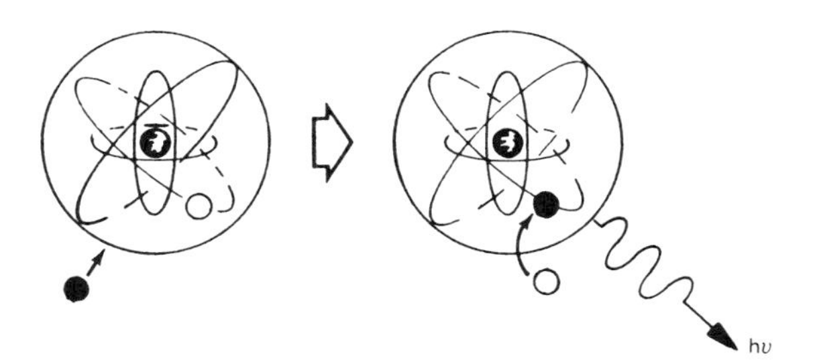

Figure 2-20. Radiative recombination

SOME OTHER COLLISION PROCESSES.

The four basic inelastic processes that were considered in the first part of this chapter – ionization and recombination, excitation and relaxation – are the inevitable ingredients of a glow discharge. We can understand the basic phenomena of a discharge by considering just these four inelastic processes and the reader may wish to omit the rest of this section on the first time through.

But as we said at the beginning of the chapter, there are many other collisional processes and some of these are important under some circumstances. We'll consider now the more relevant of these.

Dissociation

The process of *dissociation* is the breaking apart of a molecule. An oxygen molecule can be dissociated into two oxygen atoms, but an atomic gas such as argon cannot be dissociated at all.

As with the other inelastic processes we have been studying, dissociation can in principle be accomplished with any energy in excess of the dissociation threshold, i.e. the relevant bond strength in the molecule. In glow discharges, electron impact dissociation is common:

$$e + O_2 \rightarrow e + O + O$$

A normal result of dissociation is an enhancement of chemical activity since the products are usually more reactive than the parent molecule. We shall see in Chapter 7 how such enhancement of activity is used to oxidize photoresists in the *plasma ashing* process. We shall also see that dissociation may or may not be accompanied by ionization:

$$e + CF_4 \rightarrow e + CF_3 + F \qquad \text{(dissociation)}$$

$$e + CF_4 \rightarrow 2e + CF_3^+ + F \qquad \text{(dissociative ionization)}$$

There are different probabilities, and hence different cross-sections, for each of these processes.

Electron Attachment

There is a possibility that an electron colliding with an atom may join on to the atom to form a negative ion. This process is known as *electron attachment*. The noble gases, including argon, already have filled outer electron shells and so have little or no propensity to form negative ions. Halogen atoms, however, have an unfilled state in their outer electron shells; they have high electron affinities and so readily form negative ions. Figure 2-21 shows the rate of production of SF_6^- and SF_5^- negative ions by attachment as a function of the electron energy, from SF_6 gas. The production of SF_5^- in this way is known as *dissociative attachment* ($e + SF_6 \rightarrow SF_5^- + F$). It can be seen that the SF_6^- ion is formed more effectively by the attachment of electrons of almost zero energy. This is another manifestation of the problem of simultaneously satisfying energy and momentum conservation ("Recombination"). Certain interpretations are necessary to convert these production rates into collision cross-sections, but it seems (Massey 1969) that the cross-section may exceed 10^{-15} cm^2, with the resulting ion being metastable with a lifetime greater than 10^{-6} secs. Massey also cites several references to other halogen-containing gases; gases of this type are of importance in plasma etching.

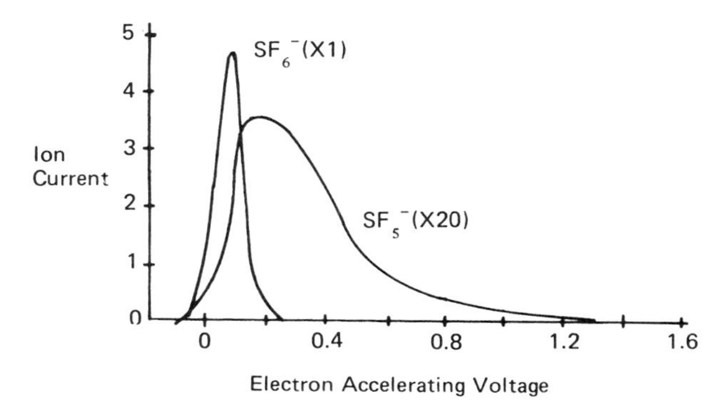

Figure 2-21. The SF_6^- and SF_5^- currents obtained by Hickam and Fox (1956) in their mass spectrometric studies of SF_6.

Ion-Neutral Collisions

Ions and neutrals can collide with each other elastically, or inelastically to either exchange charges or cause further ionization.

Charge Transfer

The probability of a collision leading to the exchange of charge, generically known as *charge transfer*, is usually greater for atomic ions moving in parent atoms (*symmetrical resonant charge transfer*), e.g.

$$A + A^+ \rightarrow A^+ + A$$

(Figure 2-22) or similarly for molecular ions moving in parent molecular gases, than in charge exchange between unlike systems, e.g.

$$B^+ + C \rightarrow B + C^+$$

which is known as *asymmetric charge transfer* and tends to be less efficient (Figure 2-23). Figure 2-24 shows both the elastic and symmetric charge transfer cross-sections for noble gas ions in their parent gases.

Symmetrical charge exchange in a field-free glow region is rather unimportant since there is, effectively, only a momentum exchange. But both types of charge transfer become important in a sheath region, where they have the effect of changing the energy distribution of ions and neutrals on the electrode (Chapter 4, "Collisions in the Sheath").

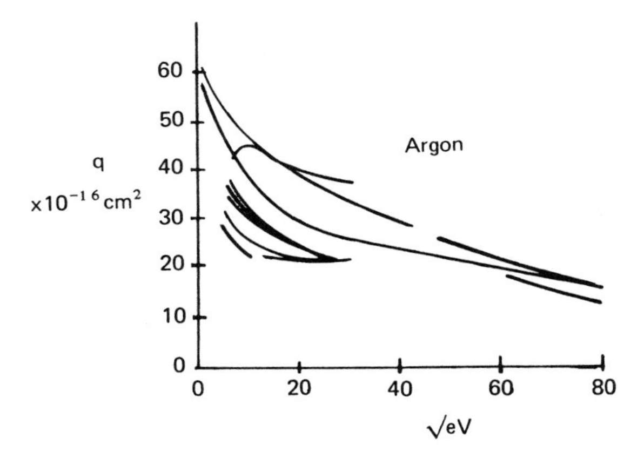

Figure 2-22. Cross-sections for resonance charge transfer of Ar^+ on Ar, as shown in McDaniel 1964; data from several authors

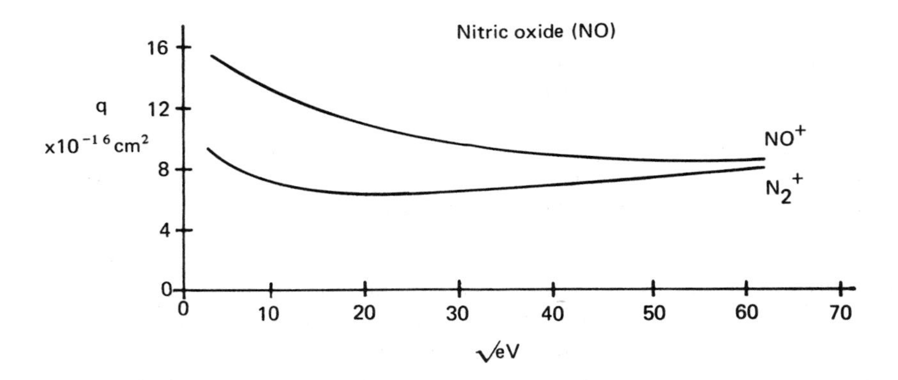

Figure 2-23. Charge transfer cross-sections for NO^+ and $N_2{}^+$ ions in nitric oxide (Stebbings et al. 1963)

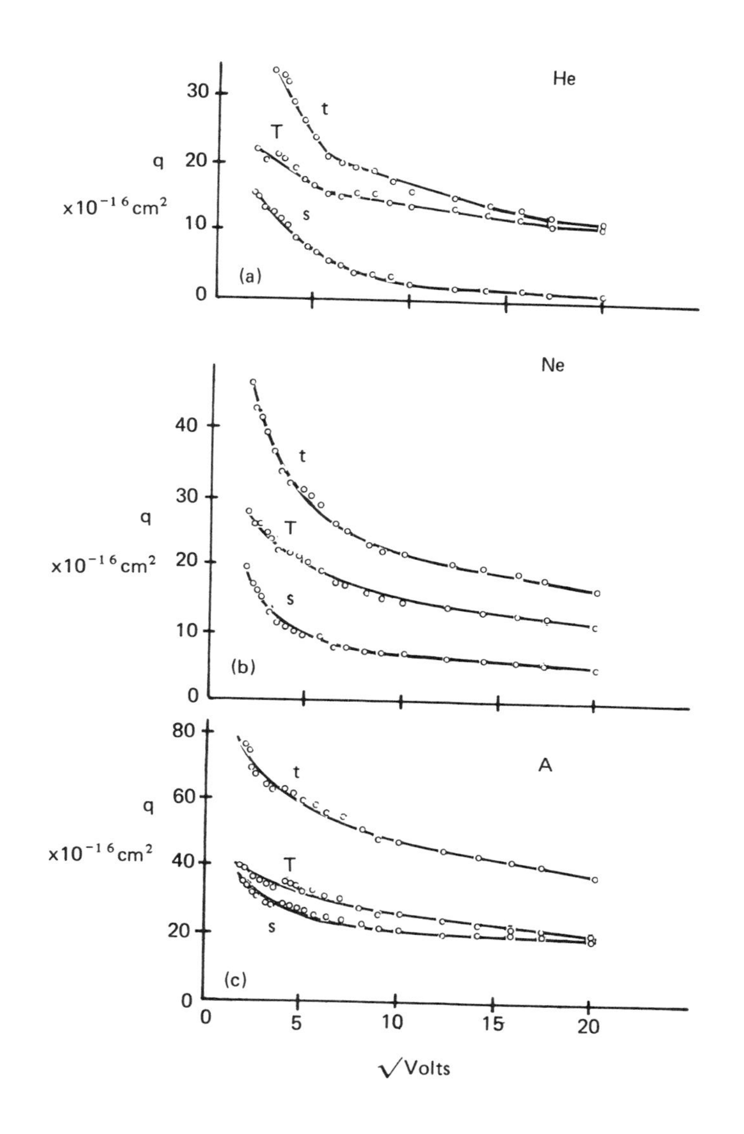

Figure 2-24. Experimental values for scattering cross-sections of ions in He, Ne and Ar; q is in units of 10^{-16} cm^2. The symbol s refers to elastic scattering, T to charge transfer, and t to the sum of s and T.
(a) He^+ on He; Cramer and Simons 1957
(b) Ne^+ on Ne; Cramer 1958.
(c) Ar^+ on Ar; Cramer 1959.

Ionization By Ion Impact

Just as ionization can be produced by photon bombardment, so it can also be produced by fast ion or fast atom bombardment, provided the incident particles have enough energy. Figure 2-25 shows the energy dependence of the ionization cross-section of several ions in their parent gases, whilst Figure 2-26 shows corresponding values for fast neutral atoms. Note that the cross-sections are of the order of 10^{-16} cm^2 (see also McDaniel 1964, p. 276). We shall consider the potential significance of these processes in glow discharges in Chapter 4.

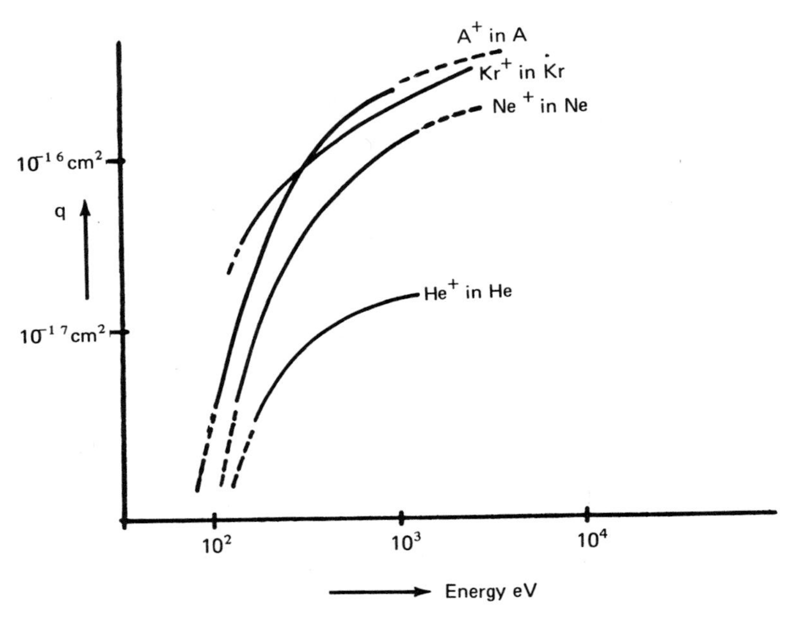

Figure 2-25. Ionization cross-sections of ions in their parent gases. Actually these values are too high because the ionizing effects of the secondary electrons produced have not been subtracted. The curve for Ar^+ is uncertain. (From von Engel 1965; includes data from Rostagni 1934, 1938 and Wien 1927)

Ion Chemistry

The subject of ion-atom interactions is also the basis of *ion chemistry*. I am not qualified to discuss this subject. It is interesting how atoms can completely change their chemical nature by becoming ionized. Argon, for example, is a noble unreactive gas because it has a closed outer shell configuration. However, on being ionized, it loses an electron and acquires the electron shell configuration of chlorine, becoming similarly reactive. This is presumably why complex ions such as ArH^+ are observed in sputtering glow discharges (Coburn and Kay 1971), being the ion equivalent of HCl.

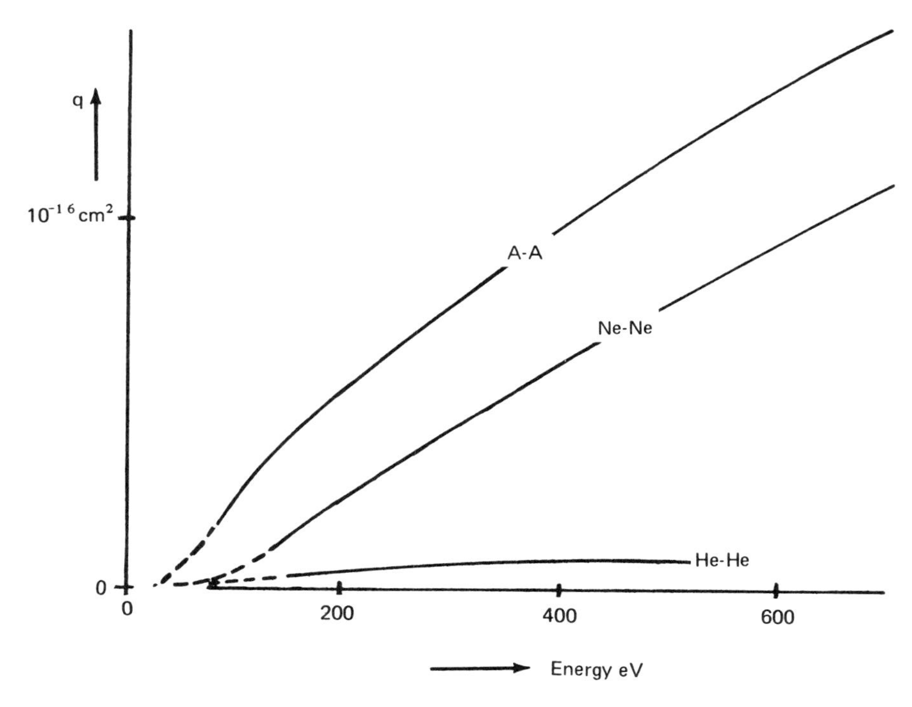

Figure 2-26. Ionization cross-section of fast atoms in their parent gas. Again secondary electron effects should be accounted for. (From von Engel 1965, including data from Rostagni 1934, 1938)

Metastable Collisions

The probability of collisions involving excited atoms depends on the density of the excited atoms, and hence on their lifetime. Some excited atoms have very long lifetimes (1 mS - 1 s, cf. 1nS for resonant states) and these are known as *metastable* excited atoms; they arise because the selection rules forbid relaxation to the ground state, or in practice, make such a transition rather unlikely. All of the noble gases have metastable states; argon has metastables at 11.5 eV and 11.7 eV.

Metastable – Neutral Collisions

When a metastable atom collides with a neutral, the neutral can become ionized if its ionization energy is less than the excitation energy of the excited atom:

$$A^* + G \rightarrow G^+ + A + e$$

This is known as *Penning Ionization*.

Coburn and Kay (1971) demonstrated the Penning ionization effect by sputtering a europium oxide target containing a small amount of iron, in both neon and argon discharges. Eu^+ (eV_i = 5.7 eV) and Fe^+ (eV_i = 7.8 eV) were observed in both gases, but O^+ (eV_i = 13.6 eV) was observed only in the neon discharge. Argon metastables are at 11.5 eV and 11.7 eV, whereas neon metastables are at 16.6 eV and 16.7 eV.

Metastable – Metastable Ionization

Lounsbury (1969) has pointed out that two metastable argon atoms, each of energy 11.55 eV, have sufficient energy that their collision could result in the ionization (threshold 15.76 eV) of one of the pair:

$$Ar^* + Ar^* \rightarrow Ar + Ar^+ + e$$

Using emission spectroscopy techniques, Lounsbury concluded that this was not a major ionization mechanism in an rf discharge. The metastable density is probably $\sim 10^{10}$ cm^{-3} (Ekstein et al. 1975). Assuming that the mean free path for argon metastable encounters is the same as for argon ground state atoms of the same density, then there will be only about 10^{11} metastable-metastable encounters/cm^3, much less than the electron impact ionization rate in a self-sustained discharge.

Electron – Metastable Ionization

Just as a ground state atom can be ionized by electron impact, so can a metastable:

$$e + Ar^* \rightarrow Ar^+ + e + e$$

The major difference between the ionization of a metastable and of a ground state is that the threshold for the latter is 15.76 eV, whilst for the former it would be only 4.21 eV, assuming an 11.55 eV metastable. So although there will usually be many fewer metastables than ground state atoms in our glow discharges, there will be many more electrons capable of ionizing the metastables than of ionizing the ground state atoms. Unfortunately, there seems to be little known about the ionization cross-section of the metastables. We shall return to consider their contribution to glow discharges in Chapter 4.

TOTAL COLLISION CROSS-SECTION

As we have seen, an electron travelling through a gas may take part in several processes: elastic scattering, excitation, ionization, recombination, attachment. There is a certain probability for each of these processes to take place, expressed as a collision cross-section. But since each cross-section is a probability, then the chance of *any* one of these processes taking place is just a sum of the individual probabilities. So there is a *total* collision cross-section which is the sum of the individual cross-sections, and is a measure of, for example, an electron being scattered elastically or inelastically. Figure 2-27 shows the total collision cross-sections for electrons in the noble gases. For electrons of 16 eV in argon, just above the ionization threshold, the total cross-section is about 23 πa_0^2, i.e. about 2 10^{-15} cm^2. So, in argon at 50 mtorr (n = 1.8 10^{15} atoms/cc), the probability of these electrons being scattered is about 0.4 per mm path length.

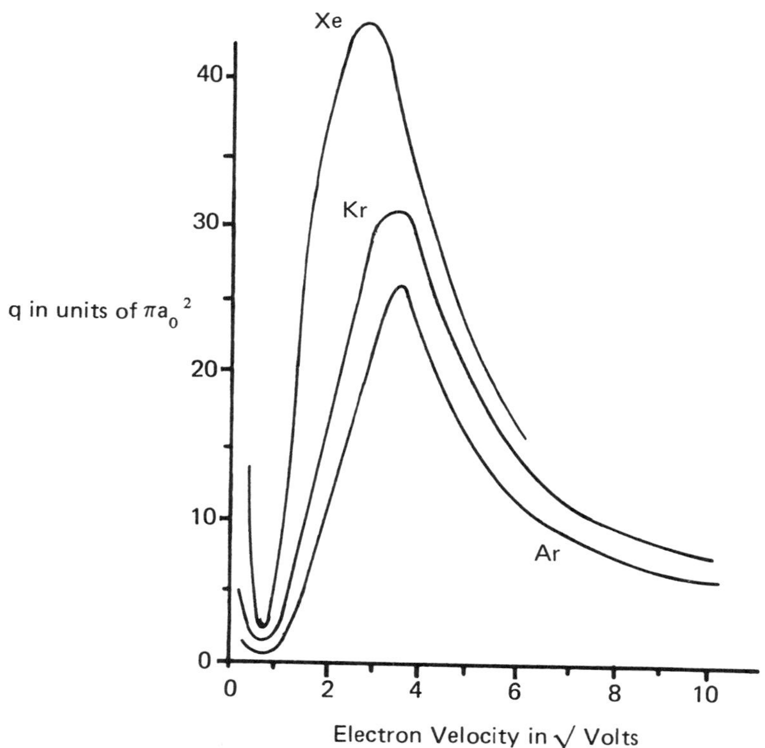

Figure 2-27. Total collision cross-section for electrons in the noble gases (Massey and Burhop 1969)

PLASMA

We have now reached the stage where we can begin to consider a simple plasma consisting of positive ions and negative electrons in a sea of neutral atoms (Figure 2-28). Ion-electron pairs are continuously created by ionization and destroyed by recombination, as we have seen above. Since these processes are always pairwise, the space occupied remains charge neutral. The electron impact process which leads to ionization is also likely to lead to excitation and the subsequent relaxation of the atom leading to photon emission is another common feature of the plasma. In the next chapter, we shall investigate some of the properties of this plasma.

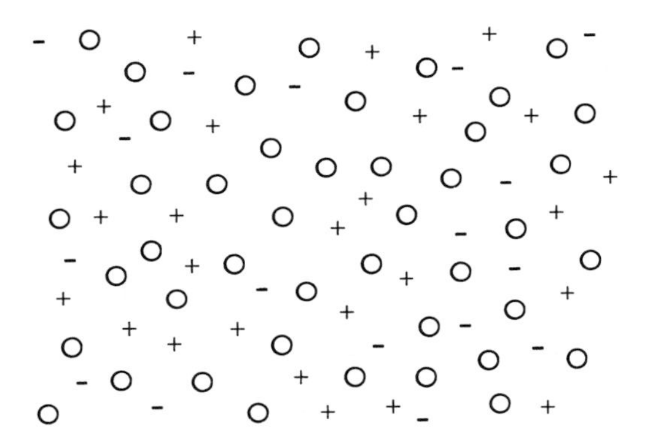

O Neutral Atoms
- Negative Electrons
+ Positive Ions

Figure 2-28. Plasma

REFERENCES

S. C. Brown, *Introduction To Electrical Discharges In Gases*, Wiley, New York and London (1966)

J. W. Coburn and E. Kay, Appl. Phys. Letters **18**, 10, 435 (1971)

W. H. Cramer and J. M. Simons, J. Chem. Phys. **26**, 1272 (1957)

W. H. Cramer, J. Chem. Phys. **28**, 688 (1958)

W. H. Cramer, J. Chem. Phys. **30**, 641 (1959)

R. D. DuBois and M. E. Rudd, J. Phys. B: Atom. Molec. Phys. **8**, 9, 1474 (1975)

A. von Engel, *Ionized Gases*, Oxford University Press, London and New York (1965)

E. W. Ekstein, J. W. Coburn, and Eric Kay, Int. Jnl. of Mass Spec. and Ion Physics **17**, 129 (1975)

J. Fletcher and I. R. Cowling, private communication (1973)

L. S. Frost and A. V. Phelps, Phys. Rev. **127**, 1621 (1962)

W. R. Harshbarger, R. A. Porter, T. A. Miller, and P. Norton, Applied Spectroscopy **31**, 3, 201 (1977)

W. M. Hickam and R. E. Fox, J. Chem. Phys. **25**, 642 (1956)

H. S. W. Massey and E. H. S. Burhop, *Electronic and Ionic Impact Phenomena*, Vol. 1, Oxford University Press, London and New York (1969)

H. S. W. Massey, *Electronic and Ionic Impact Phenomena*, Vol. II, Oxford University Press, London and New York (1969)

E. W. McDaniel, *Collision Phenomena in Ionized Gases*, Wiley, New York and London (1964)

D. Rapp and P. Englander-Golden, J. Chem. Phys. **43**, 5, 1464 (1965)

A. Rostagni, Nuovo Cim. **11**, 34 (1934)

A. Rostagni, *ibid*. **15**, 2 (1938)

B. L. Schram, F. J. de Heer, M. J. van der Wiel, and J. Kistemaker, Physica **31**, 94 (1965)

P. T. Smith, Phys. Rev. **36**, 1293 (1930)

R. F. Stebbings, B. R. Turner, and A. C. H. Smith, J. Chem. Phys. **38**, 2277 (1963)

G. L. Weissler, *Handbuch der Physik* **21**, 304, Springer, Berlin (1956)

W. Wien, *Handb. Exp. Phys*. **14**, Akad. Verlag Leipzig (1927)

I. P. Zapesochnyi and P. V. Feltsan, Optics & Spectroscopy **20**, 291 (1966)

Chapter 3. Plasmas

In Chapter 2, we introduced the idea of a *plasma* as a partially ionized gas consisting of equal numbers of positive and negative charges, and a different number of un-ionized neutral molecules. As we progress through this chapter, further requirements will be made of the gas in order to qualify it as a plasma.

In the type of plasmas discussed in this book, the degree of ionization is typically only 10^{-4}, so the gas consists mostly of neutrals. Although the Coulomb interaction between charges is both strong and long-range, it is possible to assume for an undisturbed plasma that the charges move around as free particles, since the sum of all the interactions tends to cancel, analogous to the role of a conduction electron in a solid. But also, to pursue the analogy, there are situations where the Coulomb interaction becomes dominant, as for example when the plasma is perturbed.

ELECTRON AND ION TEMPERATURES

To simplify, assume that the charged particles are singly charged positive ions and electrons. In addition, descriptions of the plasma will be made in the context of the glow discharge processes being considered. The essential mechanisms in the plasma are excitation and relaxation, ionization and recombination. To maintain a steady state of electron and ion densities, the recombination process must be balanced by an ionization process, i.e. an external energy source is required. In practice, that energy source is an electric field, which can act directly on the charged particles only. Let m_e and m_i be the masses of the electron and the ion respectively. Consider an electric field $\mathcal{E}$ acting on an initially stationary ion. The work done by the electric field, and hence the energy transferred to the ion, will be $\mathcal{E}ex$ where x is the distance travelled in time t. But:

$$x = \frac{1}{2}ft^2$$

where f is the acceleration due to the field (Figure 3-1), given by

$$\mathcal{E}e = m_i f$$

hence:

$$\text{Work done} = \mathcal{E}ex = \mathcal{E}e\frac{1}{2}\frac{\mathcal{E}e}{m_i}t^2 = \frac{(\mathcal{E}et)^2}{2m_i}$$

A similar relationship holds for the electrons, but since $m_i \gg m_e$, the action of the field is primarily to give energy to the electrons. The argument above ignored collisions, and we can always choose t to be short enough that this is so. But we have seen that, in general, collisions abound in plasmas. Electrons collide with neutral atoms and ions, but only a very small energy transfer to the heavy particle can take place; this is what the energy transfer function (q.v.) was about. In turn, the neutral atoms and ions share their energy efficiently in collision processes and likewise lose energy to the walls of the chamber. The net result is that electrons can have a high average kinetic energy, which might typically be 2 - 8 eV. The ions, which can absorb just a little energy directly from the electric field, have an average energy not much higher than that of the neutral molecules, which gain energy above the ambient only by collisions with ions (effectively) and electrons (ineffectively) and remain essentially at room temperature. We saw earlier that for the neutral gas atoms:

$$\frac{1}{2}m\overline{c^2} = \frac{3}{2}kT$$

The average energy is characterized by the kT term and although this would conventionally be measured in ergs, it is more convenient here to work in electron volts. It is useful to remember that kT has a value of 1/40 eV at 290 K,

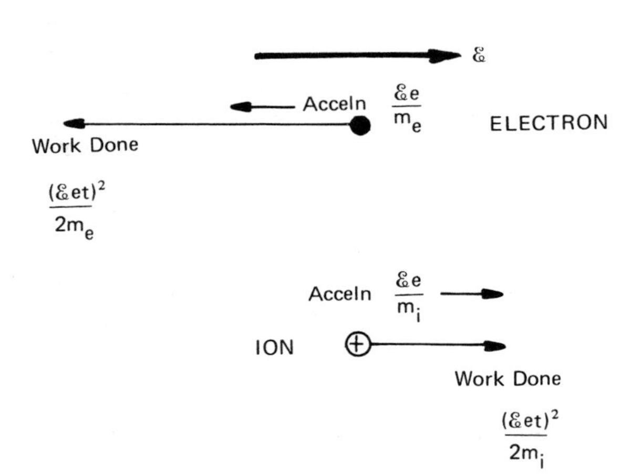

Figure 3-1. Energy transfer from the field to the electrons and ions

i.e. about room temperature. The concept of temperature applies to a random, i.e. Maxwell-Boltzmann, distribution. Can we apply this concept, then, to the energetic electrons? Based on an expectation of a large number of electron-electron collisions and other interactions, and very efficient energy sharing amongst the electrons because the energy transfer function takes all values between 0 and 1 for equal mass particles as the impact angle θ varies, a Maxwell-Boltzmann distribution seems quite reasonable. We assume this now and consider it again later. Since

$$\frac{1}{2} m_e \overline{c_e^2} = \frac{3}{2} kT_e$$

applies to electrons too, we can associate an effective temperature T_e with the electron motion. Measurements on glow discharge plasmas yield average electron energies around 2eV, which corresponds to an electron temperature of 23200 K! That doesn't mean that the containing vessel will melt, and that is because the heat capacity of the electrons is too small; we just have to think more carefully about the temperature concept. Since the ions are able to receive some energy from the external electric field, their temperature is somewhat above ambient; 500 K is representative.

PLASMA POTENTIAL

So three sets of particles exist in the plasma – ions, electrons, and neutrals – varying by mass and temperature. In addition, we saw in Chapter 1 that $\bar{c} = (8kT/\pi m)^{1/2}$, as in indicated for the typical parameters shown in Figure 3-2 based on argon. The *electron density* and *ion density* are equal (on average); this number, which is much less than the density of neutrals, is often known as the *plasma density*. The average speed of the electrons is enormous compared with those of the ions and neutrals, due to both the high temperature and low mass of the electrons.

Suppose we suspend a small electrically isolated substrate into the plasma. Initially it will be struck by electrons and ions with charge fluxes, i.e. current densities, predicted in Chapter 1 to be (Figure 3-3):

$$j_e = \frac{e n_e \overline{c_e}}{4}$$

$$j_i = \frac{e n_i \overline{c_i}}{4}$$

But $\overline{c_e}$ is much larger than $\overline{c_i}$. For the values shown in Figure 3-2,

$$j_e \sim 38 \text{ mA/cm}^2$$

$$j_i \sim 21 \ \mu\text{A/cm}^2$$

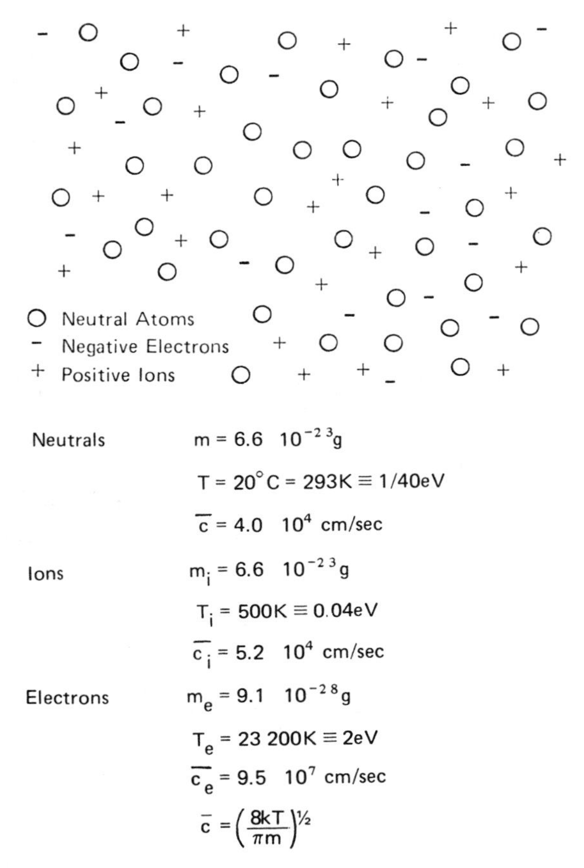

Figure 3-2. Typical parameter values for a glow discharge plasma

Since $j_e \gg j_i$, then the substrate immediately starts to build a negative charge and hence negative potential with respect to the plasma. Immediately the quasi-random motions of the ions and electrons in the region of our object, are disturbed. Since the substrate charges negatively, electrons are repelled and ions are attracted. Thus the electron flux decreases, but the object continues to charge negatively until the electron flux is reduced by repulsion just enough to balance the ion flux. We shall show shortly ("Debye Shielding") that the plasma is virtually electric field free, except around perturbations such as above, and so is equipotential. Let's call this potential the *plasma potential* V_p, also sometimes known as the *space potential*. Similarly, we can associate a *floating potential* V_f

with the isolated substrate. [In the case of a plasma container having insulating walls, these walls also require zero steady state net flux, so that *wall potential* and floating potential are related terms.] Since V_f is such as to repel electrons, then $V_f < V_p$. In the absence of a reference, only the potential difference $V_p - V_f$ is meaningful. Because of the charging of the substrate, it is as though a potential energy 'hill' develops in front of the substrate (Figure 3-4). However, it is a downhill journey for ions from the plasma to the substrate, but uphill for the electrons, so that only those electrons with enough initial kinetic energy make it to the 'top', i.e. the substrate.

SHEATH FORMATION AT A FLOATING SUBSTRATE

Since electrons are repelled by the potential difference $V_p - V_f$, it follows that the isolated substrate (assumed planar for simplicity in Figure 3-3) will acquire a net positive charge around it. This is generally known as a *space charge* and, in the context of glow discharge plasmas, forms a *sheath*. The sheath has a certain density of charges, known as the *space charge density* ρ. Poisson's equation relates variation of potential V with distance x across regions of net space charge:

$$\frac{d^2V}{dx^2} = -\frac{\rho}{\epsilon_0}$$

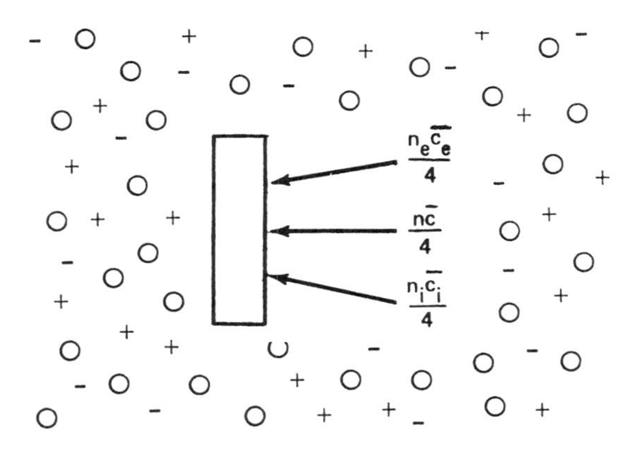

Figure 3-3. Initial particle fluxes at the substrate

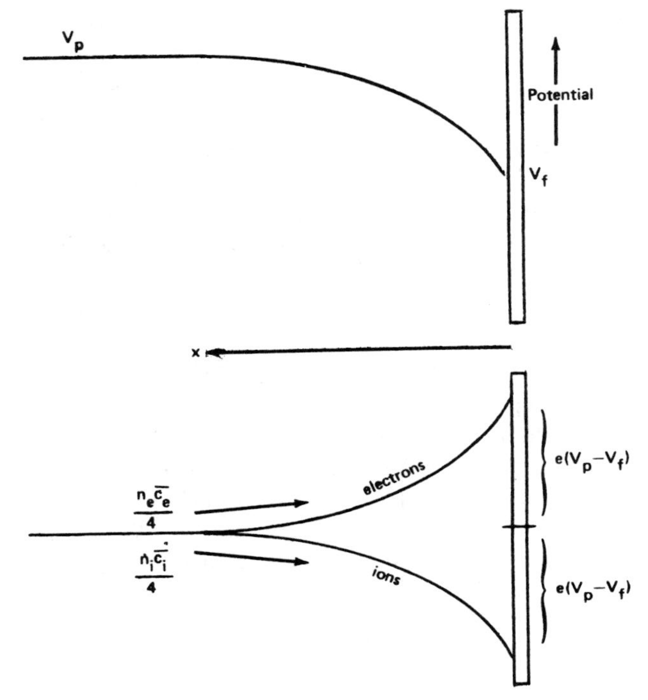

Figure 3-4. Variation of the electrical potential (upper) and of the potential energies of electrons and positive ions (lower), in the vicinity of an electrically floating substrate

This is the one-dimensional form for MKS units, where ϵ_0 is the permittivity of free space. Since electric field $\mathcal{E}$ is given by

$$\mathcal{E} = -\frac{dV}{dx}$$

then

$$\frac{d\mathcal{E}}{dx} = \frac{\rho}{\epsilon_0}$$

and this just says that the electric field across a gap changes as we go through regions of net charge, consistent with experience.

If the sheath acquires a net positive charge, it follows that the electron density decreases in the sheath – we shall obtain a quantitative expression for the decrease below. But one of the obvious features of a discharge is that it glows, and as we have already seen, this is due to the relaxation of atoms excited by electron impact. So the glow intensity depends on the number density and energy of the exciting electrons. Since the electron density is lower in the sheath, it doesn't glow as much. So we can actually see the sheath as an area of lower luminosity than the glow itself – the substrate is surrounded by a (comparatively) *dark space*, a feature common to the sheaths formed around all objects in contact with the plasma, even though the sheath thicknesses may vary greatly.

Let us now try to get an idea of the magnitude of $V_p - V_f$, which represents a barrier to electrons. To surmount this barrier, an electron must acquire $e(V_p - V_f)$ of potential energy (Figure 3-5). Hence, only electrons that enter the sheath from the plasma with kinetic energies in excess of $e(V_p - V_f)$, will reach the substrate. The Maxwell-Boltzmann distribution function tells us that the fraction n_e'/n_e that can do this is:

$$\frac{n_e'}{n_e} = \exp - \frac{e(V_p - V_f)}{kT_e}$$

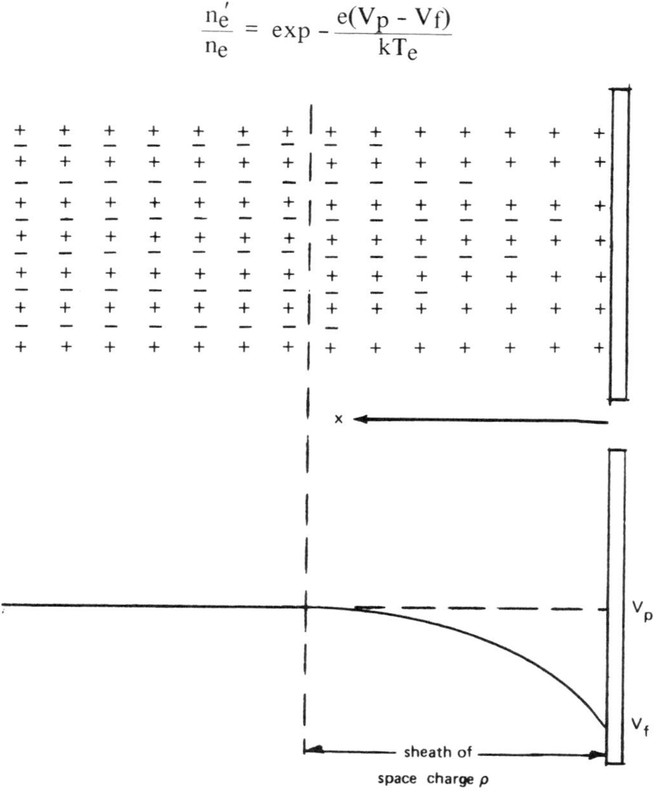

Figure 3-5. A space charge sheath develops in front of a floating substrate (upper), and establishes a sheath voltage (lower)

If the density n_e' just achieves charge flux balance at the object, then

$$\frac{n_e' \overline{c_e'}}{4} = \frac{n_i \overline{c_i}}{4}$$

One might at first think that the n_e' electrons close to the substrate would have a lower mean speed $\overline{c_e'}$ than the n_e electrons in the plasma, since the n_e' electrons suffer an $e(V_p - V_f)$ loss of kinetic energy in crossing the sheath. However, one must also bear in mind that the n_e' electrons that reach the substrate were not 'average' electrons, but had energies greater than average. In fact, the average energies of the n_e and n_e' groups of electrons are the same, i.e. they are at the same temperature. This can be shown from the Maxwell-Boltzmann distribution which, in a region of potential energy $e\phi$, becomes:

$$dn_e' = 4\pi n_e \left(\frac{m_e}{2\pi kT_e}\right)^{3/2} c_e^2 \exp -\frac{(\frac{1}{2}m_e c_e^2 + e\phi)}{kT_e} \quad dc_e$$

$$= \exp\left(-\frac{e\phi}{kT_e}\right) dn_e$$

$$\therefore \quad \overline{c_e'^2} = \frac{\int c_e^2 \exp\left(-\frac{e\phi}{kT_e}\right) dn_e}{\int \exp\left(-\frac{e\phi}{kT_e}\right) dn_e} = \overline{c_e^2}$$

Furthermore, by integration

$$n_e' = n_e \exp\left(-\frac{e\phi}{kT_e}\right) = n_e \exp -\frac{e(V_p - V_f)}{kT_e}$$

Returning to the charge flux balance equation, and substituting for n_e' and $\overline{c_e'}$, then

$$n_e \exp -\frac{e(V_p - V_f)}{kT_e}\Bigg) \frac{\overline{c_e}}{4} = \frac{n_i \overline{c_i}}{4}$$

But $n_e = n_i$ and $\overline{c} = \left(\frac{8kT}{\pi m}\right)^{1/2}$ (Chapter 1, "Mean Speed $\overline{c}$"), and so charge balance requires

$$V_p - V_f = \frac{kT_e}{e} \ln \frac{\overline{c_e}}{\overline{c_i}}$$

$$= \frac{kT_e}{2e} \ln\left(\frac{m_i T_e}{m_e T_i}\right)$$

(When we have learned a little more about sheath formation, in "Sheath Formation and The Bohm Criterion", we shall need to modify this result slightly.)

In our example (Figure 3-2), (V_p - V_f) should have a value of +15 volts, which is of the right order to agree with observation. Note the polarity, which is to make the plasma positive with respect to the floating object, and indeed, positive with respect to almost everything. The rapid motion of the electrons, relative to the ions, means they can easily move away from the plasma. But in doing so they lease the plasma more positive which hinders the escape of the negative electrons and makes the process self-limiting.

Since the charging of the floating substrate serves to repel electrons, it also . attracts positive ions. This does not increase the flux of ions, which is limited by the random arrival of ions at the sheath-plasma interface – in terms of the model in Figure 3-4, it doesn't matter how steep or high the hill is (this isn't quite true, as will be discussed later). However, the voltage across the sheath does directly influence the energy with which the ion strikes the substrate. The ion enters the sheath with very low energy. It is then accelerated by the sheath voltage, and, in the absence of collisions in the sheath, would strike the substrate with a kinetic energy equivalent to the sheath voltage.

In practice, the sheath above an electrically isolated substrate varies from one or two volts upwards. The resulting kinetic energies must be compared with interatomic binding energies in a thin film or substrate of typically 1 - 10 eV, so that it is easy to imagine that a growing thin film or an etching process on an electrically isolated surface in the plasma might be much affected by such impact.

DEBYE SHIELDING

If the numbers of ions and electrons in the plasma are equal and very large, then it is not surprising that their net Coulomb interaction with a particular charge sums to zero. But although this must be true on the average, we might expect that the instantaneous potential at a point due to some disturbance is both non-zero and time dependent. Let's consider this case (in 1 dimension, for simplicity) by assuming that the potential at $x = 0$ is ΔV_o (measured relative to the plasma), and then see how the potential $\Delta V(x)$ varies with x (Figure 3-6). In thinking about the problem assume that ΔV_o is less than V_p, i.e. more negative. Then a net positive space charge will form in front of the charged surface, since only energetic electrons can enter, as in the previous example. To a first approximation, the ion density in the sheath will be n_i as in the undisturbed plasma, since the ions are too massive to react rapidly to the space charge. This would not be true if the potential ΔV_o were maintained for a long time, but the random fluctuations that cause ΔV_o often happen on a very short time scale. And even when the potential perturbation is semi-permanent, this only serves to make n_i dependent on x, changing the argument in detail only.

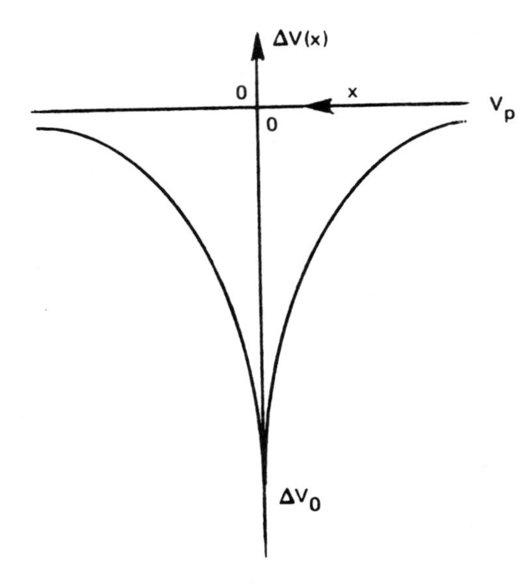

Figure 3-6. Variation of potential around a perturbation

If the electron density varies as $n_e(x)$, then Poisson's equation becomes:

$$\frac{d^2 V}{dx^2} = -\frac{e}{\epsilon_o}(n_i - n_e(x))$$

This is actually the MKS units version of Poisson's equation, where ϵ_o is the permittivity of free space, having the value $1/(36\pi\ 10^9)$ farads/metre. The purpose of using the MKS form is to avoid the cgs units of charge for the electron, and to avoid confusing myself; I always find the wrong answers when I use other units for this calculation!

Using the Boltzmann relation again:

$$\frac{n_e(x)}{n_e} = \exp -\frac{e\,\Delta V(x)}{kT_e}$$

Substituting into Poisson's equation, and remembering that $n_e = n_i$ in the undisturbed plasma, we obtain

$$\frac{d^2 V}{dx^2} = -\frac{en_i}{\epsilon_o}\left(1 - \exp -\frac{e\Delta V(x)}{kT_e}\right)$$

There is a difficulty here in that this equation can be simplified and solved when $\Delta V(x) \ll kT_e$. But in many cases that arise, this inequality does not hold for all x. However, if one solves the equation more exactly (Mitchner and Kruger 1973,

p. 133), and then numerically compares the resulting solution with that obtained using the inequality above, then apparently there is good agreement; this is because the major variation of $\Delta V(x)$ is near $x = 0$, where it varies very rapidly. So using the inequality $\Delta V(x) \ll kT_e$ to expand the exponential, then

$$\frac{d^2 V}{dx^2} \simeq \frac{e^2 n_i}{kT_e \epsilon_o} \Delta V(x)$$

This approximate differential equation has a solution (also approximate):

$$\Delta V(x) = \Delta V_o \exp - \frac{|x|}{\lambda_D}$$

where

$$\lambda_D = \left(\frac{kT_e \epsilon_o}{n_e e^2} \right)^{1/2}$$

This quantity λ_D has the dimensions of a length, and is known as the *Debye length.* The spatial dependence of $\Delta V(x)$ tells us that if the potential in the plasma is perturbed, then the plasma reacts to oppose that change. The Debye length tells us how rapidly the potential perturbation is attenuated in the plasma; over a distance λ_D the perturbation is reduced to 0.37 (1/e) of its initial value. For the example we have chosen ($n_i = n_e = 10^{10}/cm^3$ and $kT_e = 2$ eV), λ_D has the value $1.05\ 10^{-2}$ cm, or 105 μm.

Another way of regarding the Debye length concept is to say that, from the perspective of a particular charge at a particular point in the plasma, we need to consider the sum of the individual interactions with all of the other charged particles contained within a sphere centered on the particular point having a radius of 1 or 2 Debye lengths. Outside of this sphere, the detailed nature of the interaction becomes immaterial and the net interaction is zero. Hence, the unperturbed plasma is equipotential except for small fluctuating voltages which are attenuated over distances of the order of the Debye length.

One of the requirements for a collection of charged particles to be considered a plasma is that the range of these microfields must be very small on the scale of the total dimension of the plasma, i.e. $\lambda_D \ll d$ where d is the characteristic diameter of the discharge.

A similar argument to that used above could be made for the case where ΔV_o is imposed on a conducting element in the plasma, by an external source, e.g. a battery. If ΔV_o is dc or low frequency ac, then the ions around the object do have an opportunity to respond to the applied field, and n_i becomes a function of x. Nevertheless, the basis of the argument is the same, and we again come to the conclusion that the plasma attenuates voltage perturbations by forming a sheath, leaving the undisturbed region, i.e. the plasma itself, equipotential.

These screening phenomena also have a bearing on our initial assumptions about treating the plasma as a collection of three quasi-independent ideal gases. Since the plasma is equipotential, then it is also electric-field-free, so none of the constituent charged particles is subject to any externally imposed fields, except to the extent that the plasma will respond to any further applied voltages by forming a screening sheath around the relevant electrode. So the charge assembly does exhibit *collective behaviour*, a necessary criterion for its classification as a plasma. And even within the plasma, the individual charged particle interactions are important over the range of a few Debye lengths, and there is certainly a considerable and continuous energy interchange amongst the gas species; hence the gases are only quasi-independent.

Finally, we must note that the Debye shielding effect is not complete. A screening charge cloud forms around a voltage perturbation, but the resulting electric field becomes weak towards the edge of the cloud. As soon as the electrostatic potential reduces to the thermal energy of the electrons and ions, then they can escape from the charge cloud. So we come to the conclusion that the edge of the cloud is where $\Delta V \sim kT_e$, and that voltages $\sim kT_e/e$ can penetrate into the plasma. We shall see an effect of this later, in "Sheath Formation and The Bohm Criterion". Note that assuming the edge of the shielding charge cloud is where $\Delta V \sim kT_e$, contradicts the earlier assumption made in the derivation of Debye length that $\Delta V \ll kT_e$, which means that one has to be cautious in using the Debye screening length concept.

PROBE CHARACTERISTICS

Let us return to the simple plasma of Figure 3-2. Previously we considered what would happen to an electrically isolated probe placed in the plasma. Now let us pursue further what happens when that probe is maintained at a potential V set by an external power source (Figure 3-7). To make the situation more realistic, introduce a conducting wall at ground potential (0 V) to act as a reference voltage and as a return current path. The plasma potential V_p is then defined with respect to ground. The random fluxes in the plasma are $n_e\overline{c_e}/4$, and $n_i\overline{c_i}/4$ for electrons and ions respectively. We have already seen that the net flux, and hence net current, would be zero when the probe acquires a potential V_f, the floating potential. So we can begin to plot a curve of probe current density versus probe voltage (Figure 3-8). By biasing the probe negatively with respect to V_p, some electrons are prevented from reaching the probe, but the ion current density j_i remains at a value dictated by the arrival rate of ions at the edge of the sheath, and this is limited to the random flux in the discharge, i.e. $n_i\overline{c_i}/4$. If V is made very negative with respect to V_p, then the electron current would be completely suppressed. The saturation current density for negative V is then

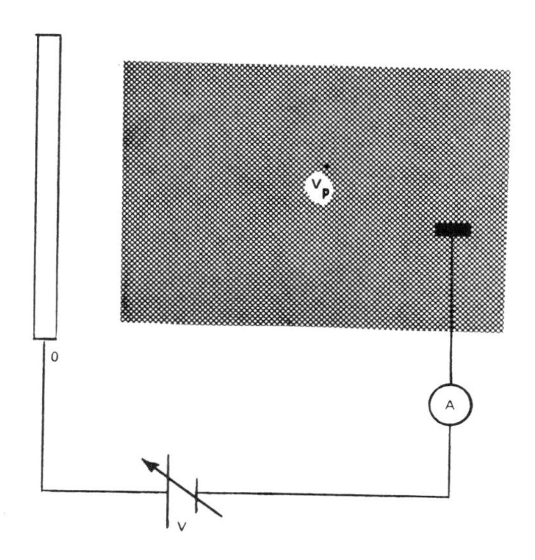

Figure 3-7. Schematic for probe measurements in a plasma

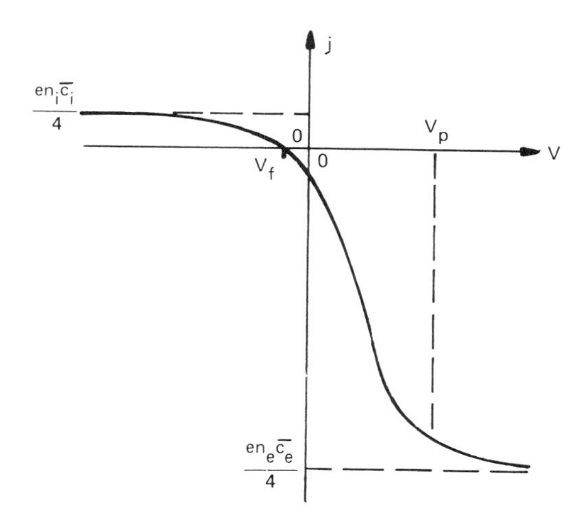

Figure 3-8. Current density-voltage characteristic of a probe

just $en_i\bar{c_i}/4$. From our earlier considerations ("Sheath Formation at a Floating Substrate"), the electron current density j_e to the probe at voltage V should follow the form:

$$j_e = \frac{en_e\bar{c_e}}{4} \exp\left(-\frac{e(V_p - V)}{kT_e}\right)$$

and hence

$$\ell n\, j_e = \ell n \frac{en_e\bar{c_e}}{4} - \frac{e(V_p - V)}{kT_e}$$

This expression is derived on the assumption that the electrons have a Maxwellian energy distribution, and it predicts that $\ell n\, j_e$ is linearly dependent on $(V_p - V)$. This prediction is substantiated by experimental results (Figure 3-9), adding credence to our initial assumption ("Electron and Ion Temperatures") that the electrons do indeed have a Maxwellian energy distribution.

The net current density to the probe, for $V < V_p$, is just the sum of j_i and j_e:

$$j = \frac{en_i\bar{c_i}}{4} - \frac{en_e\bar{c_e}}{4} \exp - \left(\frac{e(V_p - V)}{kT_e}\right)$$

By a similar argument, one would expect for $V > V_p$ that

$$j = \frac{en_i\bar{c_i}}{4} \exp\left(-\frac{e(V - V_p)}{kT_i}\right) - \frac{en_e\bar{c_e}}{4}$$

and also, since $T_i \ll T_e$, that the ion current term would rapidly go to zero as soon as V exceeds V_p, leaving the electron saturation current and a fairly well-defined V_p at the knee of the curve.

In principle, this probe technique, which was introduced by Irving Langmuir and colleagues in the 20's (Langmuir 1923, Langmuir and Mott-Smith 1924) and carries his name, should be able to give us quite simply all of the parameters of the plasma that we need to know – electron and ion temperatures, plasma density and plasma potential. But ....

Practical Complications

Unfortunately, the situation with real probe measurements is much more complex, for a variety of reasons. The effective current-collecting area of the probe is not its geometric surface area, but rather the area of the interface between the plasma and the sheath around the probe (Figure 3-10); and the thickness of the sheath, for a given plasma, is a function of the probe potential. This would not matter for a plane probe except that such a probe has ends where the problem arises again; and the relative contribution of the problem is increased because of the requirement that the probe be small, so that the probe current does not constitute a significant drain on the plasma. With a cylindrical probe, the varying sheath thickness is an even larger effect.

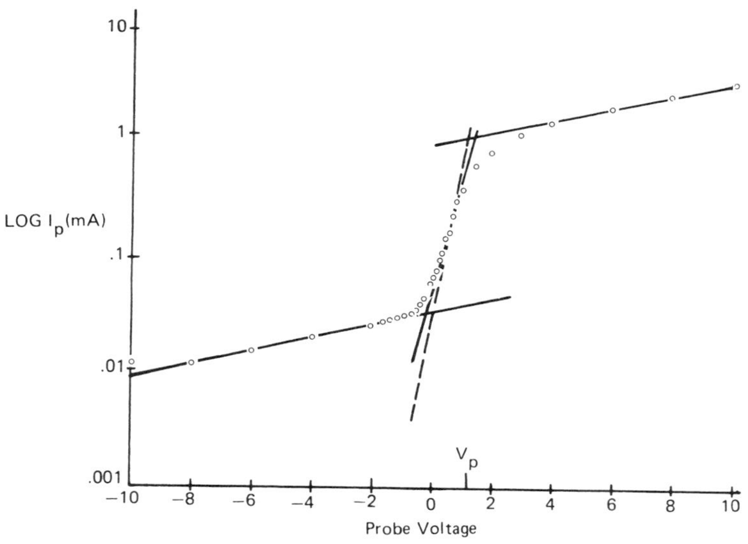

Figure 3-9. Typical probe characteristic showing a quasi-linear region where $\log j \propto V - V_S$ (Ball 1972). Tantalum target, 1000 cm². Argon discharge at 10 mtorr, 3kV and 59 mA.

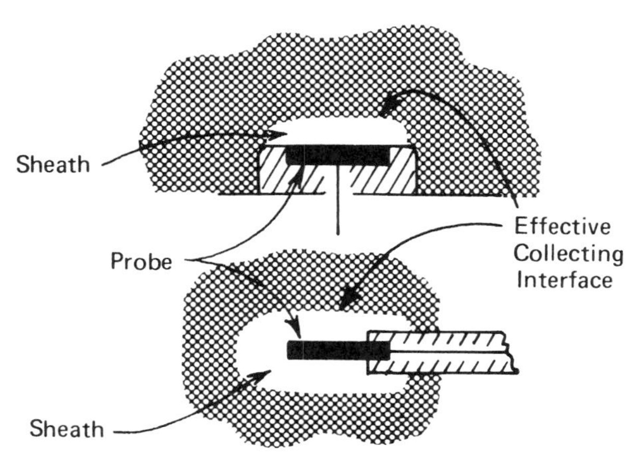

Figure 3-10. Effective current-collecting areas of probes

Two more complications are associated with additional charge generation. Secondary electrons (Chapter 4, "Secondary Electron Emission") may be generated at the probe due directly to the impact of ions, electrons, and photons or the the heating effects caused by such impact, giving rise to additional current flow; electron impact ionization may occur in the sheath, again enhancing current flow.

Yet one more problem concerns the tendency of charged particles to take up orbital paths around the probe, further influencing the probe characteristics. Even our assumption that the ion current density at the edge of the sheath is equal to the random density $en_i\bar{c}_i/4$, turns out to be incorrect, as we shall see later. And in the glow discharges used in sputtering and plasma etching, there are additional difficulties due to directed high energy electrons which flow through the plasma.

All of these effects, and others that exist, add considerable complexity to proper interpretation of probe data. The reader is referred to one of the many reviews of probe techniques, such as those of Chen (1965), Laframboise (1960), Swift and Schwar (1970), and Loeb (1961), or more recently to articles referring specifically to sputtering discharges by Clements (1978), Thornton (1978), and Eser et al. (1978).

Positively Biased Probes

Another probe effect is quite difficult to deal with: as soon as the probe potential approaches the plasma potential, the electron current density to the probe should approach the saturation value, $en_e\bar{c}_e/4$. But even with a tiny probe, the actual current drain can easily become a serious drain on the plasma, causing a significant perturbation, at least for glow discharge processes, which are of rather low density. This current drain can be limited by minimizing the size of the probe, but the following example shows that a *very* small probe is required. Use the typical plasma parameters shown in Figure 3-2 and a total current of 10 mA. Let us estimate a tolerable electron current drain of 1 mA. Since the random electron current density is 38 mA/cm^2 ("Plasma Potential"), 1 mA would be drawn by a collection area of 2.6 10^{-2} cm^2. Imagine a thin cylindrical wire probe 0.25 cm in length; such a collection area would correspond to a cylinder radius of 166 μm. But this radius corresponds to the sum of the probe and sheath radii (Figure 3-10) and the sheath itself is going to be ~ 1 Debye length, which alone is 105 μm for our example ("Debye Shielding")!

The effect of attempting to draw too much electron current from the plasma is illustrated in Figure 3-11 where the probe circuit of Figure 3-7 is redrawn along with the discharge circuit. The electron current to the probe is in addition to the electron current to the anode. So either the ion current to the cathode must increase or the electron current to the anode must decrease. Under normal circumstances where the probe circuit supplies very little power to the discharge, the latter dominates. A decrease in the electron current to the cathode is accomplished by an increase in the plasma potential, causing more electron retardation in the anode sheath. One arrives at the same result by arguing that the probe starts to drain the plasma of electrons, leaving it space charge positive so that the plasma potential has to rise; or by arguing that the probe becomes the new anode as soon as its potential exceeds that of the original anode and that the plasma potential is determined by the anode potential and the need to maintain current continuity in the circuit. Coburn and Kay (1972) have encountered just this difficulty of not being able to find a small enough probe for sputtering discharges, and, using an independent technique to determine plasma potential based on measuring the energy distribution of ions accelerated across a sheath, have found that application of positive probe voltages serves only to increase the plasma potential, in agreement with the above argument.

So we are left with the conclusion that, at least in the rather tenuous discharges of sputtering and plasma etching, the plasma potential will be the most positive potential in the system. This becomes increasingly true with increasing size of the perturbing electrode.

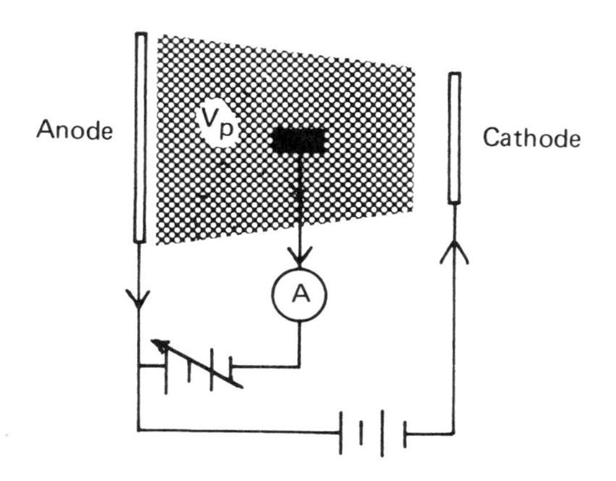

Figure 3-11. Schematic of probe and discharge circuits

SHEATH FORMATION AND THE BOHM CRITERION

Earlier in this chapter, in the section on "Plasma Potential", we calculated the random ion current density $n_i\bar{c}_i/4$ which flows in the plasma and found that it had a value of 21 μA/cm² for a typical plasma of density 10^{10}/cm³ and ion temperature 500K. We further reasoned, in "Sheath Formation at a Floating Substrate", that the ion current density to any object more negative than the plasma potential should be equal to the random ion current density. In that section, the substrate was electrically floating so that the net current flow was zero. However, it is a simple matter to extend the arguments given there to include the case where there is a net ion current to the object, and one would still expect to find a current density of 21 μA/cm². But if we measure the current density at the target in a dc sputtering glow discharge, as in Figure 4-1 of Chapter 4, we find that the current density is larger, of the order of a few tenths of a milliamp per square centimetre. Although we shall learn in the next chapter that some of this latter current is due to the emission of electrons from the target, there is apparently a discrepancy in these two values of current density. Although the ion temperature which we used to derive $\bar{c}_i$ was only an estimate, this estimate can't be far out, and anyway $\bar{c}_i$ varies as the square root of the ion temperature, which is rather a weak dependence. So the reason for the discrepancy must lie elsewhere.

The problem turns out to be due to an oversimplification of the model for the sheath. We had assumed that the sheath terminated at the plane where the ion and electron densities became equal, to become an undisturbed plasma again (Figure 3-5). In fact, between these two regions there is a quasi-neutral *transition region* of low electric field (Figure 3-12), and the effect of this region is to increase the velocity of ions entering the sheath proper. The existence of this velocity change was demonstrated by Bohm (1949) and the resulting criterion for sheath formation has come to be known as the *Bohm sheath criterion*, and is demonstrated as follows:

In Figure 3-12, we assume a monotonically decreasing potential V(x) as ions traverse the positive space charge sheath; x = 0 corresponds to the boundary between the two regions so that $n_i(0) = n_e(0)$, i.e. space charge neutrality at x = 0. We also assume that the sheath is collisionless and the consequent absence of ionization ensures that the ion current $e\, n_i(x)\, u(x)$ is constant.

Conservation of energy for the ions requires that

$$\tfrac{1}{2} m_i u(x)^2 = \tfrac{1}{2} m_i u(0)^2 - e[V(x) - V(0)]$$

$$\therefore \quad u(x) = \left(u(0)^2 - \frac{2e[V(x) - V(0)]}{m_i} \right)^{1/2}$$

and

$$n_i(x) = \frac{n_i(0)u(0)}{u(x)} = n_i(0)\left(1 - \frac{2e[V(x) - V(0)]}{m_i u(0)^2}\right)^{-1/2}$$

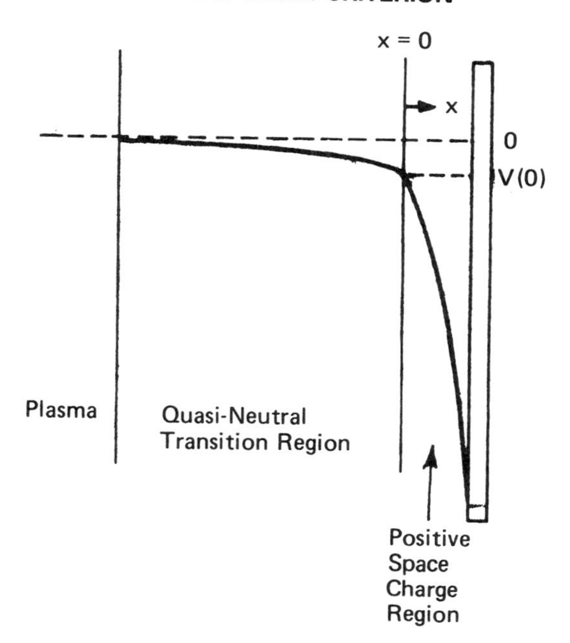

Figure 3-12. Potential variation near a negative electrode. Density $n_i(x)$ and potential $V(x)$ at $x \geq 0$

By the Boltzmann relation for the electrons

$$n_e(x) = n_e(0) \exp \frac{e[V(x) - V(0)]}{kT_e}$$

Poisson's equation is then

$$\frac{d^2\phi}{dx^2} = \frac{e}{\epsilon_o}(n_e(x) - n_i(x))$$

$$= en_e(0)\left(\exp \frac{e[V(x) - V(0)]}{kT_e} - \left(1 - \frac{2e[V(x) - V(0)]}{m_i u(0)^2}\right)^{-1/2}\right)$$

But if this is to be a positive space charge sheath, then d^2V/dx^2 must be negative for all $x > 0$ (and zero for $x = 0$)

i.e. $$\left(1 - \frac{2e[V(x) - V(0)]}{m_i u(0)^2}\right)^{-1/2} > \exp \frac{e[V(x) - V(0)]}{kT_e}$$

Squaring and inverting, then

$$\exp - \frac{2e[V(x) - V(0)]}{kT_e} > 1 - \frac{2e[V(x) - V(0)]}{m_i u(0)^2}$$

We now restrict out attention to the beginning of the space charge sheath where V(x) - V(0) is very small compared to kT_e so that we can expand and approximate the exponential thus:

$$1 - \frac{2e[V(x) - V(0)]}{kT_e} > 1 - \frac{2e[V(x) - V(0)]}{m_i u(0)^2}$$

i.e.

$$u(0) > \left(\frac{kT_e}{m_i}\right)^{1/2}$$

This says that the ion velocity on entering the sheath must be greater than $(kT_e/m_i)^{1/2}$, i.e. is determined by the electron temperature, which is a rather peculiar result and demonstrates how the ion and electron motions are coupled. Chen (1974) demonstrates (Figure 3-13) that the physical significance of the criterion is that the acceleration of ions in the sheath and repulsion of electrons there, both of which decrease the relevant particle volume densities, must be such that the ion density decreases less rapidly than the electron density across the sheath. This is equivalent to the requirement that $d^2 V/dx^2$ is negative, and it is clear from Figure 3-13 that this requirement is most stringent at the beginning of the sheath where V(x) - V(0) is very small, as we had assumed.

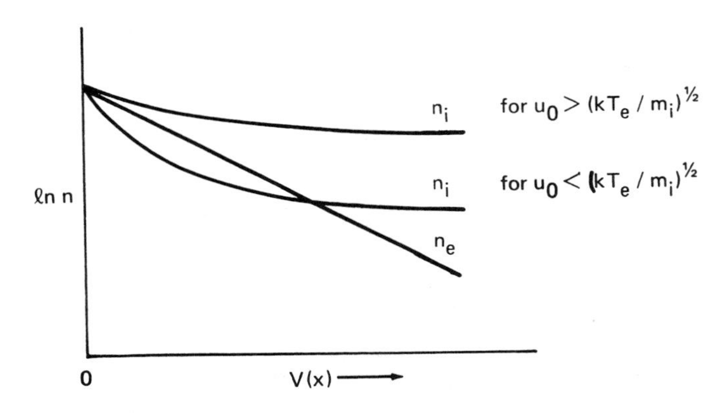

Figure 3-13. Variation of ion and electron density with potential V(x) in a sheath, for two cases: u_0 greater than and u_0 less than the critical velocity $(kT_e/m_i)^{1/2}$. From Chen 1974.

How do the ions acquire this velocity? There must be an electric field across the transition region so as to give the ions a directed velocity of u(0) towards the electrode. If we assume that the ion temperature is negligibly small so that the random motion of the ions can be neglected, then since the potential at the boundary is V(0) with respect to the plasma,

$$\tfrac{1}{2} m_i u(0)^2 = eV(0)$$

$$\therefore \quad V(0) = \frac{m_i u(0)^2}{2e} = \frac{m_i kT_e}{2e\, m_i} = \frac{kT_e}{2e}$$

The existence of a field in the transition region does not contradict our earlier claim that the plasma is equipotential, since that claim was qualified then to the extent that voltages of the order of kT_e/e could 'leak' into the plasma, and here we see an example of this.

We can pursue the exercise further to calculate the ion flux at the sheath boundary. Since the potential there is V(0) with respect to the plasma in which the electron density is n_e, then using the Boltzmann relation again,

$$\begin{aligned} n_e(0) &= n_e \exp - \frac{V(0)}{kT_e} \\ &= n_e \exp - \tfrac{1}{2} \\ &= 0.6\, n_e \end{aligned}$$

since $V(0) = kT_e/2$. But $n_e(0) = n_i(0)$, and so the ion flux is given by

$$n_i(0)\, u(0) = 0.6\, n_e \left(\frac{kT_e}{m_i}\right)^{1/2}.$$

Substituting in the values from Figure 3-2 again, we obtain an ion current density of 0.2 mA/cm², which is more like reality. However, this derivation is still not quite realistic since it assumes that the ion temperature is zero, which is never so; and that there are no collisions in the sheath, which is not true for the sheaths that form in front of our glow discharge cathodes, although it is reasonably true for the much thinner sheaths that form in front of low voltage anodes and probes. We also know that the cathode current depends on the cathode voltage in practice, and the ion current expression derived above does not explicitly include the electrode voltage except to the extent that the electrode voltage does control the electron temperature and plasma density, as we shall see when we explore how discharges are maintained, in the next chapter.

The Floating Potential – Again

The effect of the Bohm criterion is to increase the ion flux to any object negatively biased with respect to the plasma. In particular it will change the ion flux to a floating substrate. We had calculated the floating potential earlier in the chapter, and apparently we must now change this to allow for this changed ion flux. Using a similar derivation to before, the criterion for net zero current becomes:

$$\left(n_e \exp - \frac{e(V_p - V_f)}{kT_e}\right) \frac{c_e}{4} = n_i \, 0.6 \left(\frac{kT_e}{m_i}\right)^{1/2}$$

$$\therefore \quad V_p - V_f = - \frac{kT_e}{e} \, \ell n \, 2.4 \left(\frac{kT_e}{m_i}\right)^{1/2} \left(\frac{\pi m_e}{8kT_e}\right)^{1/2}$$

$$= \frac{kT_e}{2e} \, \ell n \left(\frac{m_i}{2.3 \, m_e}\right)$$

In our example (Figure 3-2), $V_p - V_f$ should have a value of 10.4 V compared with 15 V as derived earlier. The larger ion flux requires a larger electron flux for current neutrality, and so a smaller electron retarding potential. The logarithmic dependence minimizes the change in potential due to the increased ion flux.

PLASMA OSCILLATIONS

One might at first think that in a dc plasma, all parameters would be time independent. This is not the case. Although the electrons and ions are in equilibrium as a whole, this is only the average result of the many detailed interactions. If a plasma, or even a small section of it, is perturbed from neutrality for any reason, then there will be large restoring forces striving to re-establish charge neutrality. Because of the large mass difference between ions and electrons, it will be the electrons which will first respond to the restoring forces. We shall find that these restoring forces are proportional to displacement, which is just the condition for *oscillations*.

Electron Oscillations

The frequency of oscillation can be found in the following way. Consider a slab of plasma of thickness L and density n (Figure 3-14a) and then suppose that all the electrons are displaced a distance Δ along the x axis by some external force (Figure 3-14b). The regions between $x = 0$ and $x = L$, and for all $x < -\Delta$, will remain space charge neutral. However, the electrons between $x = -\Delta$ and $x = 0$ will give rise to a space charge there. By Poisson's equation, since the electron density is n,

$$\frac{d\mathcal{E}}{dx} = \frac{ne}{\epsilon_0}$$

Integrating,

$$\mathcal{E} = \frac{ne\Delta}{\epsilon_o}$$

But the action of the field $\mathcal{E}$ is to exert a restoring force on the electron:

$$m_e \frac{d\Delta}{dt} = -e\mathcal{E}$$

$$= -\frac{ne^2}{\epsilon_o}\Delta$$

When released, the inertia of the electrons will cause them to overshoot their original positions, and they will continue to describe motion determined by the same equation, where Δ now becomes a function of time.

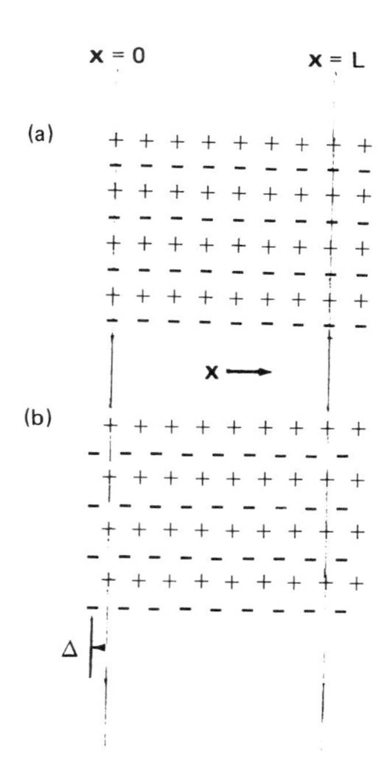

Figure 3-14. When a uniformly distributed plasma (a) is displaced (b), oscillations may result

This is just an equation of simple harmonic motion of angular frequency ω_e given by

$$\omega_e = \left(\frac{ne^2}{m_e \epsilon_o}\right)^{1/2}$$

which corresponds to a *plasma frequency* of 8.98 10^3 $n_e^{1/2}$ Hz, where n_e is the density per cm^3. So for a plasma density of $10^{10}/cm^3$, the plasma frequency is 9 10^8 Hz, which is very much higher than the 13.56 Mhz with which rf plasmas are usually driven (see Chapter 5).

The period of these oscillations, about 1 nS in our example, tells us the response time of the plasma to charge fluctuations. Since the frequency is determined by the interaction between the ions and the electrons, it is not too surprising to find that ω_e is related to the Debye length λ_D.

$$\lambda_D \, \omega_e = \left(\frac{\epsilon_o \, kT_e}{ne^2}\right)^{1/2} \left(\frac{mne^2}{m_e \, \epsilon_o}\right)^{1/2} = \left(\frac{kT_e}{m_e}\right)^{1/2} \simeq \overline{c_e}$$

This relationship enables us to give some more physical meaning to λ_D and ω_e (Mitchner and Kruger 1973). In the derivation of the plasma frequency, the time required for the electron displacement Δ to build up would be about $\Delta/\overline{c_e}$. This displacement would be impeded if the response time $1/\omega_e$ of the electrons was shorter than $\Delta/\overline{c_e}$. Therefore regions of disturbance will be restricted to a distance Δ given by

$$\frac{\Delta}{\overline{c_e}} \simeq \frac{1}{\omega_e}$$

or

$$\Delta \simeq \frac{\overline{c_e}}{\omega_e} \simeq \lambda_D$$

This is consistent with our earlier picture of λ_D as the extent of deviation from charge neutrality.

Alternatively, the relationship $\lambda_D \, \omega_e \simeq \overline{c_e}$ says that the electrons can move a distance of about λ_D in a time of $1/\omega_e$. This means that if the plasma is disturbed by an electromagnetic wave of angular frequency ω, then the plasma electrons can respond fast enough to maintain neutrality if $\omega < \omega_e$. So ω_e is the minimum frequency for propagation of longitudinal waves in the plasma.

In our simple derivation, the resulting oscillation was stationary, but we ignored the thermal random motion of the electrons, and when this is taken into account it can be shown that disturbances can be propagated as waves. In fact a plasma is very rich in wave motion. This gives a means for propagating energy through the plasma. There can also be energy interchange between these plasma waves and fast electrons which travel through the plasma. We shall need to consider these energy exchanges in Chapter 4.

$m\frac{d^2x}{dt^2} = -eE$; $P = -enx$ WHERE $P \rightarrow D = \epsilon_0 E + P$

Ion Oscillations

Just as the electrons could oscillate in the plasma, so also can the ions. The mass of the ions ensures that their oscillations are so slow that the electrons can maintain thermal equilibrium. The ion frequency is more complex to find than the electron frequency; although we were able to ignore the ion motion when deriving the electron frequency, we cannot ignore the electron motion when deriving the ion frequency. However, in the case where T_e is large, the ion frequency simplifies to the same form $\omega_i = (ne^2/m_i\epsilon_o)^{1/2}$ as the electron plasma oscillation frequency. For our example, this would amount to 3.3 Mhz. In the more general case, these low frequency oscillations occur with frequencies between zero and a few megahertz. They can be observed (Pekarek and Krejci 1961) as *striations* in the positive column of dc glow discharge tubes (our glow discharge processes don't usually have positive columns – see Chapter 4). When the ion frequency is low enough, these striations can be observed with the naked eye as slow moving or even stationary regions of higher optical emission intensity.

AMBIPOLAR DIFFUSION

Finally, there is a topic which we won't be using much in this book, but it does play a significant role in plasmas and therefore we need to know of the concept.

Whenever there is a concentration gradient of particles, the random motion of the particles results in a net flow down the gradient. This is the phenomenon of *diffusion*. The resulting ion and electron current densities in the presence of a diffusion gradient dn/dx (assumed in one dimension for simplicity) can be written:

$$j_e = -e D_e \frac{dn_e}{dx}$$

$$j_i = -e D_i \frac{dn_i}{dx}$$

D_e and D_i are the diffusion coefficients of the electrons and ions respectively. It is possible to show that the diffusion coefficient and mobility μ (the drift velocity in unit electric field) are related by temperature:

$$\frac{D}{\mu} = \frac{kT}{e}$$

This is *Einstein's relation*. We already know that the mobility of the electrons is very much greater than that of the ions, and therefore the electron diffusion coefficient will be very much greater than the ion diffusion coefficient. One might expect as a result that, in a region of concentration gradient, the electrons would stream out very much faster than the ions. This is initially true, but the exodus of the electrons leaves the rest of the plasma more positive and sets up a restraining electric field $\mathcal{E}$ which grows large enough to equalize the diffusion rates of the ions and electrons.

In the presence of both the resulting electric field $\mathcal{E}$ and the diffusion gradient, the resulting ion and electron densities can be written as follows:

$$j_i = e\, n_i\, \mu_i\, \mathcal{E} - e\, D_i \frac{dn_i}{dx}$$

$$j_e = -e\, n_e\, \mu_e\, \mathcal{E} - e\, D_e \frac{dn_e}{dx}$$

where μ_i and μ_e are the mobilities of the ions and electrons respectively. The equalization of diffusion rates is achieved by putting $j_i = j_e$ in our current flow equations. Since n_i and n_e (and hence their concentration gradients) are closely equal throughout the main body of the plasma, equating the current densities yields the following result for $\mathcal{E}$:

$$\mathcal{E}\, n_e\, (\mu_i + \mu_e) = (D_i + D_e) \frac{dn_e}{dx}$$

Substituting this value of $\mathcal{E}$ back, we obtain the following expression for the current flow of ions and electrons:

$$j_i = j_e = \left(\frac{D_e\, \mu_i + D_i\, \mu_e}{\mu_i + \mu_e} \right) \frac{dn_e}{dx}$$

So the collective behaviour of the ions and electrons causes them to move with the *same* diffusion coefficient. This is the phenomenon of *ambipolar diffusion*, which will apply to all motion within the plasma.

The current density formulation that we have used in this derivation, is borrowed from solid state physics where mobilities, which actually imply collision-dominated motion, are relevant. Using mobilities in the present application is stretching things rather, but we can take care of this by making the mobilities dependent on the electric field and plasma conditions. But again we run into the problem that simple concepts become rather complex when the details are considered.

REFERENCES

D. J. Ball, J. Appl. Phys. **43**, 7, 3047 (1972)

D. Bohm, in *The Characteristics of Electrical Discharges in Magnetic Fields*, ed. A. Guthrie and R. K. Wakerling, McGraw Hill, New York and London (1949)

F. F. Chen, in *Plasma Diagnostic Techniques*, ed. R. H. Huddlestone and S. L. Leonard, Academic Press (1965)

F. F. Chen, *Introduction to Plasma Physics*, Plenum Press, New York and London (1974)

R. M. Clements, J. Vac. Sci. Tech. **15**, 2, 193 (1978)

J. W. Coburn and E. Kay, J. Appl. Phys. **43**, 12, 4966 (1972)

E. Eser, R. E. Ogilvie, and K. A. Taylor, J. Vac. Sci. Tech. **15**, 2, 199 (1978)

J. G. Laframboise, Univ. of Toronto, Inst. Aerospace Studies, Report No. 100 (1966)

I. Langmuir, General Electric Review **26**, 731 (1923)

I. Langmuir and H. Mott-Smith, ibid **27**, 449 (1924)

I. Langmuir and H. Mott-Smith, ibid **27**, 538 (1924)

I. Langmuir and H. Mott-Smith, ibid **27**, 616 (1924)

I. Langmuir and H. Mott-Smith, ibid **27**, 762 (1924)

I. Langmuir and H. Mott-Smith, ibid **27**, 810 (1924)

L. B. Loeb, *Basic Processes of Gaseous Electronics*, U. of California Press, Berkeley (1961)

M. Mitchner and Charles H. Kruger Jr., *Partially Ionized Gases*, Wiley, New York and London (1973)

L. Pekarek and V. Krejci, Czech. J. Physics **B-11**, 729 (1961)

J. D. Swift and M. J. R. Schwar, *Electrical Probes for Plasma Diagnostics*, Iliffe, London (1970)

J. A. Thornton, J. Vac. Sci. Tech. **15**, 2, 188 (1978)

Chapter 4. DC Glow Discharges

So far we have been dealing with a rather idealized homogeneous plasma with a well-defined potential and density, and with constituent particles in equilibrium motion characterized by relevant temperatures. The glow discharges which we're using only approximate this condition, for various reasons which we shall be discussing. Nevertheless, many of the plasma concepts are of great utility in helping us to derive some understanding and control of glow discharge processes, even on a semi-quantitative basis. Amongst sputtering and plasma etching folks, the words 'plasma' and 'glow discharge' tend to be used synonymously – to the horror of plasma physicists, I'm sure! One can get into semantic discussions and argue that some discharges are plasmas with two or three different groups of electrons each with a well-defined temperature. That argument could probably be extended indefinitely. So let's accept that our glow discharges are certainly not ideal plasmas, and keep this in mind when we lapse into glow discharge – plasma synonyms.

One of the complicating factors in trying to understand glow discharges is that most of the literature, particularly the 'classical' literature of the 1920's and 30's, deals with dc discharges; whereas practical plasma processes are more usually rf excited. And, as we said above, none of our practical glow discharges are truly plasmas. This gives then, in a sense, a choice: we can either pursue some plasma physics rather exactly, and then find that it does not entirely apply to our systems; or we can follow some simpler, if not always entirely accurate, models which convey the physical ideas rather well and, in the event, are probably just as accurate. In the present book I have opted for the latter.

Before commencing battle, I would recommend reading a delightful history of gaseous electronics by Brown (1974). Prof. Brown tells, for example, the story of the unfortunate pioneer Hittorf who laboured week after week, gradually extending the length of a thin glass discharge tube to try to discover the length of the positive column. Eventually the tube ran back and forth across Hittorf's laboratory. At this stage, a frightened cat pursued by a pack of dogs came flying through the window . . . "Until an unfortunate accident terminated my experiment", Hittorf wrote, "the positive column appeared to extend without limit."

ARCHITECTURE OF THE DISCHARGE

We could make a dc glow discharge by applying a potential between two electrodes in a gas; Figure 4-1 shows the resulting current density j flowing due to the application of a dc voltage V between a chromium cathode and a stainless steel anode, in argon gas at two different pressures. Each electrode was 12.5 cm diameter, and the electrodes were 6.4 cm apart. Most of the space between the two electrodes is filled by a bright glow known as the *negative glow*, the result of the excitation and subsequent recombination processes we discussed in Chapter 2. Adjacent to the cathode is a comparatively dark region known as the *dark space*. This corresponds to the sheath formed in front of the cathode; there is a similar sheath at the anode, but it is too thin to clearly see.

In this chapter, we shall be looking at dc discharges. These are somewhat easier to begin to analyze than rf discharges, although they are still extremely complex and we certainly don't understand all the details. Fortunately, much of what we learn can also be applied to rf systems.

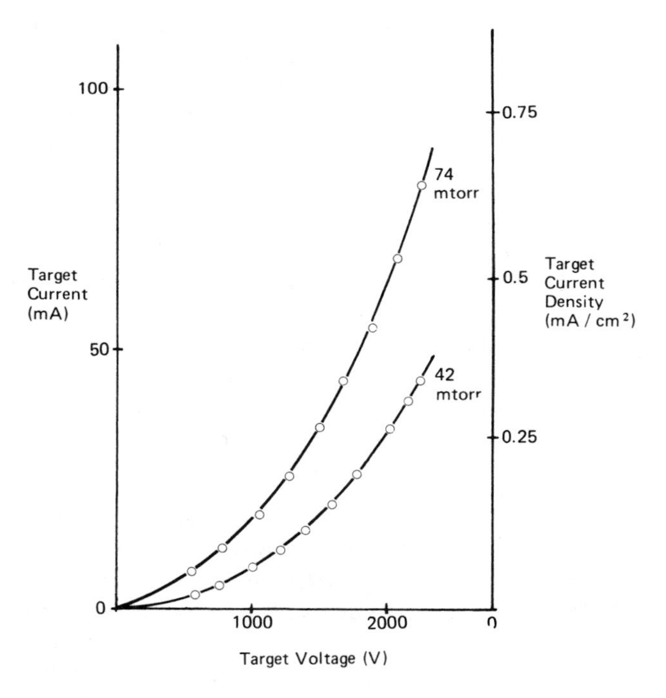

Figure 4-1. I-V characteristics for chromium sputtering in argon (Chapman 1975)

Many textbooks show a whole series of glowing and dark spaces in dc discharges. Figure 4-2 is from Nasser (1971); virtually the same figure appears in Cobine (1958), von Engel (1965) and doubtless many other texts. The *positive column* is the region of the discharge which most nearly resembles a plasma, and most of the classic probe studies have been made on positive columns. It is found that, when the two electrodes are brought together, the cathode dark space and negative glow are unaffected whilst the positive column shrinks. This process continues so that eventually the positive column, and then the Faraday dark space, are 'consumed', leaving only the negative glow and dark spaces adjacent to each electrode. This last situation is the usual case in glow discharge processes (Figure 4-3), where the inter-electrode separation is just a few times the cathode dark space thickness. The minimum separation is about twice the dark space thickness; at less than this, the dark space is distorted and then the discharge is extinguished.

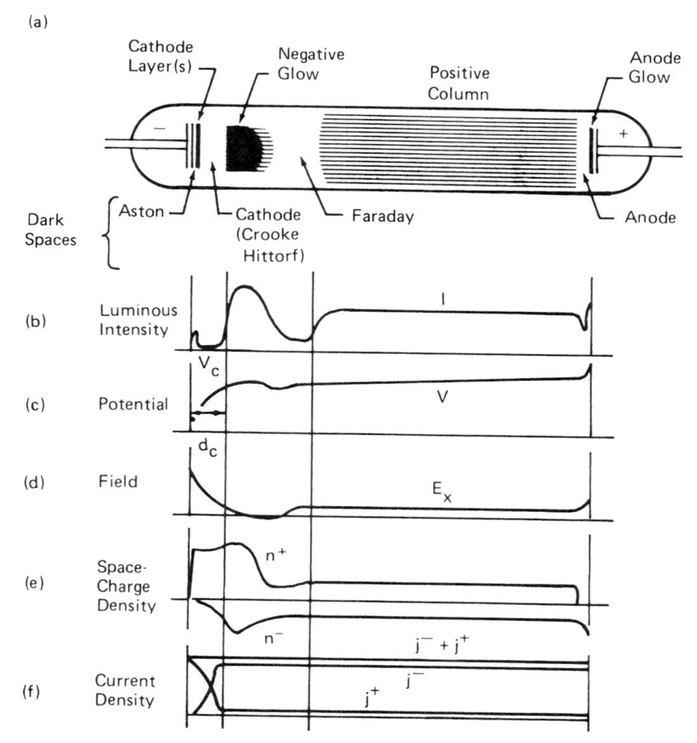

Figure 4-2. The normal glow discharge in neon in a 50 cm tube at p = 1 torr. The luminous regions are shown shaded (Nasser 1971). The abnormal glow would be somewhat different, although the glowing and dark regions would look the same

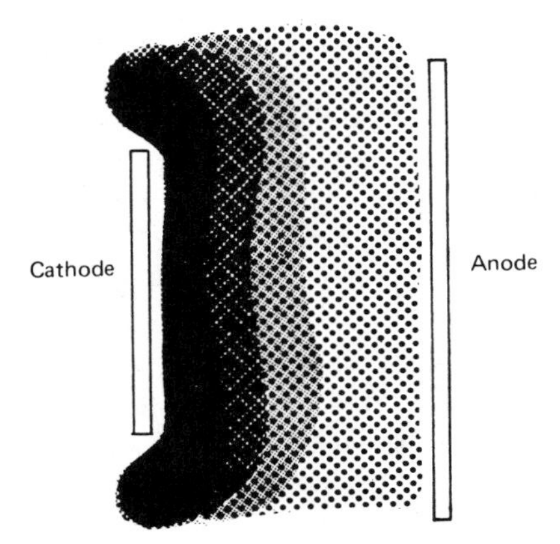

Figure 4-3. DC glow discharge process

Since current must be continuous in a system, it is clear that the currents at the two electrodes must be equal. In this particular system, the only other grounded electrode was remote from the discharge and had a small surface area; thus, the current densities at the chromium cathode and stainless steel anode were approximately equal. Take a typical datum point, which might be 2000V and 0.3 mA/cm^2 at 50 mtorr. This represents an electron current density to the anode that is much smaller than the random current density $\frac{1}{4}en\overline{c_e}$ and so there must be a net decelerating field for electrons approaching the anode, i.e. the plasma is more positive than the anode. But there is still some electron current flowing, so apparently the anode is more positive than floating potential. We earlier calculated a 'reasonable' floating potential 15V less than the plasma potential, and this is consistent with commonly found values of $V_p \sim +10V$ (with respect to a grounded anode) in dc sputtering systems.

The plasma is virtually field-free, as we saw earlier, so the plasma has the same potential V_p adjacent to the sheath at the cathode. But the cathode has a potential of -2000V, so the sheath voltage is $-(2000 + V_p)$, i.e. -2010V in our example (Figure 4-4).

Notice some peculiarities about this voltage distribution:

1. The plasma does *not* take a potential intermediate between those of the electrodes, as might first be expected. This is consistent with our earlier contention that the plasma is the most positive body in the discharge.
2. The electric fields in the system are restricted to sheaths at each of the electrodes.
3. The sheath fields are such as to repel electrons trying to reach either electrode.

All of these peculiarities follow from the mass of the electron being so much less than that of an ion. The third, in particular, is illustrative of the role played by electrons in a discharge.

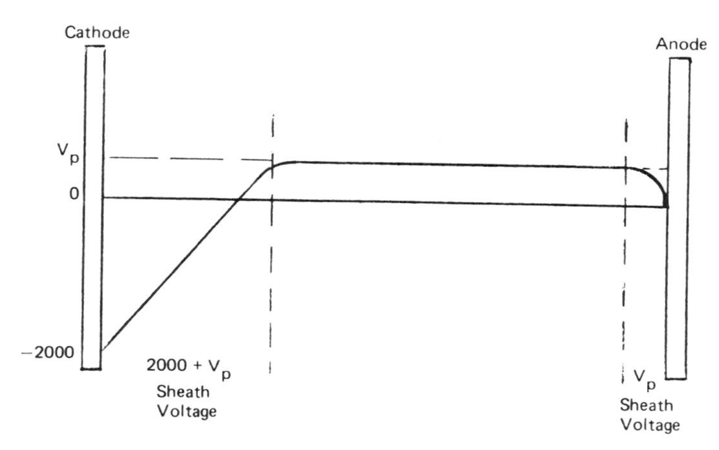

Figure 4-4. Voltage distribution in a dc glow discharge process

MAINTENANCE OF THE DISCHARGE

How is this glow discharge sustained? Electrons and ions are lost to each of the electrodes and to all other surfaces within the chamber. The loss processes include electron-ion recombination (which takes place primarily on the walls and anode due to energy and momentum conservation requirements, as we saw in Chapter 2), ion neutralization by Auger emission at the target, and an equivalent electron loss into the external circuit at the anode. To maintain a steady state discharge, there must be a numerically equal ion-electron pair generation rate; i.e. there must be a good deal of ionization going on in the discharge.

There is also a considerable energy loss from the discharge. Energetic particles impinge on the electrodes and walls of the system, resulting in heating there; this energy loss is then conducted away to the environment. So another requirement for maintaining the discharge is that there is a balancing energy input to the discharge.

How are these ionization and energy requirements satisfied? The simplest answer is that the applied electric field accelerates electrons, so that the electrons absorb energy from the field, and that the accelerated electrons acquire sufficient energy to ionize gas atoms. So the process becomes continuous. But that's a very simple answer, and raises various other questions. Where does most ionization take place, and what are the major processes involved? Can the model of the discharge that we've been developing account for the amount of ionization required? To what extent is the dc discharge like the plasma of Chapter 3?

In trying to decide where most ionization occurs, the glow region must be an obvious candidate. In chapter 2, we saw that ionization and excitation are rather similar processes. Their thresholds and cross-section energy dependences are not so different, so that for electrons with energies well above threshold, ionization and excitation will be achieved in a rather constant ratio; as the electron energy decreases towards threshold, then excitation will occur in an increasing proportion since it has a lower threshold. So we would expect that excitation, and subsequent emission from de-excitation, will always accompany ionization – at least for the glow discharges we're considering. Hence the choice of the glow region as the prime candidate for the main ionization region. But if we look in the literature, then we often find descriptions of glow maintenance that rely entirely on ionization in the cathode sheath region. So apparently there is some disagreement over this matter.

In the rest of this chapter, we shall be examining a practical dc discharge in some more detail, and we shall do this by dividing the discharge into three regions: the cathode region, the glow itself, and the anode region. We shall be looking not only at the ionization question raised above, but also at practical matters such as charge exchange collisions in the sheath which have the important effect of controlling the energy of bombarding ions at the cathode – important in practical applications. But before looking at these three regions, we shall discuss the phenomenon of *secondary electron emission* that takes place at cathode, anode, and walls.

SECONDARY ELECTRON EMISSION

When a particle strikes a surface, one of the possible results is that an electron is ejected. The number of electrons ejected per incident particle is called the *secondary electron coefficient* or *yield*. Secondary electron emission is observed for bombardment by ions, electrons, photons and neutrals (both ground state and metastable); each will have a different coefficient and a different energy dependence.

Electron Bombardment

The emission of electrons due to electron impact has been closely studied because of its importance in valves, cathode ray tubes, and electron multipliers. By looking at the energy dependence of the emitted electrons (Figure 4-5), it appears that some of the bombarding electrons are elastically or inelastically scattered, and that some *'true' secondaries* are also emitted. The 'true' secondaries are frequently, but not always, more numerous than the scattered primaries. Electron bombardment processes will be significant at the anode and at walls; there is no electron bombardment at the cathode. The yield due to electron impact is usually given the symbol δ, which depends on the energy of the bombarding electron, and is typically unity for clean metals (Figure 4-6). However,

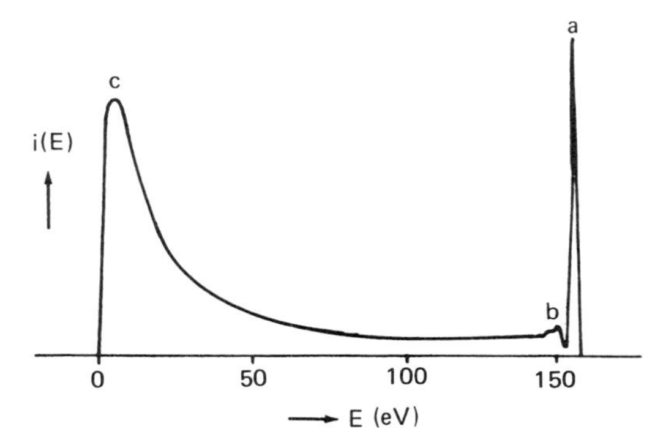

Figure 4-5. The energy distribution of secondary electrons emitted by silver (Rudberg 1930, 1934); a – elastically reflected primaries, b – inelastically reflected primaries, c – 'true' secondaries

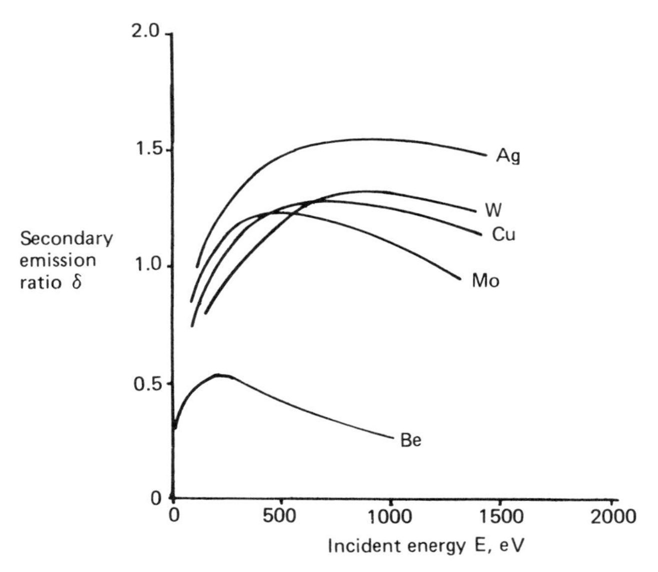

Figure 4-6. Secondary emission coefficient δ of different metals as a function of the energy of incident electrons (Hemenway et al. 1967)

δ is also strongly dependent on the presence of contamination or surface adsorbed layers, and is higher for insulating materials. Table 4-1 gives the maximum yield values δ_m and corresponding bombardment energies, and the *unity points* (one electron out for one electron in, therefore no net charging) for a number of materials. In glow discharge processes, we have to deal with electron bombardment at low energies of a few eV (and also some by high energy electrons – see later) so we would really like some δ data at correspondingly low energies, but it doesn't seem to be too readily available.

Ion Bombardment

The corresponding secondary electron emission coefficient for ion bombardment is given the symbol γ_i. Some values for γ_i are shown in Tables 4-2 and 4-3. The energy dependence of γ_i for noble gas ions on tungsten and molybdenum is shown in Figure 4-7, and for various other ion-metal combinations in Figure

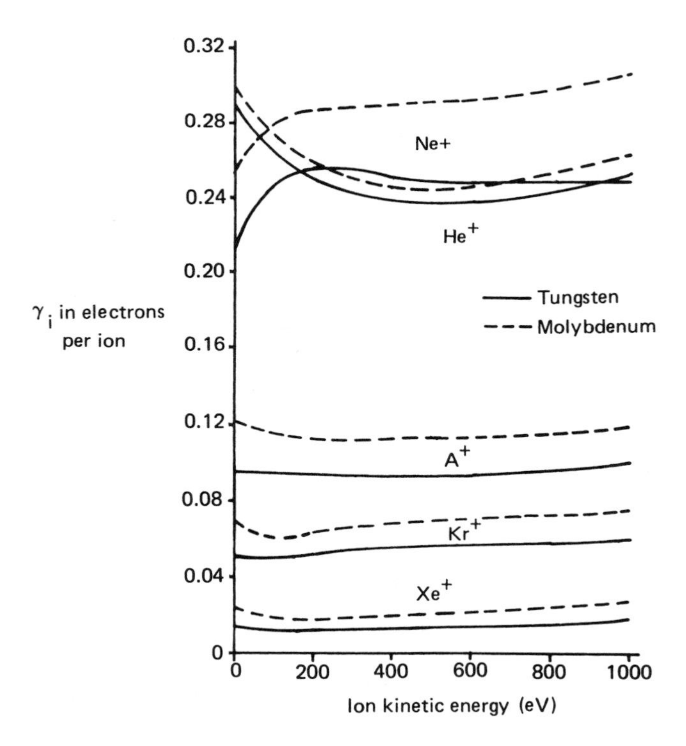

Figure 4-7. Secondary electron yields γ_i for noble gas ions on atomically clean tungsten and molybdenum (Hagstrum 1956b)

Table 4-1 Secondary Electron Emission Coefficients δ

	$\delta = 1$ at		δ_{max}	Energy (eV) for δ_{max}
	V eV	V′ eV		
Ag			1.5	800
Al			1.0	300
Au			1.5	800
C	160	~ 1000	1.3	600
Cu	> 100		1.3	600
Fe			1.3	350
Ge			1.1	400
K			0.7	200
Li			0.5	85
Mo	140	1200	1.3	350
Na			0.8	300
Pt			1.6	800
Pt			1.8	800
Pt	150	> 2000	1.8	800
Si			1.1	250
W			1.5	500
Zn	100	400	1.1	200
NaCl			6	600
NaCl	~ 20	1400	6–7	600
MgO (vacuum cleaved)			21	1100
MgO			2.4–4	400
MgO	< 100	> 5000	7.2	1100
Pyrex glass	30–50	2400	2.3	300–400
Soda glass	30–50	900	~ 3	300
Oxide cathode BaOSrO	40–60	3500	5–12	1400
ZnS		6000–9000		
Ca tungstate		3000–5000		

Two and three different sets of data (independent sources) are shown for Pt, NaCl, and MgO with considerable disagreement in the case of MgO (Lye 1955, Von Engel 1965, and Johnson and McKay 1953, respectively)

Data From:

McKay 1948
Bruining 1954
Woods 1954
Hachenberg & Brauer 1959
Von Engel 1965
Dekker 1963
Lye 1955
Johnson and McKay 1953, 1954
Copeland 1931

Table 4-2 Secondary Electron Coeficients γ_I for Argon Ion Impact

	Ion Energy		
	10 eV	100 eV	1000 eV
Mo	0.122	0.115	0.118
W	0.096	0.095	0.099
Si (100)	0.024	0.027	0.039
Ni (111)	0.034	0.036	0.07
Ge (111)	0.032	0.037	0.047

Data From:
Hagstrum 1956a, 1956b, 1960
Takeishi & Hagstrum 1965
Carlston et al 1965

Table 4-3 Values of γ_i From Metals for Slow (Sic) Ions

Metal	Ar	H_2	Air	N_2	Ne	
Al	0.12	0.095	0.021	0.10	0.053	
Ba	0.14	.....	0.100	.....	0.14	
C	.....	0.014				
Cu	0.058	0.050	.....	0.025	0.066	
Fe	0.058	0.061	0.015	0.020	0.059	0.022
Hg	.....	0.008	0.020			
K	0.22	0.22	0.17	0.077	0.12	0.22
Mg	0.077	0.125	0.031	0.038	0.089	0.11
Ni	0.058	0.053	0.019	0.036	0.077	0.023
Pt	0.058	0.020	0.010	0.017	0.059	0.023
W	.....	.....	.....	.....	.....	0.045

From Knoll et al. (1935); reported in Cobine (1958)

4-8. The yield is again very dependent on the condition of the surface: Figure 4-9 shows how γ_i depends on the crystal face exposed and Figure 4-10 shows how the yield of polycrystalline tungsten decreases from the clean metal value on exposure to nitrogen, reaching a new quasi-steady state after about 10 minutes, coinciding with the completion of the first monolayer coverage of the nitrogen. (Note also in Figure 4-10 that the ion bombardment energy is only 10 eV, so that γ_i is still quite high in this case, even at such low ion energies). The effect of surface contamination is again shown in Figure 4-11, this time for argon ion bombardment of tungsten. Figure 4-12 shows similar effects due to other gas adsorptions on tantalum and platinum.

These variations of yield γ_i with surface condition are quite important in dc sputtering where the magnitude of the yield plays a role in determining the V-I characteristics of the discharge. A sputtering target is immediately contaminated on exposure to the atmosphere, commonly with the formation of an oxide surface layer on metal targets. When the target is subsequently sputtered, there is a period when the V-I characteristic is continuously changing as the surface layer is removed.

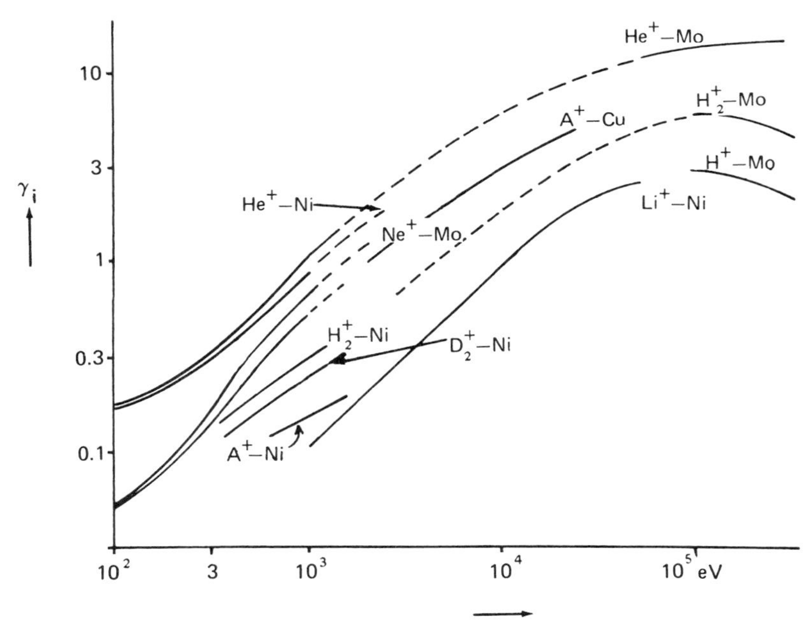

Figure 4-8. Secondary emission coefficient γ_i for ions of energy K falling on the surface of various substances, from von Engel (1965). References: Rostagni (1938), Healea and Houtermans (1940), Hill et al. (1939)

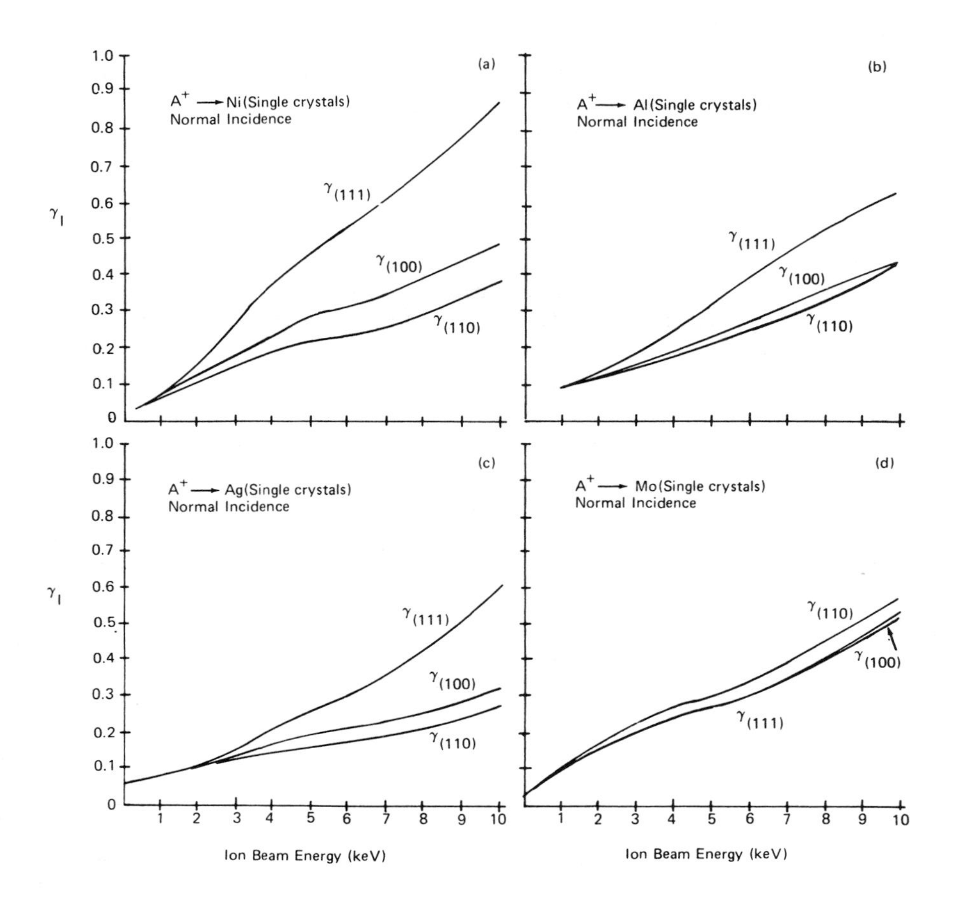

Figure 4-9. Variation of γ_i with ion energy for Ar^+ bombardment of (111), (100) and (110) surfaces of (a) nickel, (b) aluminium, (c) silver and (d) molybdenum (from Carlston et al. 1965)

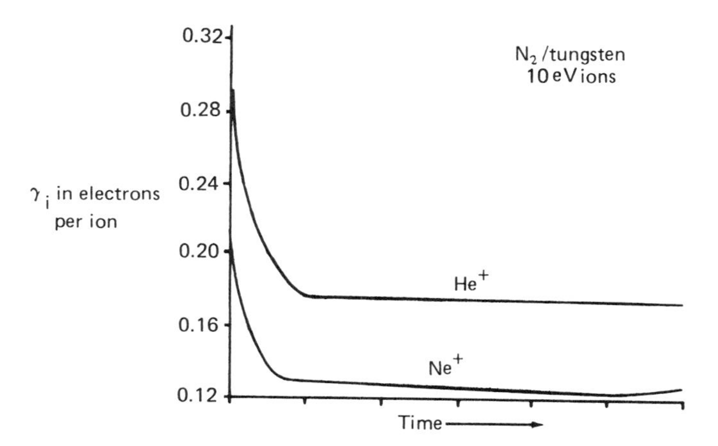

Figure 4-10. Secondary electron yields γ_i for He^+ and Ne^+ ions, as a monolayer of nitrogen forms on tungsten. The break in the plot represents the completion of the first monolayer (from Hagstrum 1956c)

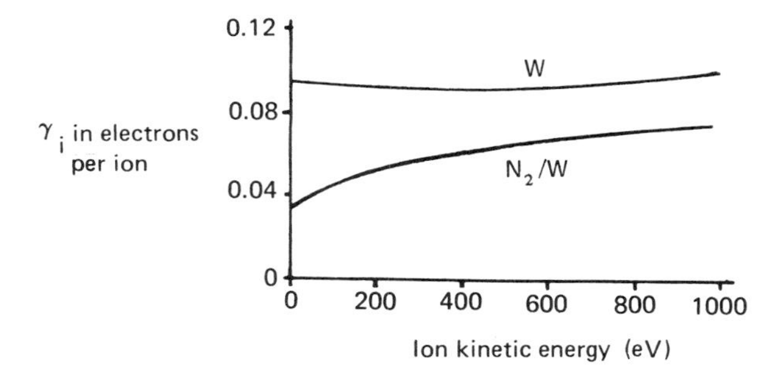

Figure 4-11. Secondary yield γ_i for argon ions on clean tungsten and on tungsten covered with a monolayer of nitrogen (from Hagstrum 1956c)

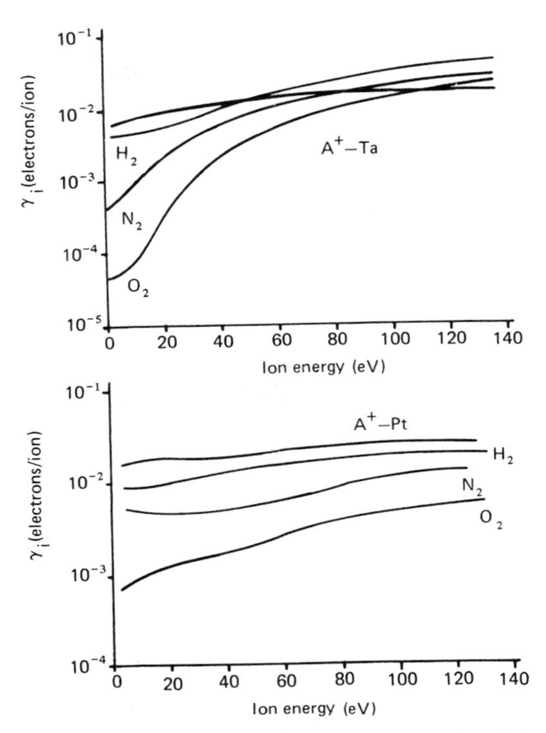

Figure 4-12. Secondary electron yields for Ar^+ ions on outgassed tantalum and platinum, and on these metals after treatment with hydrogen, nitrogen and oxygen (Parker 1954); n.b. logarithmic vertical axis. (From McDaniel 1964)

We have so far been dealing with pure metals having γ_i much less than unity. Insulators generally have much larger values, but there is a problem in obtaining accurate yield values due to the charging of the insulator. Some alloys also have large yields (Figure 4-13) which make them suitable for use as electron multipliers in dynode arrays.

We saw earlier that many of the secondary electrons emitted due to electron impact had rather low energies of a few eV. The same is true when the impacting particles are ions. Figure 4-14 shows the yields for various 40 eV noble gas ions. The dependence of the secondary electron energy distribution on the energy of the incident ion is rather weak (Figure 4-15), so that all emitted electrons have initial energies ~ 5 - 10 eV.

The interested reader is referred to McDaniel (1964), from which much of the data shown here has been taken, for a more thorough review of secondary electron emission due to ion impact.

In glow discharges, ion energies on targets and substrates range from a few eV up to a few hundred eV, and so the secondary electron yield data over the corresponding range are the most useful for the present investigation.

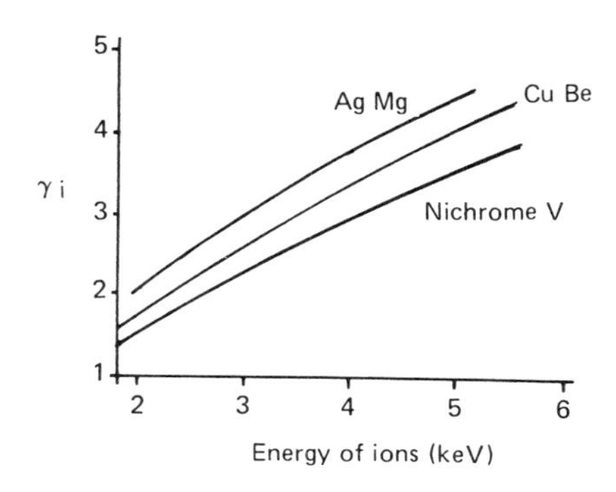

Figure 4-13. Electron yields for Ar^+ ions on Ag-Mg, Cu-Be and Nichrome V alloys (from Higatsberger et al. 1954)

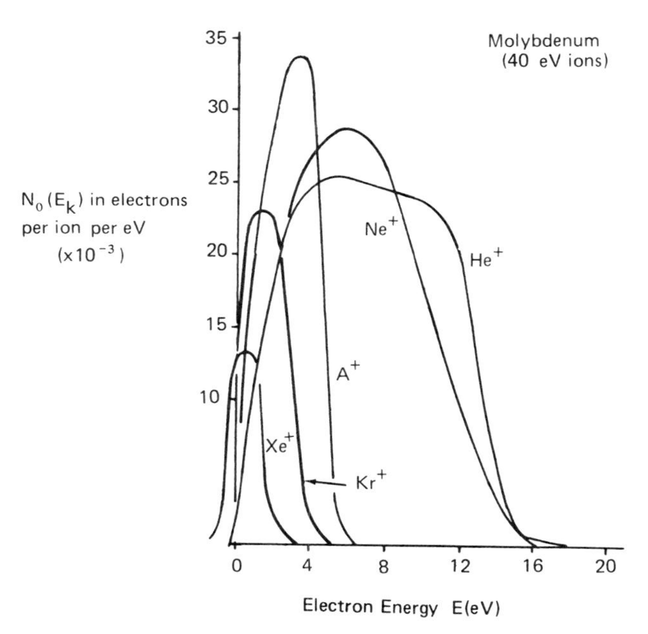

Figure 4-14. Energy distributions of secondary electrons ejected from Mo by 40 eV ions of the noble gases (Hagstrum 1956b)

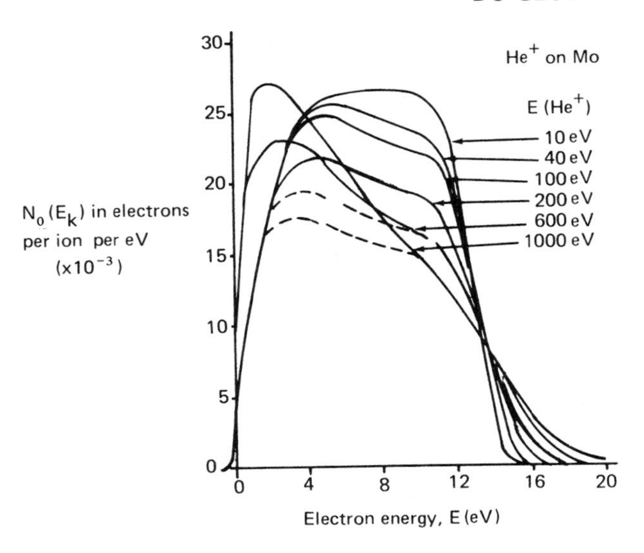

Figure 4-15. Energy distributions of secondary electrons ejected from Mo by He^+ ions of various energies (Hagstrum 1956b)

Neutral Bombardment

In the sheath at an electrode, energetic ions frequently collide with neutrals either elastically or with charge exchange (see Chapter 2) in either case giving rise to energetic neutrals. If sufficiently energetic, these neutrals can cause secondary electron emission. Figure 4-16 shows the yields for argon ions and argon neutrals on molybdenum. It appears that there is a potential energy component for the ions only. Unfortunately, there is rather little of this data available; Figure 4-16, if typical, suggests that electron emission due to neutrals is rather unimportant in glow discharge processes where neutral energies are a few hundred eV at most.

In Chapter 2, we saw that there are likely to be long-lived metastable neutrals, particularly in noble gas discharges. Although these metastables cannot be accelerated by electric fields, being neutral, they will receive energy by collision with energetic ions, the energy transfer function making this an efficient process. Since the metastables have some potential energy, they will presumably be somewhat more effective in producing secondary electrons than their corresponding ground state parents. There seems, however, to be rather little quantitative information available.

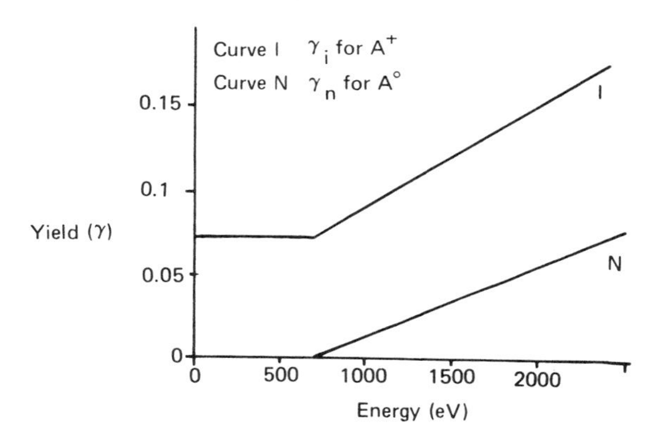

Figure 4-16. Secondary electron emission as a function of energy for argon ion and neutral atom bombardment of molybdenum (from Medved et al. 1963)

Photon Bombardment

The ejection of electrons due to photon bombardment is well-known, and is usually referred to as *photoemission*. For pure metals, the photoelectric yield γ_p depends on the work function ϕ of the metal, with a threshold for emission of $hc/\lambda = e\phi$. The photoelectric yields for most pure metals are only 10^{-4} to 10^{-3} electrons per photon in the visible to near ultraviolet frequencies, largely because the photon is usually efficiently reflected, except at very short wavelengths where a corresponding increase in photoelectric yield is seen, as in Figure 4-17. There doesn't seem to have been much consideration of the effect of photons in sputtering and plasma etching glow discharges. It does seem that, under the right circumstances, photoelectric yields can be as large as ion yields, and certainly there are believed to be strong photon effects in rather specific cases such as hollow cathode sources. Holmes and Cozens (1974) propose a contribution from photoelectric emission in their rather high current density mercury discharge (in which they also make the rather interesting observation of a pressure gradient near the target, believed to be due to the strong ion flux there). But on the whole, the effects of photoelectric emission and photoionization in glow discharges are not well understood.

Summary

Electrons can be emitted from solid surfaces due to the impact of ions, electrons, neutrals, and photons. Some of the processes are well understood, at least for clean metals. The situation for insulators and contaminated surfaces is much less

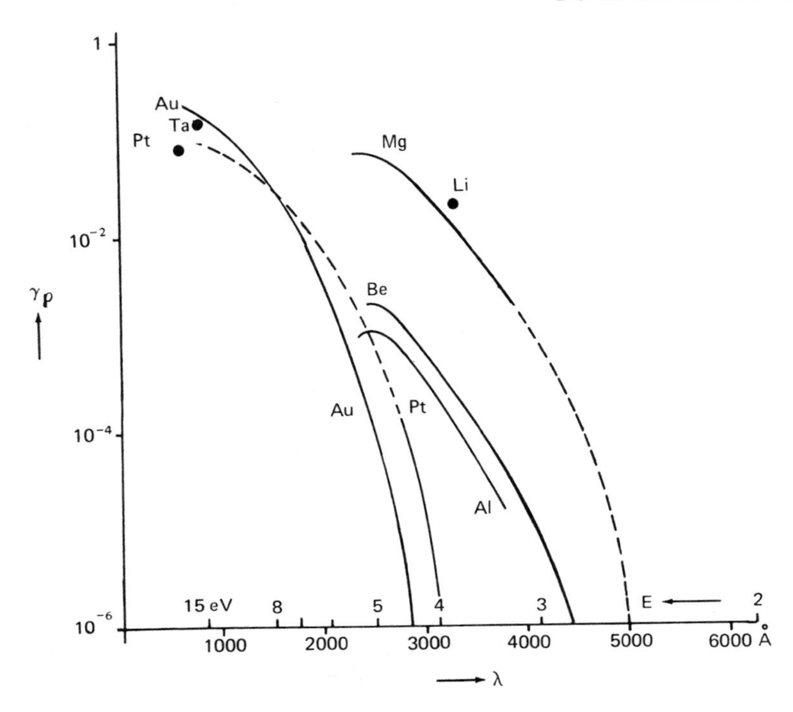

Figure 4-17. Photoelectric yield γ_p as a function of the wavelength λ of the incident light (energy E of quantum) for various substances (von Engel and Steenbeck 1932, Kenty 1933, Stebbins 1957, Wainfan et al. 1953). 2537 Å light yields $\gamma_p \sim 10^{-4}$ for borosilicate glass, 6 10^{-4} for soda glass; light of $\ll$ 1250 Å gives $\gamma_p \sim 10^{-3}$ for borosilicate glass (Rohatgi 1957). From von Engel (1965)

clear; experiments are complicated by the resultant charging of the surface. Harold Winters has pointed out to me that some of the literature on secondary electron emission, particularly the earlier literature and including some shown here, is likely to be erroneous, often because of the experimental difficulties encountered. For a discussion of modern measurement techniques, and an illustration of the importance of surface condition, see Sickafus (1977). Similar problems exist in measurements of both δ and γ. Theoretical considerations lead one to expect that γ will be independent of ion energy below 500 eV; the incoming ion is neutralized by an electron from the target, which then may Auger-emit another electron, so that the potential energy of the ion is important rather than its kinetic energy. This explanation is consistent with some of the data shown. The data of Hagstrum and colleagues is well regarded.

These processes are important in glow discharge processes because each of them can contribute electrons to the discharge and help to counter electron loss processes. Since the plasma is more positive than the potential of any surface in the discharge, the action of the sheath is to accelerate electrons from the surface into the glow, giving both electrons and energy to the discharge.

Our practical processes result in surface bombardment energies from a few eV up to several hundred eV or even a few thousand eV and we need, therefore, to consider secondary electron data over this range. Ion bombardment will clearly be of importance at the cathode of a dc discharge, and both electron and ion bombardment at the anode. The importance of metastable and ground state neutrals, and of photons, has to be further assessed.

The detail of the loss processes for electrons and ions at electrodes and walls is complicated by secondary electron emission from those surfaces. When we have previously looked at currents to surfaces, e.g. in "Sheath Formation at a Floating Substrate" in Chapter 3, we have tacitly ignored the effects of secondary emission, which would change the net current to a surface or modify its floating potential, for example.

THE CATHODE REGION

The type of dc discharge used in glow discharge processes is known as an *abnormal glow discharge*. At lower applied voltages and consequent lower currents, a discharge can result which is characterized by constant voltage and constant current density. This is a *normal glow discharge*. More power applied to the system is manifested by an increase in the size of the region of the cathode carrying current (j and V remaining constant) until the whole cathode is utilised, at which stage the discharge becomes abnormal. We shall not consider normal discharges further in this book.

The cathode plays an important part in dc sputtering systems because the sputtering target actually becomes the cathode of the sputtering discharge. The cathode is also the source of secondary electrons, as we have seen, and these secondary electrons have a significant role both in maintaining the discharge and in influencing the growth of sputtered films.

When the formation of sheaths was being considered in Chapter 3, we made the assumption that there were no collisions in the sheath. Many books and papers on plasma physics are concerned specifically with *collisionless plasmas*, but this is because most current interest is in plasmas which have very high temperatures of many keV, and these are essentially collisionless; such plasmas are of interest in fusion. 'Our' plasmas are very different and do have lots of collisions, both in the sheaths and in the glow. In a moment we shall look at some of these collision processes.

As already pointed out, in trying to understand the mechanisms by which a discharge is sustained, it is clearly necessary to account for all the recombination and energy loss processes which occur (Figure 4-18). We could simplify the situation for analysis purposes by considering a discharge between very large electrodes close together, which is usually the case in high pressure planar diode plasma etchers (see Chapter 7) and some sputter deposition systems (see Chapter 6). Unfortunately I don't have any quantitative data for this dc situation, but the data in Figure 4-1 should be reasonably representative.

To return to our example in "Architecture of the Discharge", a current density of 0.3 mA/cm^2 means that net currents of 1.9 10^{15} $ions/cm^2$ and 1.9 10^{15} $electrons/cm^2$ are flowing each second to the cathode and anode respectively. The ion flux at the anode should also be about 1.9 $10^{15}/cm^2$ sec, as we discussed in Chapter 3. So if we ignore the small electron current at the cathode due to secondary electron emission, and ion-electron recombination at the walls and in the gas volume, then we need an ion-electron pair production rate of at least 3.8 10^{15} ions per second for each cylinder of discharge emanating perpendicularly from the cathode and having 1 cm^2 cross-sectional area.

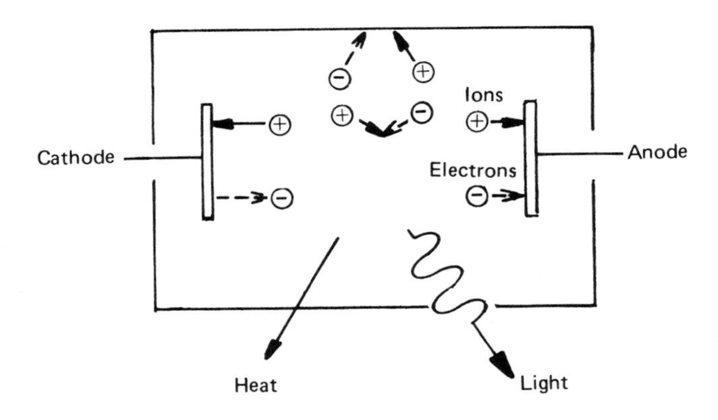

Figure 4-18. Discharge loss processes

Ionization In The Sheath

Electron Impact Ionization

Some descriptions of the glow discharge process rely on ionization caused by secondary electrons from the target as they are accelerated across the dark space (Figure 4-19). This can be modelled by considering the amount of ionization caused by a flux $N_e(x)$ electrons passing through a thin slab of thickness Δx located x from the cathode (Figure 4-20). The density of neutrals is n and the ionization cross-section (assumed energy-independent for simplicity) is q.

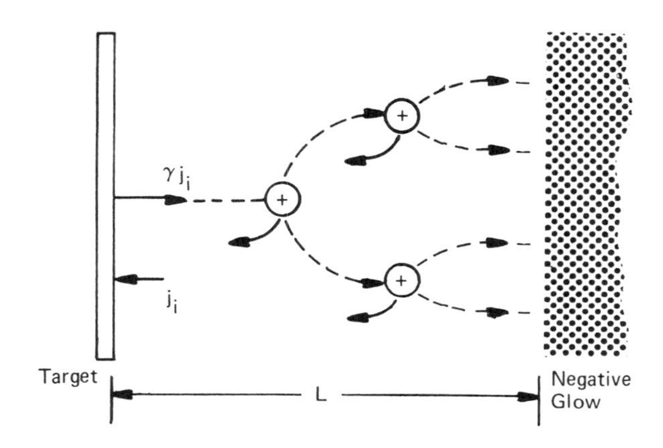

Figure 4-19. Ion pair production in the dark space

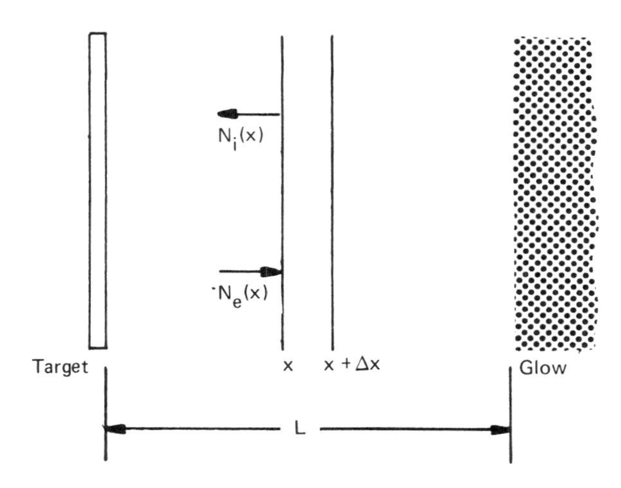

Figure 4-20. Analysis of ion pair production in the dark space

Number of ionizing collisions = $N_e(x)nq\,\Delta x$

$$\therefore \frac{dN_e(x)}{dx}\,\Delta x = N_e(x)nq\,\Delta x$$

$$\therefore \int \frac{dN_e}{N_e} = \int nq\,dx$$

$$\therefore N_e(x) = N_e(O)\exp nqx$$

So each electron that leaves the target is multiplied by exp nqL by the time it reaches the edge of the dark space. The electric field in this region is strong enough that the major part of the electron travel will be straight across the dark space along the field lines.

Let's obtain an idea of the magnitude of this electron multiplication for the practical conditions under consideration. In Chapter 2, "Ionization", we found that the maximum ionization cross-section for electrons in argon is $2.9\ 10^{-16}$ cm^2 for 100 eV. Davis and Vanderslice (1963), whose work on collisions in the sheath we shall be considering shortly, found a sheath thickness of 1.3 cm for a discharge voltage of 600V in argon at 60 mtorr, for which $n = 2.1\ 10^{15}$, using a Kovar alloy cathode. These figures put an upper limit on electron multiplication of $\exp(2.1\ 10^{15} \times 2.9\ 10^{-16} \times 1.3) = 2.2$.

For each ionization, a new ion is formed as well as a new electron. For each electron that leaves the target, (exp nqL - 1) ions will be formed. For each ion that strikes the target, γ secondary electrons will be emitted, where γ is the sum yield for all of the various processes (see "Secondary Electron Emission"). Hence, each ion that strikes the target will lead to the generation of $\gamma(\exp nqL - 1)$ ions within the dark space. The yield γ is unlikely to exceed 0.2 for most metals, and this suggests an ion production rate of 0.24 ions per ion; remember that this is an upper limit based on the use of the maximum cross-section for ionization in argon.

Ion Impact Ionization

There may be other ionization mechanisms in the sheath. In Chapter 2, we saw that photoionization and ion impact on neutrals were both possible ionization mechanisms. I don't know the photon fluxes to be able to assess photoionization, although these could presumably be obtained with optical emission spectroscopy. But we can make an estimate of ion impact ionization. In the same way that we estimated electron impact ionization earlier, we can use the exp nqL expression to estimate ion impact ionization. Using the same example, n will be $2.1\ 10^{15}$ cm^3 and L will be 1.3 cm, as before. From Figure 2-25 (Chapter 2), we find a cross-section q of about $5\ 10^{-17} cm^2$ for ions with mean energies of a hundred eV or so. So the ion multiplication factor will be about exp

$(2.1\ 10^{15}\ x\ 5\ 10^{-17}\ x\ 1.3) = 1.15$. Since secondary electron effects were ignored in Figure 2-25, this will be an overestimate (as was our earlier value for electron impact ionization). We therefore have an ion production rate by this process of 0.15 ions per ion compared with an equivalent of 0.24 ions per ion for electron impact ionization; although the electron cross-section is larger, there are fewer electrons. Both are maximum possible values, not actual; in reality, the contributions may be much smaller, particularly from the ions.

Sheath Ionization – Conclusion

An ion production rate of 1 ion per ion in the sheath would be adequate to maintain the ion flux to the cathode. But according to our analysis, this would be achieved by electron impact ionization only if L = 2.9 cm (for $\gamma = 0.2$) or L = 3.9 cm (for $\gamma = 0.1$), and there would not be such errors in the measurement of L. A production rate of 1 ion per ion would also be achieved, for the given value of nqL, if $\gamma = 0.8$. This also is unlikely, although it is true that our working figure of $\gamma = 0.1$ is based largely on ion impact secondary electron values, and we should add the effects of bombardment by fast neutrals, metastable and photons, so $\gamma = 0.8$ isn't out of the question. On the other hand, the q value we used was the maximum possible, and so values for ion production would be considerable overestimates.

Our finding of a certain amount of ionization by ion impact does not really change the situation since the maximum possible ion multiplication by this means was only 1.14.

Our general conclusion is therefore that there is some ionization in the sheath, but very probably not enough to maintain the ion flux to the target. In the next two sections on "Charge Exchange in the Sheath" and "Generation of Fast Electrons", we shall present some further experimental evidence to add credence to this conclusion, so that although none of the evidence presented is really conclusive by itself, the overall weight of evidence is quite convincing.

The inadequacy of sheath ionization becomes even further apparent when we remember that a similar amount of ionization is required to account for the ion current to the anode. Ions produced in the cathode sheath certainly cannot travel to the anode because of the polarity of the cathode sheath field. So we need a large source of ionization in either the negative glow or in the anode sheath. But the latter is so much thinner than the cathode sheath that any significant amount of ionization there is immediately ruled out. Which leaves the glow.

Charge Exchange in the Sheath

An ion arriving at the interface between the glow and a sheath has a kinetic energy that is negligible compared with most sheath voltages (see Chapter 3, "Sheath Formation and the Bohm Criterion"). In the absence of collisions, the ion would accelerate across the sheath, losing potential energy as it does so, and would hit the electrode with an energy equivalent to the sheath voltage. But the ion usually does collide, with or without the exchange of charge (Figure 4-21) (see Chapter 2, "Ion-Neutral Collisions"). This effect is important in glow discharge processes because it modifies the energy distributions of particles striking the electrodes and substrate.

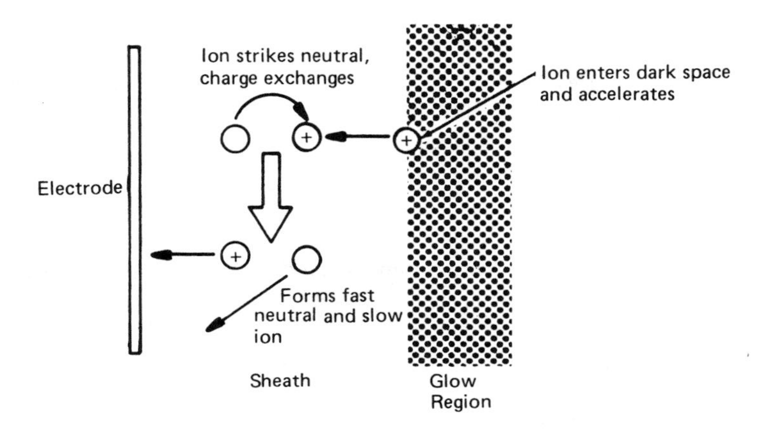

Figure 4-21. Charge exchange in an electrode sheath

Studies of the energy distributions of ions striking an electrode have been made by a number of authors; the work most relevant to glow discharge processes is by Davis and Vanderslice (1963). The apparatus they used is shown in Figure 4-22. Some of the ions striking the cathode pass through a tiny hole into a much lower pressure region where they are energy analyzed and then mass analyzed. The energy distribution of ions striking the cathode would be influenced not only by charge exchange but also by ionizing collisions in the sheath, and so could give us information about the latter. The results obtained by Davis and Vanderslice for Ar^+ ions in an argon discharge are shown in Figure 4-23; these results are consistent with their model which is based on the following assumptions:

1. All ions originate in the negative glow or very close to it. The model assumes little or no ionization in the sheath, and then uses the predictions of the model to test this assumption.
2. The dominant collision process is of symmetrical charge transfer ($Ar^+ + Ar \rightarrow Ar + Ar^+$) with the new ion formed starting at rest, and then accelerating

in the sheath field. There is no net change in ion flux, which therefore remains constant across the sheath.

3. The charge exchange cross-section is independent of energy, which is an approximation over the range used.
4. The electric field across the sheath decreases linearly to zero at the dark space – negative glow interface.

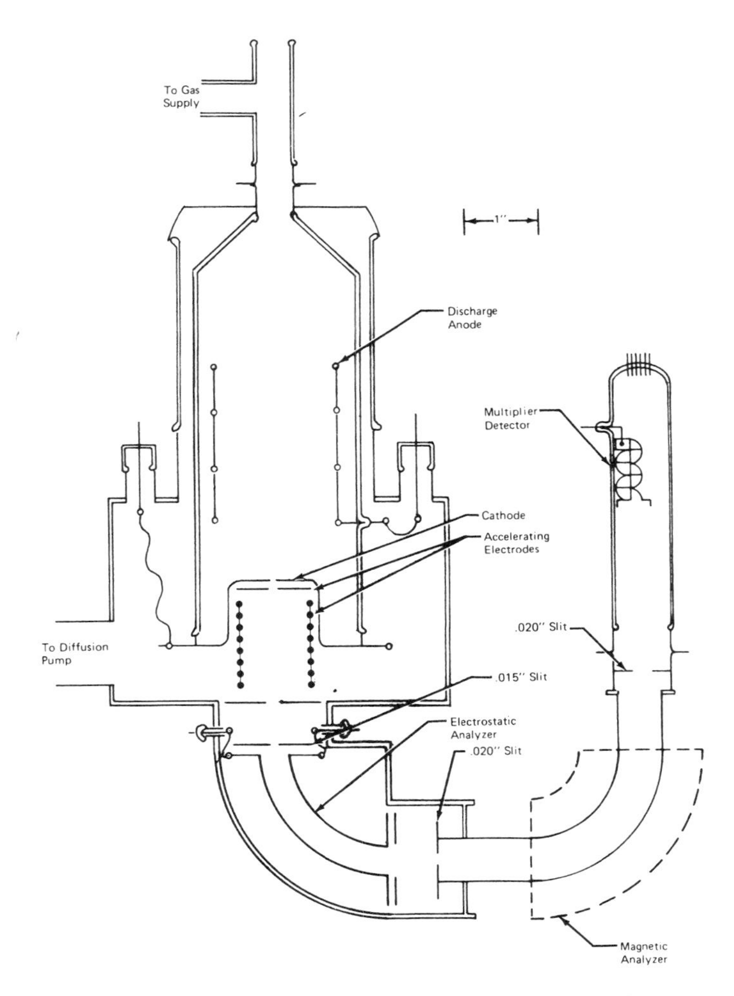

Figure 4-22. Experimental apparatus for the energy and mass analysis of ions bombarding the cathode in a dc discharge (Davis and Vanderslice 1963)

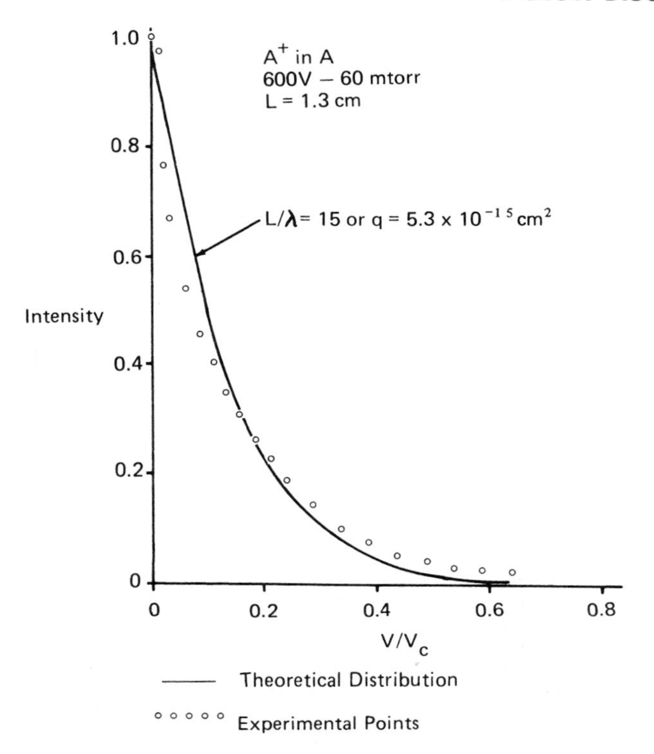

Figure 4-23. Energy distribution for Ar^+ from an argon discharge (Davis and Vanderslice 1963)

Using these assumptions and the parameters shown in Figure 4-24, Davis and Vanderslice obtained the theoretical distribution

$$\frac{V_c}{N_o}\frac{dN}{dV} = \frac{L}{2\lambda}\left(1 - \frac{V}{V_c}\right)^{-\frac{1}{2}} \exp - \frac{L}{\lambda}\left[1 - \left(1 - \frac{V}{V_c}\right)^{\frac{1}{2}}\right]$$

where V_c is the target voltage, and dN is the number of ions arriving with energies between eV and eV + edV; L is the dark space thickness and λ is the charge exchange mean free path. When $\lambda \ll L$, this distribution function reduces to

$$\frac{V_c}{N_o}\frac{dN}{dV} = \frac{L}{2\lambda} \exp\left(- \frac{L}{2\lambda}\frac{V}{V_c}\right)$$

The result for argon shows reasonable agreement with this model; in Figure 4-23, the open circles are experimental results and the solid line is a best fit from the

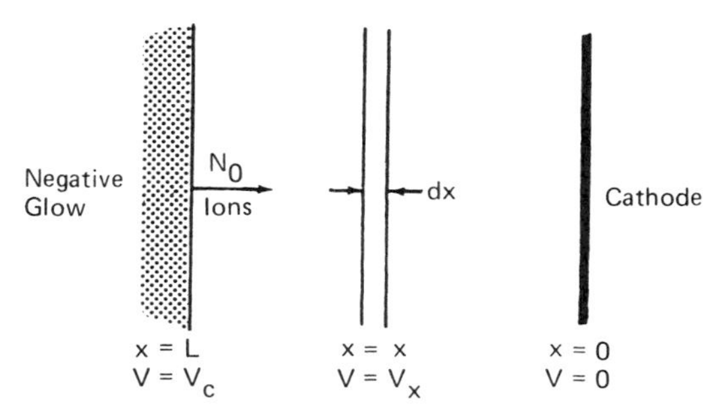

Figure 4-24. Model used to derive energy distributions (Davis and Vanderslice 1963)

theoretical expression, giving a cross-section of 5.3 10^{-15} cm^2 in reasonable agreement with other published values. Note that:

- The effect of gas pressure on the energy distribution is found to be small, if the discharge voltage is held constant. This is a result of the pressure – dark space thickness product being fairly constant for a dc discharge, so that the average number of collisions per ion in traversing this distance is reasonably constant.
- Increasing the target voltage (at constant pressure) causes the dark space to decrease in thickness, so that a relatively larger proportion of high energy ions will reach the cathode.
- Reduction of the collision cross-section also causes a larger proportion of high energy ions. Figure 4-25 is for A^{++} ions in argon, where a significant number of A^{++} ions (cross-section 7 10^{-16} cm^2) apparently traverse the sheath without collision.

The results of Davis and Vanderslice, which are confirmed by the later experiments of Houston and Uhl (1971), are used to illustrate the effect of charge exchange in limiting the energy of ion bombardment at the cathode. The good agreement between the theoretical and experimental results is also taken as confirmation that there is little or no ionization in the sheath.

However, I wonder how much of this agreement is fortuitous. Certainly there should be some ionization in the sheath, as we showed in the previous section, and this should have an effect on the energy distribution by generating ions in the sheath. Another questionable assumption in the Davis and Vanderslice model is that of a linear field variation in the sheath. From Poisson's equation, we know that such a variation is synonymous with uniform net positive space charge in the sheath; if $d\mathcal{E}/dx$ is proportional to x, then $\rho/\epsilon_0 = d^2\mathcal{E}/dx^2 =$ constant. Several authors refer to findings of linear field variations in the sheath.

Apparently the original findings are due to Aston (1911) who observed the deflection of a beam of electrons fired across the sheath. There are also some comments of findings of departures from linearity towards both interfaces of the sheath. If all ionization was in the glow, and ions entering the sheath were then accelerated freely, the ion density would decrease towards the cathode. The effect would be greatest in a collisionless sheath, and would be reduced by charge exchange collisions. If a linear field existed in the sheath, the potential would vary as x^2. Ingold (1978) has pointed out that, in practice, the potential variations of $x^{4/3}$ and $x^{3/2}$ given by the free-fall and high pressure versions, respectively, of the Child-Langmuir space charge equation (see "Space Charge Limited Current") are similar enough that they might be interpreted as a linear field variation. So it may be that the experimental evidence for the linear field is not accurate enough to differentiate between the various possibilities; alternatively the results obtained may not apply to our discharges.

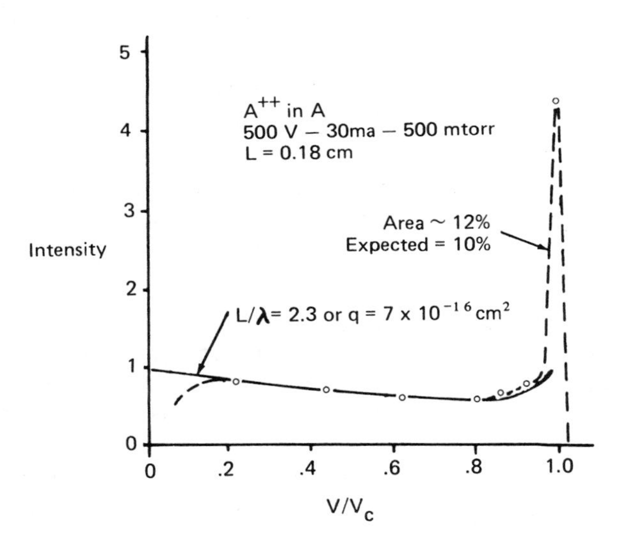

Figure 4-25. Energy distribution for Ar^{++} from an argon discharge. Dashed line and circles are experimental values while full line is the calculated distribution for L/λ = 2.3. The area of the peak represents those ions with the full cathode fall potential (Davis and Vanderslice 1963)

Generation Of Fast Electrons

If the secondary electrons emitted from the cathode were accelerated across the cathode sheath without making any collisions, then they would reach the edge of the sheath and enter the negative glow with an energy equivalent to the voltage drop across the sheath (give or take the energy of emission, and the sheath edge potential due to the Bohm criterion). Conversely, collisions in the sheath, including ionizing collisions, would attenuate this energy and also introduce a distribution of electron energies.

Brewer and Westhaver (1937) examined the energy distribution of electrons passing through a perforation in an aluminium electrode. The electrons were then deflected by a magnetic field and observed by fluorescence on a suitable screen. When the anode was at the sheath-glow interface, there was no change in the spot size or shape indicating that the electrons were still monoenergetic. When the anode was moved back into the glow region, the spot on the screen started to lengthen, implying electron energy inhomogeneity.

Voltages of 400 V to 30 000 V were used in these experiments. For 1000 V electrons, it was estimated that a change of 100 V could be readily detected. The conclusion was that less than 2 ions per electron were formed in the sheath. Although not specifically stated, it was implied that nitrogen was used for this work at pressures between about 0.1 - 4 torr.

It is not clear in Brewer and Westhaver's paper whether they used their technique to measure the actual energies of the electrons or only to measure the energy spread. It is also unclear how they moved their anode to the sheath-glow interface without grossly perturbing the discharge. Finally, I wonder how the motion of electrons was affected by the travel from the anode slit to the fluorescent screen 9 cm away, since there is no indication that differential pumping was used to reduce the pressure and eliminate collisions in this region.

In another group of experiments described in the same paper, Brewer and Westhaver measured the length of the negative glow (their discharge could be up to 40 cm long, unlike our applied discharges) for discharges in helium, hydrogen, argon and nitrogen. They obtained very good agreement with a theory of Lehmann (1927) for the range of fast electrons from the sheath, with the implication that the glow resulted from these fast electrons. Brewer and Westhaver again concluded that a large number, if not most, of the electrons entering the glow had an energy corresponding to the voltage across the sheath.

These experiments are further evidence that there is not a great deal of ionization in the sheath. However, they have a further and more practical significance: the fast electrons entering the glow have quite a small cross-section due to their velocity (as was seen in Chapter 2, and is rationalized in "The Glow Region") and as a result a significant number of these electrons hit the anode with substantial energies. In Chapter 6, "Life on the Substrate", we shall present experimental evidence for this phenomenon, which applies to rf as well as dc discharges, and see how it can influence thin film growth in a sputtering system.

Space Charge Limited Current

We still have some inconsistencies to eliminate. One of these is that the cathode sheath length in the examples earlier in this chapter was about 1 cm, which is typical of the values found, whereas earlier we had calculated sheath thicknesses characterized by a Debye length of the order of 100 μm.

Before we can remove this inconsistency, we need to understand the phenomenon of *space charge limited current*. We shall see that this does apply to the sheath regions of glow discharges, but we shall initially introduce the idea in relation to the emission of electrons from a heated filament in high vacuum, for simplicity.

Collisionless Motion

Figure 4-26 shows a heated wire filament emitting electrons to a positively biased anode distance d away. In Chapter 6, "Some Other Sputtering Configurations", we shall see how such *hot filament systems* are used for both sputtering and plasma etching applications. But in the present illustration, high vacuum is used to avoid ionization and thus restrict the system to a single charge carrier – the electron. The electron emission from a heated filament is given by the Richardson-Dushman equation:

$$j = AT^2 \exp - \frac{e\phi}{kT}$$

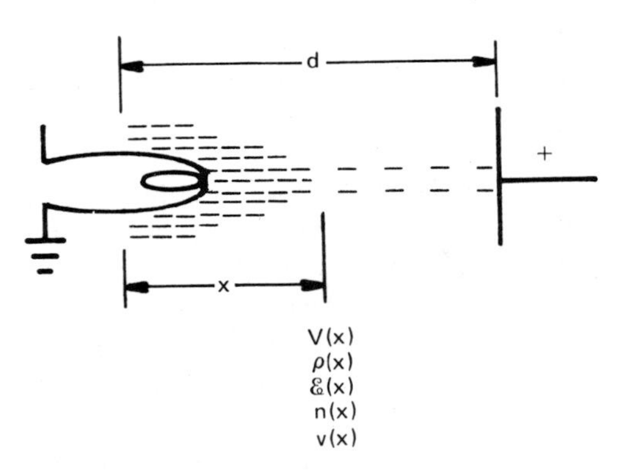

Figure 4-26. Space charge limited current from a heated filament

where ϕ is the *work function* of the filament material and A is a constant. In practice this value is not usually attained because an electron leaving the filament meets the strong Coulomb repulsion of the electrons which left previously, i.e. the actual emission current is limited by the space charge of the electrons. The limitation is overcome to some extent by applying an electric field to move the electrons away from the filament and reduce the space charge there.

At x (Figure 4-26), let the potential, electric field, electron density, electron velocity be V, $\mathcal{E}$, n_e and v respectively, where each is a function of **x**. Assuming a constant cross-section of the electron flux, j will be constant across the gap; m is the mass of the charge carrier, in this case the electron. For a single carrier:

$$j = nev \tag{1}$$

To find v, then by energy conservation

$$\tfrac{1}{2} mv^2 = eV$$

$$\therefore \quad v = \left(\frac{2eV}{m}\right)^{\frac{1}{2}} \tag{2}$$

To find n, then by Poisson's equation

$$\frac{d^2 V}{dx^2} = -\frac{\rho}{\epsilon_o} = \frac{ne}{\epsilon_o}$$

This cannot be integrated directly because n is a function of x. But using equations (1) and (2) to express n in terms of V,

$$\frac{d^2 V}{dx^2} = \frac{j}{\epsilon_o}\left(\frac{m}{2e}\right)^{\frac{1}{2}} V^{-\frac{1}{2}}$$

$$\therefore \frac{dV}{dx}\frac{d^2 V}{dx^2} = \frac{j}{\epsilon_o}\left(\frac{m}{2e}\right)^{\frac{1}{2}} V^{-\frac{1}{2}} \frac{dV}{dx}$$

Integrating,

$$\frac{1}{2}\left(\frac{dV}{dx}\right)^2 = \frac{j}{\epsilon_o}\left(\frac{m}{2e}\right)^{\frac{1}{2}} 2V^{\frac{1}{2}}$$

The integration constant is removed since, if more electrons are being emitted than manage to reach the electrode, then the field is about zero at x = 0 (depending on the emission velocity of the electrons). Rearranging this and integrating once more,

$$V^{-1/4}\,dV = \left(\frac{4j}{\epsilon_0}\right)^{1/2}\left(\frac{m}{2e}\right)^{1/4} dx$$

$$\frac{4}{3}\,V^{3/4} = \left(\frac{4j}{\epsilon_0}\right)^{1/2}\left(\frac{m}{2e}\right)^{1/4} x$$

Again the integration constant is eliminated since $V = 0$ at $x = 0$. This equation is put into its more usual form by squaring and rearranging:

$$j = \frac{4\epsilon_0}{9}\left(\frac{2e}{m}\right)^{1/2}\frac{V^{3/2}}{x^2}$$

Note also that $V \propto x^{4/3}$, so that $\mathcal{E} = dV/dx \propto x^{1/3}$. This is the high vacuum version of the *Child-Langmuir* space charge limited current equation. It applies for all values of x, including d, the full extent of the voltage sheath. Note that it applies to a single charge carrier under collisionless conditions, so that the energy conservation equation can be used. It can clearly be used for any charge carrier by suitable choice of m.

In our example of thermionic emission, by increasing the voltage V we would eventually reach the saturation current limitation imposed by the Richardson-Dushman equation. Current could be increased further only by raising the filament temperature, as shown in Figure 4-27. So space charge limitation applies only in the absence of a more stringent limitation such as the supply of charge carriers.

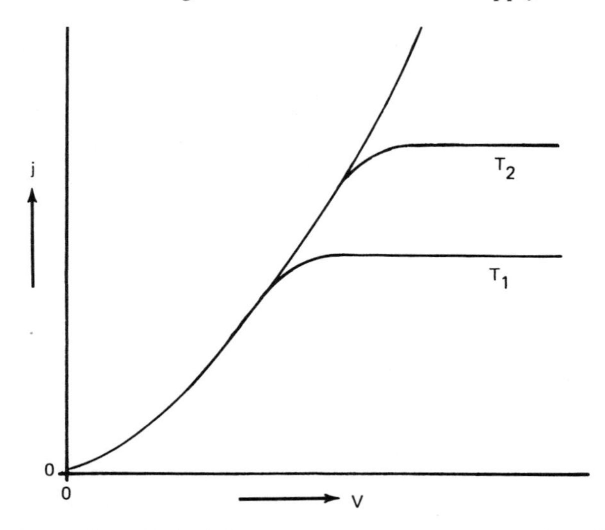

Figure 4-27. Space charge limited electron emission current versus voltage for two filament temperatures $T_2 > T_1$

Mobility Limited Motion

An electron travelling through a metal makes so many collisions that its *drift velocity* is quite small compared with its thermal velocity. We say that it is *mobility limited*. The drift velocity is proportional to the electric field; the constant of proportionality is the *mobility* μ, which we have already encountered in Chapter 3, "Ambipolar Diffusion".

To some extent, the same concept can be applied to gases, particularly in situations where the motion of the charge carriers is dominated by collisions. We can derive a space charge limited current equation for this case too, by substituting the drift velocity $V = \mu\mathcal{E}$ for the carriers instead of the velocity acquired by free fall. We would expect the resulting current to be smaller than for collisionless travel since it is now more difficult for the charge carriers to accelerate. Using a similar derivation,

$$j = nev = ne\mu\mathcal{E}$$

$$\frac{d^2V}{dx^2} = \frac{d\mathcal{E}}{dx} = \frac{ne}{\epsilon_0}$$

$$= \frac{j}{\mu\epsilon_0}\,\frac{1}{\mathcal{E}}$$

$$\therefore \tfrac{1}{2}\mathcal{E}^2 = \frac{jx}{\mu\epsilon_0}$$

$$\mathcal{E} = \left(\frac{2jx}{\mu\epsilon_0}\right)^{1/2} = \frac{dV}{dx}$$

$$\therefore V = \frac{2}{3}\left(\frac{2j}{\epsilon_0\mu}\right)^{1/2} x^{3/2}$$

and

$$j = \frac{9\epsilon_0\mu}{8}\frac{V^2}{x^3}$$

This is the mobility limited version of the Child-Langmuir equation, sometimes known as the high pressure version, though somewhat misleadingly. Note that $V \propto x^{3/2}$, so that $\mathcal{E} \propto x^{1/2}$.

Application to Glow Discharge Sheaths

Which one of these space charge limited current equations apply to the sheaths in our discharges, if either?

The first problem is that the equations were derived for single charge carriers, and we have two – electrons and ions (and even more if multiple ions are included). Actually this isn't much of a problem because the electrons accelerate away from the sheath so rapidly that they produce a very small space charge density. However, this assumption of negligible electron density would not be true if there were copious ionization in the sheath.

Let's see what order of current densities are predicted by the two space charge equations. We'll use again the example from the data of Davis and Vanderslice – a 600 V sheath of thickness 1.3 cm, in argon so that m is 6.6 10^{-26} kg. Substituting these values into the collisionless Child-Langmuir equation, we obtain a value of 75 $\mu A/cm^2$. This seems quite low, at the bottom end of the values obtained in sputtering systems. But 600 V is quite a low cathode voltage for a dc sputtering system. Unfortunately Davis and Vanderslice do not report the current they obtained for this condition, but they do for another situation – 30 mA current from a 500 V sheath of thickness 0.18 cm at 500 mtorr. For these conditions the high vacuum current would be 2.9 mA/cm^2 ; since their target was 4.5 cm diameter, their actual current density was 1.9 mA/cm^2. The difference could well have been due to the charge exchange collisions in the sheath.

To use another example, Güntherschulze (1930) reports values for a helium discharge with an iron cathode, equivalent to a dark space thickness of 0.64 cm at 1 torr for a voltage of 1000 V. The high vacuum space charge density should then be equal to 2.1 mA/cm^2, which compares very well with the measured value of 2 mA/cm^2. The good agreement may be fortuitous, although the charge exchange cross-section for He^+ in He is several times lower than the equivalent figure for argon. And returning to argon, we should note that a 1000V sheath of thickness 1 cm would give a current density of 0.27 mA/cm^2 ; all of these values are consistent with observed sputtering values. We cannot expect to achieve very precise values of the space charge limited current because of the difficulties involved in assessing L, as will become more apparent in the next section. However, it does seem that the observed cathode currents are almost as large as the values predicted by the collisionless Child-Langmuir equation. This implies either that the saturation value of ion current from the glow has not been reached, or that the sheath thickness adjusts itself to extract precisely the saturation current. It would be difficult to test this in a diode discharge because increasing the cathode voltage would increase the power input to the discharge. The high voltage probe characteristics might be more illuminating. Tisone and Cruzan (1975) have measured the target voltage and sheath thickness for a target immersed in a hot filament discharge (see Chapter 6). They obtained rather good agreement with a $V \propto x^{4/3}$ relationship. It seems as though the sheath thickness is determined by the ion production rate in the glow and by the space charge limitation, at least in this case.

A second implication of the small differences between the free fall current limit and the measured values is that there are not many collisional processes in the sheath involving ions. This is further evidence that there is not much ionization in the sheath.

By definition, any motion in the sheath that is not free-fall is mobility limited, though not generally with the simple field-independent mobility μ assumed in the derivation of the mobility limited space charge equation. In Appendix 4, there are several sets of data relating to the drift velocity and mobility of ions and electrons in argon. You can see that these are plotted against $\mathcal{E}/p$. This is common practice when looking at conduction in gases at lower fields and higher pressures; even then μ is very dependent on $\mathcal{E}$, as can be seen from the data presented. Note that $\mathcal{E}/p$ is typically around a few volts/cm torr in these examples (although up to 100 and 240 volts/cm torr in two untypical cases). By comparison, our earlier example of a 500 V sheath of thickness 0.18 cm at 500 millitorr corresponds to $\mathcal{E}/p$ values increasing from about 0 at the sheath-glow interface up to 11 100 V/cm torr at the cathode, if we follow the assumptions of Davis and Vanderslice. Obviously we can 'predict' the observed values by suitably choosing μ, which in this example would need to be 446 cm^2/volt sec. If we guess, from Figure 4-23, at an average argon ion arrival energy at the cathode of 100 eV, then this is equivalent to a velocity of 2.1 10^6 cm/sec. The field at the cathode in this example is predicted to be 5.6 10^3 V/cm, so this gives a crude estimate of μ equal to 375 cm^2/volt sec. Mobility figures obtained in these two ways are virtually forced to agree, but the consistency is encouraging. The main point, however, is that these mobility figures are more than two orders of magnitude higher than equivalent figures obtained for conventional mobility limited situations. We can therefore conclude that ion motion in the sheaths of our dc discharges is much closer to free fall than conventional mobility limitation.

Finally, we should note that since the product of sheath thickness and pressure in dc systems is observed to be constant, then reducing the operating pressure will not significantly change the number of collisions in the sheath. By the same token, neither will increasing the pressure, and ion motion will remain closer to free-fall that mobility limited. Hence the earlier comment that the title of 'high pressure space charge equation' for the mobility limited situation was rather misleading. We can change the situation in rf systems which retain sheath thicknesses of about 1 cm even when the pressure is reduced down to 1 millitorr. At such a low pressure, collisions in the sheath become very unlikely and motion becomes essentially free-fall, albeit modulated by the applied rf.

Structure of the Cathode Sheath

We are now ready to account for the large difference between the Debye length and typical cathode sheath dimensions.

The Debye length was introduced in Chapter 3 by considering the space charge sheath formed around a perturbation in the discharge. In the subsequent derivation of the potential distribution around the perturbation, we assumed that the ion density remained constant at its unperturbed value. But as we have just seen, a large semi-permanent negative potential causes the formation of a positive space charge sheath of varying density. This sheath may be as much as a few cm thick. Our final sheath model (Figure 4-28) therefore has 3 regions:

- a quasi-neutral 'pre-sheath' in which ions are accelerated to satisfy the Bohm criterion, as discussed in Chapter 3.
- a region of the extent of a few Debye lengths in which the electron density rapidly becomes negligible.
- a region of space charge limited current flow, which would be of zero electron density in the absence of secondary electron emission from the target, and in practice is not so different because of the rapid acceleration of the electrons.

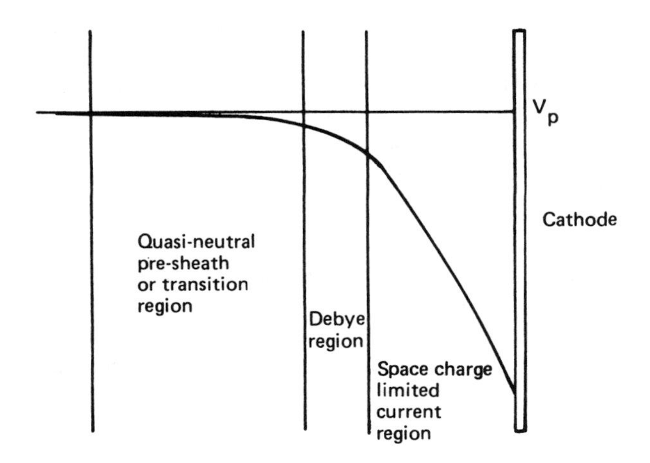

Figure 4-28. Regions of a cathode sheath

Of course, these divisions are in our minds only. A difficulty in experiments on sheath thicknesses is of trying to decide where the edge of the sheath is. Practically, people generally look for the change of luminous intensity due to de-excitation, either with a travelling microscope or an emission spectrometer with spatial resolution. But a change in intensity need not necessarily coincide with the boundary of the sheath as we have defined it. Fortunately, since the average thermal velocity of excited atoms will be about 5 10^4 cm/sec, at least we don't generally have to worry about atoms moving appreciably between excitation and relaxation, which might not be the case for excited ions in the sheath or excited sputtered atoms, which have greater than thermal energies.

THE ANODE REGION

Structure of the Anode Sheath

In Chapter 3, we saw how a small sheath must be set up in front of the anode, of sufficient magnitude to repel some of the random flux ¼ $n_e \overline{c_e}$ of electrons and reduce the current density at the anode to a more practical value. Our model of the sheath was essentially the same as that in front of a floating substrate (Chapter 3, "Sheath Formation at a Floating Substrate") except that the sheath voltage isn't as large at the anode. Later in Chapter 3, we needed to involve a pre-sheath or transition region to satisfy the Bohm criterion, and we expect this to apply to the anode too. The anode sheath is found to be so thin, usually about an order of magnitude less than the cathode sheath, that it should be essentially collisionless – and in particular not a source of ionization, which was tenuous even in the much thicker cathode sheath.

The anode sheath won't be very different from that in our derivation of Debye shielding. The Bohm criterion requires the ions to enter the sheath with an energy of about kT_e/e, and they then accelerate through the anode sheath to reach energies of 10 - 15 eV. The energy increase of a factor of 3 - 10 is equivalent to a velocity increase of $\sqrt{3}$ - $\sqrt{10}$, and an inverse change in ion density. The main point is that the ion density is not far from the uniform density assumed in the Debye sheath derivation, and does not vary anywhere near as much as in the cathode sheath. At the same time, the sheath voltage is small enough that the electron density does not go to zero as in the cathode sheath. The net result is that the anode sheath consists primarily of a pre-sheath and a Debye-like region.

Secondary Electron Emission

Unlike our simple model, in reality there is secondary electron emission from the anode. With the usual polarity of the anode sheath, these electrons are accelerated back into the glow, acting as a source of both electrons and energy to the glow. The anode is bombarded by ions, photons and electrons. Most of these come from the glow, except for the fast electrons which are generated in the cathode sheath; many of these travel through the glow without making many collisions and strike the anode with considerable energy. These fast electrons are responsible for a significant power input to the anode.

As well as the fast electrons, there are slower electrons from the glow. The coefficient δ for electron bombardment tends to be larger than the coefficient γ_i for ion bombardment, so there are a significant number of electrons emitted and injected back into the glow, where we expect them to have an effect on the generation of the glow. Gillery (1978) has observed a considerable change in the V - I characteristics of his glow as substrates pass in front of his sputtering target. There is the implication in his work that it is the fast secondaries from the target which are having the greatest effect.

Note that another effect of secondary electron emission is to invalidate our earlier calculations of floating potentials and anode sheath potentials, which are therefore only approximately correct.

Space Charge Limited Anode Current

In our earlier attempts to consider the effect of space charge in limiting current flow, we were able to assume that there was a single charge carrier, or at least reasonably so. But in the anode sheath, the electron density is not necessarily insignificant, particularly if secondary electron emission from the anode is included. Therefore our existing collisionless form of the Child-Langmuir equation appears to be inadequate for the anode sheath, and a two carrier model is required. Testing of a model would be difficult since sheath voltage and thickness would both be small and subject to measurement error.

Polarity of the Anode Sheath

I have so far argued that the polarity of the anode sheath is such that the plasma potential will always be more positive than the anode. This is not always the case. Two reasons for the polarity would be

- high secondary electron coefficient at the anode
- physically small anode.

To treat the first of these, suppose that the electron and ion fluxes to an anode are j_e and j_i respectively. If the secondary electron coefficient for electron bombardment is δ, and if we add together the effects of ions and photons so that the secondary electron flux due to these bombardments is γj_i, then the net electron flux to the surface will be $j_e - (\delta j_e + j_i + \gamma j_i)$. Clearly this expression would be negative for values of $\delta > 1$. Of course the actual net current must be an electron current. The discharge achieves this by reducing the anode sheath voltage so that j_e (and also δj_e) increases. By continuing this process, the sheath voltage reduces and then *changes polarity*. At this stage j_e has reached its random value and no longer increases, whilst the ions encounter a retarding field at the anode so that j_i decreases; the ejected secondary electrons now encounter a repulsive force so that the less energetic of the δj_e and γj_i electrons return to the anode. It would be interesting to run a discharge using an anode with a large δ to assess the practical extent of this sheath field reversal.

A second reason for polarity change at the anode sheath is anode size. For a given discharge of given total current, the current density at the anode would have to increase as the anode size is decreased. This would be achieved by decreasing the voltage of the anode sheath so that fewer electrons are repelled. The electron current could increase in this way until it reaches saturation when the anode is at plasma potential. Further increases in net electron current are then achieved by reducing the ion current, i.e. the sheath polarity reverses.

Main Effects in the Anode Region

The polarity of the anode sheath is usually such as to accelerate secondary electrons from the anode back into the glow, and also to accelerate ions from the glow onto the anode and onto any substrate there. Although the sheath is too thin to be a likely source of ionization, the accelerated secondary electrons act as both an electron source and an energy source to the glow. The sheath has to rely on the glow as an ion source. Since there appears to be rather little ionization in the cathode sheath and even less in the anode sheath, the ion fluxes at each electrode are of similar magnitude.

THE GLOW REGION

And so we come to the glow region. Although the glow is an ionized gas of approximate charge neutrality, it certainly isn't the uniform isotropic plasma described in Chapter 3. The main reason for this is the beam of fast electrons entering the glow region from the cathode sheath; these penetrate into and through the sheath and make it very anisotropic. People refer to three groups of electrons in the glow:

- *primary electrons* which enter from the cathode sheath with high energy; the name is slightly confusing because these same electrons were secondary electrons emitted from the target.
- *secondary electrons* of considerably lower energy; these are the product electrons of ionizing collisions or primaries which have lost much of their energy.
- *ultimate electrons* which have become thermalized to the plasma temperature; these ultimate electrons have the highest density. In a low current density neon discharge at 1 torr, Francis (1956) reports densities of $5\ 10^6$, $5\ 10^7$, and $4\ 10^9$ for the primary, secondary, and ultimate electrons, respectively.

One might at first think that the primary electrons would soon lose their directionality and energy. However, as we saw in Chapter 2, there is a tendency for cross-sections to decrease with increasing energy at high energies. The rationale for this is illustrated in Appendix 4 for electron interactions; all interactions in plasmas are fundamentally electronic in nature. Another consequence of the weak interaction is that of *forward scattering*, i.e. the incident particles are not deflected much from their initial path.

So there is a good chance of fast electrons passing through the glow and colliding with the anode. We shall see experimental evidence for this in Chapter 6, in a sputtering application. Since the electron collision cross-section continues to decrease with increasing energy, there comes a point where increasing the voltage across a diode system has little effect – the electrons pass right through. This is the phenomenon of the *runaway electron*, and is encountered in very high temperature plasmas as an obstacle to heating (i.e. putting energy into) the plasma. This is not really a problem in our cold plasmas, although the seeds of the problem are evident.

In a long glow discharge, where there is room for a positive column to develop, then the energy of the primary electrons can be attenuated before they reach the positive column. As a result the positive column is much more like an idealized plasma, which has made it a popular testing ground for probe theories. The negative glow, with its three groups of electrons and anisotropic nature, is obviously a more difficult region to deal with. Some folks have used *directional probes* to try to distinguish between the various groups of electrons (Fataliev et al. 1939, Polin and Gvozdover 1938, Pringle and Farvis 1954), but there seem to be problems of interpretation. There are two-temperature models of the glow, pertaining to the secondary and ultimate groups of electrons, and probe measurements to substantiate a Maxwell-Boltzmann energy distribution for each of these groups. Ball (1972) has observed such two-temperature distributions in a dc sputtering discharge.

Ionization in the Negative Glow

In the next few sections we shall be considering the contributions of the various ionization mechanisms that can exist in the glow region. Remember that, using the example in "Architecture of the Discharge", we need an ionization rate of at least $3.8\ 10^{15}$ ions/second for each cm^2 of cathode to maintain an argon discharge of 0.3 mA/cm^2 at 2000V.

By Fast Electrons

The fast electrons entering the glow will obviously cause some ionization. Figure 2-8 shows the energy dependence of the ionization cross-section in argon. The ionization rate will be nq per cm per electron. Although we should really consider the energy dependence of q, let's use an average value of $1.3\ 10^{-16}\ cm^2$, which is half of the maximum value and should be reasonable over the energy range considered. In contrast to the situation in the cathode sheath, we should not use the multiplicative (exponential) version of the ionization rate here because the electron produced by the ionization will have an energy of only a few eV. In the absence of the large field of the cathode sheath, this slow electron will not immediately produce further ionization.

Assuming a value of 0.1 for γ, the electron current at the edge of the glow will be 0.03 mA in our example, equivalent to an electron flux of $1.9\ 10^{14}$ electrons/cm^2 sec. At a pressure of 50 mtorr, the ionization rate per cm^3 per sec will be

$$1.9\ 10^{14}\ (3.54\ 10^{16}\ x\ 50\ 10^{-3})\ 1.3\ 10^{-16},$$

which is $4.3\ 10^{13}$ ions/cm^3 sec. Even allowing for a glow length of 5 cm, this figure is too low, by a factor of at least 20, to sustain the discharge. This large difference could not be sensibly accounted for by underestimation of γ or of electron multiplication in the cathode sheath. So we conclude that the fast electrons do not directly cause enough ionization to sustain the glow.

By Thermal Electrons

Electrons just above threshold have a smaller ionization cross-section than the fast electrons from the cathode, so how can the slow electrons cause much ionization?

Figure 4-29 is an electron energy diagram of the discharge, redrawn from Figure 4-4. Assume that the electrons are thermalized with a Maxwell-Boltzmann distribution around the electron temperature T_e. Figure 4-29 tells us that an electron in the plasma needs an energy of eV_p to reach the anode and $e(2000 + V_p)$ to reach the cathode. The probability of the former is $\exp(-eV_p/kT_e)$, which has a small but finite value of $7\ 10^{-3}$ for $V_p = 10$ V and $kT_e = 2$ eV. The probability of the electron returning to the cathode is virtually zero – less than 10^{-99}, which is as far as my calculator goes! The net result is that electrons become trapped in the glow region, generally being reflected at the interfaces with the electrode sheaths, including the sheath at the wall, before eventually managing to overcome the anode barrier. So the effective path length is increased as necessary to maintain the ion and electron densities by electron impact ionization.

Let us see if the amount of ionization in the glow is adequate to maintain the glow. The temperature of the electrons in the glow is typically around 2 eV – 8 eV, which is not adequate to ionize argon, which has a first ionization potential of 15.7 eV. But of course, in a Maxwell-Boltzmann distribution, some particles

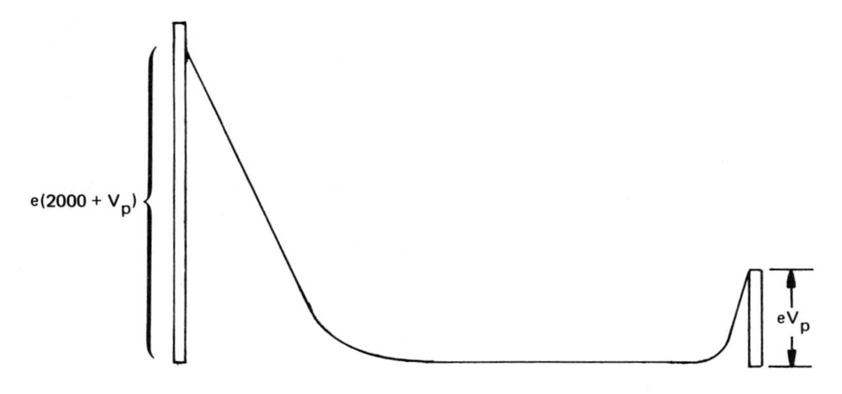

Figure 4-29. Electron energy diagram for the dc glow discharge

have energies far in excess of the mean. In Chapter 1, we saw that the speed distribution of such a gas is given by:

$$\frac{dn}{n} = 4\pi \left(\frac{m}{2\pi kT}\right)^{3/2} c^2 \exp \frac{-mc^2}{2kT} \quad dc$$

Since the kinetic energy E can be written in terms of speed c as

$$E = ½ mc^2$$

then

$$c = \left(\frac{2E}{m}\right)^{½}$$

and so

$$dc = ½ \left(\frac{2E}{m}\right)^{-½} \frac{2dE}{m}$$

Hence we can derive an energy distribution:

$$\frac{dn}{n} = \frac{2}{\pi^{½}} \frac{1}{(kT)^{3/2}} E^{½} \exp - \frac{E}{kT} \quad dE$$

perhaps more conveniently written as

$$\frac{dn}{n} = \frac{2}{\pi^{1/2}} \left(\frac{E}{kT}\right)^{½} \exp - \left(\frac{E}{kT}\right) d\left(\frac{E}{kT}\right)$$

This version of the Maxwell-Boltzmann distribution function tells us how the number of particles is distributed as a function of their energy E with respect to the temperature T of the distribution. The expression can be used in various ways, for example to calculate the fraction $f(E > E_o)$ of electrons in a Maxwellian distribution of temperature T_e, that have an energy greater than E_o:

$$f(E > E_o) = \frac{2}{\pi^{1/2}} \int_{E_o/kT_e}^{\infty} \left(\frac{E}{kT_e}\right)^{1/2} \exp\left(\frac{-E}{kT_e}\right) d\left(\frac{E}{kT_e}\right).$$

Some values obtained for this integral are shown in Table 4-4. The fraction of electrons which are energetic enough to excite argon (threshold 11.56 eV) or ionize argon (threshold 15.76 eV) are also calculated for a number of electron temperatures.

Table 4-4

The table shows the fraction $f(E > E_o)$ of particles having an energy greater than E_o in a Maxwellian distribution of temperature T_e.

E_o	kT_e	E_o/kT_e	$f(E > E_o)$
		0	1.0
		0.5	$8.2\ 10^{-1}$
		1	$5.9\ 10^{-1}$
		2	$2.7\ 10^{-1}$
		3	$1.2\ 10^{-1}$
		4	$4.8\ 10^{-2}$
		5	$1.9\ 10^{-2}$
		7	$3.0\ 10^{-3}$
		10	$1.8\ 10^{-4}$
11.56 eV	0.25 eV	46.2	$7.0\ 10^{-20}$
argon	0.5	23.1	$5.4\ 10^{-10}$
excitation	1.0	11.6	$4.0\ 10^{-5}$
threshold	2.0	5.78	$9.5\ 10^{-3}$
	4.0	2.89	$1.3\ 10^{-1}$
	8.0	1.45	$4.2\ 10^{-1}$
15.76	0.25	63.0	$4.1\ 10^{-27}$
argon	0.5	31.5	$1.4\ 10^{-13}$
ionization	1.0	15.8	$6.9\ 10^{-7}$
threshold	2.0	7.88	$1.3\ 10^{-3}$
	4.0	3.94	$5.1\ 10^{-2}$
	8.0	1.97	$2.8\ 10^{-1}$

But this tells us only the proportion of electrons *capable* of ionization. We can calculate the *rate* of creation of ion-electron pairs by using the cross-section data referred to in Chapter 2. We saw that the cross-section q(E) was a function of energy. This cross-section, as defined earlier, gives an ion pair production rate nq(E) per electron per centimetre path length of the electron. For our present purposes, it is more useful to know the rate per unit time, i.e. in c centimetres at an electron speed of c. We can therefore write the rate of ion pair production per unit volume of the plasma per unit time as:

$$\text{Ion production rate} = \int_{0}^{\infty} n\, q(E)\, c\;\; dn_e(E)$$

since $dn_e(E)$ is the number of electrons having energies between E and E + dE, with corresponding speeds varying between c and c + dc, and n q(E) is the probability per unit length of forming an ion from a volume density n of gas atoms.

Examining the function within the integral, we know that q(E) is zero for all energies up to the ionization threshold eV_i. The integrand then begins to take nonzero values with q(e) monotonically increasing and $dn_e(E)$ monotonically decreasing. The cross-section q(E) can be written as $a\,(E - eV_i)^b\; \pi\, a_o^{\;2}$ for all values of E greater than the ionization threshold eV_i, by making a power curve fit to the data of Rapp and Englander-Golden (1965) discussed in Chapter 2. The speed term can be written as $(2E/m)^{1/2}$, and $dn_e(E)$ is just the Maxwell-Boltzmann distribution function. Since there is no ionization below the threshold, the integral becomes

$$\int_{eV_i}^{\infty} n\, a(E - eV_i)^b\;\; \pi\, a_o^{\;2} \left(\frac{2E}{m}\right)^{1/2} n_e\, \frac{2}{\pi^{1/2}} \left(\frac{E}{kT_e}\right)^{1/2} \exp\left(\frac{-E}{kT_e}\right) \frac{1}{kT_e}\; dE$$

where

n = number/cc = p (torr) x 3.54 10^{16}

a = 0.125 for argon

b = 1.077 for argon

E is in eV

eV_i = 15.7 eV for argon

$\pi a_o^{\;2}$ = 8.82 10^{-17} cm^2

$\left(\frac{2E}{m}\right)^{1/2}$ is in cm/sec

= $E^{1/2}$ 5.93 10^7 cm/sec, E in eV

n_e = plasma density/cc

$$\frac{2}{\pi^{1/2}} = 1.13$$

The integral is thus

$$2.09\ 10^8\ p\ a\ n_e \int_{\frac{eV_i}{kT_e}}^{\infty} (E - eV_i)^b\ E^{1/2} \left(\frac{E}{kT_e}\right)^{1/2} \exp\left(\frac{-E}{kT_e}\right) d\left(\frac{E}{kT_e}\right)$$

This can be evaluated numerically, for example with a programmable calculator, and some values thus obtained are shown in Table 4-5. For the sake of illustration, these are based on a plasma density of 10^{10}/cc in argon at 50 mtorr, but note that the ion pair production rate is proportional to both p and n_e, so that the corresponding rates under other conditions can readily be assessed. Interestingly, the proportion of electrons having a specific energy seems to decrease with energy at about the same rate as the cross-section increases. As a result, ionization is not confined to the group of electrons just above threshold, as one might at first guess.

Table 4-5 Ion Pair Production Rates in Argon

p = 50 mtorr, $n_e = 10^{10}$/cc.

kT_e	Production Rate (per cc per sec).
0.25 eV	$5.2\ 10^{-11}$
0.5	$3.5\ 10^{3}$
1.0	$3.7\ 10^{10}$
1.5	$9.2\ 10^{12}$
2.0	$1.6\ 10^{14}$
2.5	$9.0\ 10^{14}$
3.0	$3.0\ 10^{15}$
4.0	$1.4\ 10^{16}$ *
8.0	$2.0\ 10^{17}$ *

*These values are based on an integration up to 100 eV using the same power curve fit to the ionization cross-section data. The resulting values are too large, but by less than a factor of 2.

Remembering that the minimum ionization rate required to sustain our discharge is 3.8 10^{15} ions per sec per cm^2 of the target, then if the length of the glow in the example is 5 cm, this corresponds to an ionization rate of 7.6 10^{14}/sec cm^3. From Table 4-5, this can apparently be achieved by a Maxwellian distribution of 10^{10} electrons/cm^3 with a temperature of 2.5 eV, which is a realistic figure for our discharges. And a small increase in electron temperature would provide enough ionization to account for wall losses, too. These calculations therefore suggest that the negative glow could provide enough ionization to sustain the discharge.

By Ions

One of the ionization mechanisms in the cathode sheath that we considered was of ionization by ion impact on neutrals, and it seemed as though there could be a small contribution. By contrast, the energy of ions in the glow will be very low, with an average ion temperature of less than 1000 K. And even for the very few ions with energies above the ionization threshold, the relevant cross-section will be much less than 10^{-18} cm^2, as can be seen in Figure 2-25. So ion impact can be completely ignored as an ionization source in the glow.

Of Metastables

A metastable argon atom has an excitation energy of 11.56 eV (or more) which is only 4.2 eV below the ionization energy. So the metastable can be ionized by a much larger proportion of the electrons in the glow than can a ground state atom. Since the metastable has already been excited by some energy input, this is known as a two step ionization process. Although there are many fewer metastables than ground state atoms, perhaps this is offset by the larger number of electrons which could ionize the metastables.

To make this calculation, we need to know the density of metastables and their ionization cross-section. Neither of these is well-known, so some guesswork is required. Eckstein et al. (1975) have measured the density of metastable neon atoms (16.62 eV and 16.72 eV) in a neon rf sputtering discharge at 20 mtorr, and found values of 10^{10} - 10^{11} cm^{-3}. We won't be wildly wrong if we guess at 10^{11} cm^{-3} for the argon metastables in our dc sputtering example. There is even less information about the ionization cross-section of the metastables, so let's assume that it has the same value as the maximum cross-section of 2.6 10^{-16} cm^2 for the ground state atom, but is energy independent above the threshold of 4.2 eV.

The rate of ionization by this process can be calculated in a similar way to that used for the ionization of ground state atoms. The technique that is shown in Appendix 4 can be used for any process with a constant cross-section above a given threshold. Some rates of electron impact ionization of argon metastables, for the assumed values of n^* and q and various conditions typical of the practical glow discharge, are shown in Table 4-6. If we then compare these values with the rates of ground state ionization at equivalent electron temperatures, we find that around 2 - 4 eV the ionization rate of metastables is smaller, but not much smaller. The apparent reversal of roles at low temperatures is primarily due to the assumption of constant metastable density. John Coburn has pointed out to me that the metastable atom might well have a much larger cross-section than its ground state partner; this could make the ionization of metastables comparable to that of ground states for electron temperatures of 2 - 4 eV, although it would still appear to be inadequate above 4 eV. We would be unwise, therefore, to ignore the ionization of metastables in the glow as a possible source of ion-electron pairs.

Table 4-6

Electron impact ionization of metastable argon atoms, assuming:

metastable density n^* — 10^{11} cm^{-3}
electron density n_e — 10^{10} cm^{-3}
ionization cross-section q — $2.6\ 10^{-16}$ cm^2
ionization threshold — 4.2 eV

Electron Temperature	Ionization Rate per cm^3 per second
1.0	$1.30\ 10^{12}$
2.0	$8.97\ 10^{12}$
2.5	$1.32\ 10^{13}$
3.0	$1.71\ 10^{13}$
3.5	$2.07\ 10^{13}$
4.0	$2.40\ 10^{13}$
5.0	$2.97\ 10^{13}$
6.0	$3.45\ 10^{13}$
8.0	$4.26\ 10^{13}$

Summary

From the calculations in the last few sections, it appears as though the main source of ionization in the discharge is by electron impact ionization of ground state argon atoms in the negative glow, with possible additional contributions from electron impact ionization and ion impact ionization in the cathode sheath, and from ionization of metastables in the glow. But these calculations were based on certain assumptions involving the electron distribution and the electron temperature, so let's examine those assumptions further.

The Electron Energy Distribution

We have been using the Maxwell-Boltzmann distribution to represent the energy distribution of electrons in the glow, or rather of those electrons that are not primary electrons from the cathode. The Maxwell-Boltzmann distribution applies to an assembly of particles in complete thermal equilibrium, for example the atoms in an ideal gas. By contrast, the electrons in the glow are in a non-equilibrium situation. The slower electrons make elastic collisions only, whereas electrons with energies above the excitation and ionization thresholds are liable to lose a large fraction of their energy by the corresponding inelastic processes. Fast electrons are also lost rapidly by diffusion to the walls, and recombination there. As a result, there is a transfer process of electrons from high energy to low energy states. So we might expect to have fewer electrons with high energies than the Maxwellian distribution predicts.

Druvestyn and Penning (1940) have tried to be more realistic by considering the motion of electrons in a weak electric field, such as that existing in the glow. The distribution which is so obtained, known as the *Druvestyn distribution*, when compared with a Maxwellian distribution predicts more electrons with energies around the average energy but many fewer electrons with energies greater than a few times average. However, their derivation still ignores inelastic collisions. Thornton (1967) has discussed how this model has been developed. Druvestyn and Penning (1940) and a later more detailed analysis by Holstein (1946), introduce a constant inelastic cross-section above threshold and this serves to further reduce the number of electrons with energies above threshold. Barbiere (1951) has included the velocity dependence of the elastic collision cross-section into his analysis. He shows that this has a very significant effect also in reducing the number of high energy electrons in argon because the Ramsauer effect causes the argon elastic collision cross-section to increase with increasing electron energy, in contrast to helium where it decreases.

The analyses above lead one to expect almost no energetic electrons at all. But experimentally the glow discharge electron distribution is found to be much more Maxwellian than it should be, based on these analyses. This is known as *Langmuir's Paradox*, and it was 30 years after Langmuir's original work that the resolution of the paradox began to be clarified, and I believe that the clarification is incomplete even now.

The analyses which produced this dilemma were based on the assumptions that the glow electrons gain their energy from the weak electric field across the glow, and that there is no energy interchange amongst the electrons. To treat the latter assumption, Thornton (1967) quotes the work of Dreicer (1960), who considered the effect of electron-electron and electron-ion interactions (see Appendix 4) on the electron distribution (actually in hydrogen gas). Dreicer concludes that these interactions can have a major effect in restoring the distribution to Maxwellian, but only at higher degrees of ionization ($> 10^{-2}$) than are encountered in our glow discharges ($\sim 10^{-4}$), so it still leaves us with a deficit of high energy electrons.

Any significant departure from a Maxwellian electron energy distribution will render invalid all the calculations we have made based on that distribution. In particular, it will enormously reduce the amount of ionization produced by the tail of the distribution in the glow and raise once again the question of how the glow is sustained.

In the rest of this chapter, I shall attempt to show that the main energy input to the electrons in the glow is from the fast electrons from the cathode rather than from the weak electric field across the glow, and that there are more energy interchange processes in the glow to be considered. An understanding of these sections is not essential before reading about the practical processes in Chapters 6 and 7, and a cursory reading may be adequate first time through.

Energy Dissipation in the Discharge

In order to clarify a couple of terms that I shall use, consider one of the 'water splash' rides that one sees at fairgrounds (Figure 4-30a). Having been mechanically raised, the boat accelerates rapidly down a ramp so that it acquires kinetic energy as it loses potential energy; let's say that a lot of kinetic energy is *generated* in the ramp. The boat then hits the water and is quickly slowed down as its energy is *dissipated* by transfer to the water. Note that no energy is generated in the water trough since it is level.

Let's see if we can apply some of these ideas to the discharge. There are three regions to consider: the sheaths at cathode and anode, and the glow itself. Using the values in our example again, we need at least $3.8\ 10^{15}$ ions produced per second per cm^2. Each ionization step requires a minimum energy of 15.7 eV, whether by one-step or two-step processes. The minimum energy consumption is therefore $3.8\ 10^{15}\ x\ 15.7$ eV/sec cm^2, which is equal to $3.8\ 10^{15}\ x\ 15.7\ x$ $1.6\ 10^{-19}$ joules/sec cm^2, or 9.6 mW/cm^2. In practice, the electron energy loss per ionization is more than 30 eV since the collision products also have some kinetic energy, and there will be many more ionizing collisions than $3.8\ 10^{15}$/ cm^2 sec to account for wall losses. There will also be further energy losses due to the inelastic collisions producing excitation. If, as seems likely, most ionization occurs in the glow, then the power consumption there (requiring an equivalent amount of dissipation) will be at least 9.6 mW/cm^2, and probably several times this value. The glow region is rather like the water trough in the analogy, except that the glow does have some electric field across it. We have already seen that the glow should be equipotential within a few kT_e/e, and this appears to be consistent with measurement; Brewer and Westhaver (1937) found values of just a few volts. Let's assume 10 V across the glow. The current through the glow in our example is 0.3 mA per cm^2 of the target. These values give a power generation in the glow of 3 mW/cm^2, considerably less than even the very minimum value of 9.6 mW/cm^2 which must be dissipated there.

Where does this energy come from? The main power generation in the discharge is in the cathode sheath, and amounts to 2010 x 0.3 mW/cm^2, i.e. 603 mW/cm^2. Most of this goes into kinetic energy of ions and subsequently into heating of the cathode. We won't be far wrong by assuming a collisionless sheath and a secondary electron coefficient of $\gamma = 0{\cdot}1$, so that 10% of the current is carried by electrons. In the absence of collisions, these electrons enter the glow with a kinetic energy equivalent to the cathode sheath voltage, and so inject 60 mW/cm^2 of power into the glow, notably adequate to account for the ionization required with power to spare. The excess power is consistent with the observation that some fast electrons lose very little or no energy in the glow and hit the anode at high velocity. We shall see some evidence of this in Chapter 6, when we look at sputtering. It's as though the water trough in our analogy was not completely efficient in arresting the motion of the boats, so that some boats hit the end wall with considerable velocity even in the presence of a 'braking' hill (Figure 4-30b). I now understand my fear of such amusements! Notice the similarity between Figures 4-29 and 4-30b.

Energy Transfer Amongst the Discharge Electrons

The calculations and experimental evidence we have presented so far could be made consistent if the electrons from the sheath act as an energy source to the glow region. But how is this energy transferred?

Inelastic Collisions of Fast Electrons

Some of the fast electrons make ionizing collisions. As a result, they produce a second electron with a few eV of energy and also slow down due to the energy loss. Because of the energy dependence of the ionization cross-section, their propensity for further ionization increases. The deceleration of these electrons is increased because they also excite atoms, sometimes simultaneously with ionization. Of these processes, only the production of second electrons by ionization directly adds to the energy of the glow electrons. The photons resulting from ex-

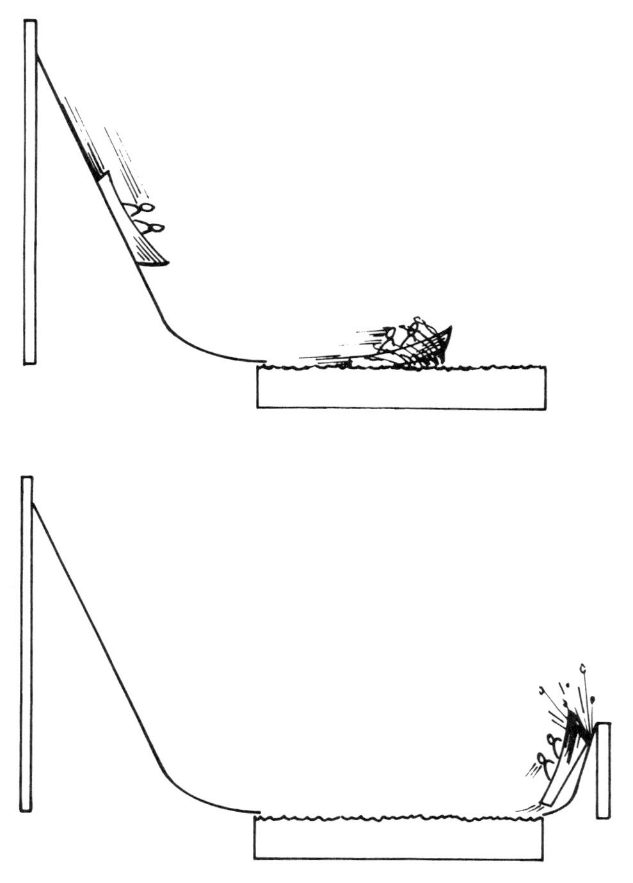

Figure 4-30. The water splash

citation probably don't do much in the gas phase because the cross-sections are too low, but they may cause secondary electron emission from the chamber walls.

In an earlier section on "Ionization in the Negative Glow", we calculated a rate of 4.3 10^{13} ionizing events per sec per cm travel of 0.03 mA of fast electrons entering the glow. Allowing a loss of 30 eV for each of these collisions, power dissipation in the glow will be 0.21 mW/cm^2 per cm length, or 1 mW/cm^2 if we include a glow length of 5 cm. This value is a significant addition to the comparable figure of 3 mW/cm^2 generated in the glow, but is still far short of requirements.

Note that the figure of 0.2 mW/cm^2 per cm is in good accord with the results of Brewer and Westhaver (1937) and Lehmann (1927). According to the latter, the range of 700 V electrons in argon at 1 torr is 5 cm, with an inverse dependence on pressure that would imply a range of 100 cm at 1 mtorr. A current of 0.03 mA at 700 V would have an initial energy of 21 mW, and if this is attenuated at a rate of 0.2 mW/cm, the resulting range would be close to 100 cm. On further thought, this is probably where the value of 30 eV per ionizing collision came from!

To return to the problem of transferring energy to the glow, the values we have obtained suggest the loss of energy by the fast electrons due to inelastic processes alone is inadequate to develop the power dissipation required in the glow.

Electron-Electron Collisions

We have already mentioned, in "The Electron Energy Distribution", the subject of electron-electron collisions. These are a potential source of energy exchange since the equal masses involved maximize the energy transfer function, and the Coulomb interaction between them is quite long-range, leading to a large effective cross-section. In principle, the cross-section would be infinite since the Coulomb interaction is, too. But we have to remember that in a plasma, the collective behaviour of electrons and ions causes electric fields to be *screened*, as discussed in Chapter 3 in connection with the Debye length λ_D. When we examine the electron-electron interactions in more detail, as in Appendix 4, we find a collision frequency of 1.3 10^5 per second, very comparable to the value of 4 10^5 per second for atom-atom collisions in an ideal gas, derived in Chapter 1, "Collision Frequency". However, the individual collisions are so weak that the energy transfer is insignificant for a 100 eV electron, amounting to only 2 10^{-4} eV/cm. The energy transfer is inversely proportional to the energy (see Appendix 4), but even for a 1 eV electron, the loss due to collisions with other electrons is only 1.4 10^{-2} eV/cm. On the other hand, the energy transfer is almost proportional to the electron density, so if the electron density were 10^{14} cm^{-3}, the transfer would be about 20 V/cm for a 10 eV electron, i.e. strong interaction, consistent with the results of Dreicer (1960) discussed in "The Electron Energy Distribution".

Interactions With Plasma Waves

The electron-electron collision analysis discussed above was based on the summation of individual pair interactions, and the analysis ignored any collective behaviour. Is this reasonable?

We have already briefly referred to some experiments by Langmuir (1925); he made some probe measurements on a *hot filament discharge*, in which electrons are thermionically emitted from a heated cathode and enable a discharge to be sustained with a few tens of volts. Using the probe technique discussed in Chapter 3, Langmuir identified three groups of electrons in the discharge:

- *primary electrons* from the cathode, which retain practically all the momentum acquired by acceleration across the cathode sheath, and hence are directional.
- *secondary electrons*, moving in random directions with a Maxwellian distribution about a temperature proportional to the primary beam energy (200 000 K for 100 eV primaries). This group includes primaries which have lost most of their energy and electrons emitted from ionizing collisions.
- *ultimate electrons*, which were the most numerous, $\sim 10^3$ times the density of the primaries and secondaries, with a Maxwellian distribution of energies around 1 - 3 eV. These were assumed to be secondary electrons which had lost most of their energy to join the ultimate group.

We have previously mentioned these three groups of electrons. What is more relevant in the present context is the energy spread of the primary electrons that Langmuir observed. At low discharge currents the electrons were quite monoenergetic with a spread of about 2 volts for a 50 V beam. However, when the current was raised, there were electrons with energies both greater and less than the interelectrode potential; for example, for a beam current of 10 mA, the primary beam energy was about 47 volts, with a spread of ± 10 V. Langmuir referred to this as the phenomenon of *high scattering*; Tonks and Langmuir (1929) subsequently found high frequency oscillations in the discharge – the *plasma oscillations* discussed in Chapter 3.

Langmuir's approach was refined, and the connection between high scattering and the oscillations established, by the experiments of Merrill and Webb (1939). They used an indirectly heated cathode to avoid magnetic field effects, and a probe which could be moved by very small increments. This enabled them to discover that, although oscillations existed everywhere, they had a sharp peak in the glow, in a very localized region a few tenths of a millimetre deep. The 'high scattering', observed as a velocity modulation about the initial primary electron energy, appeared in a distinctly separate region about 0.5 mm nearer to the cathode. The reason for these separate locations can now be understood in terms of the operation of a *klystron*. The modulation of the electron velocity in the region of high scattering introduces many different electron velocities, but does not change the current in that region. However, the faster electrons now begin to catch slower electrons so that the phenomenon of *electron bunching* appears – for the same reason that public transport buses come in threes. The bunching is observed as oscillations in current (which is what the probe was looking for) and will reach a maximum some distance 'downstream' from the region of velocity modulation, as observed by Merrill and Webb. (In the klystron, velocity modulation of a beam of electrons is produced by applying a small high frequency voltage modulation via a resonant cavity. As a result, electron bunching occurs, and the current oscillations produced further down the beam are used, with a suitably placed second cavity, to induce a power modulation in an external impedance. The power delivered to the external impedance comes primarily from the kinetic energy of the electron beam, so that power amplification from the input modulation to the output modulation has occurred, i.e. the klystron is a high frequency amplifier).

In the klystron, and in the plasma too, the fast electrons do not overtake the slower electrons. This is because the bunching causes an increased negative space charge which repels and decelerates the fast electrons as they try to overtake the slower electrons. The net result is that the electrons in the beam vibrate about the positions they would have occupied in the unmodulated beam, at the plasma frequency, as discussed in Chapter 3; at the same time, the whole electron beam moves at the original velocity. So we have *space charge waves* moving through space.

Wehner (1950) turned Merrill and Webb's findings, in a somewhat different arrangement, into a practical device – the *plasma oscillator*. This oscillator is very much like a klystron, except that it uses the plasma itself to generate the oscillations rather than an external source.

An objection to the Merrill and Webb experiments was that the probe perturbed the plasma – the standard objection to probes. Cannara and Crawford (1965) carried out similar experiments on a hot filament discharge, using an electron beam rather than a probe. The thin beam is fired across the discharge, and the resulting deflection is used to determine the electric fields in the discharge. Their results essentially confirm the earlier work, and Cannara and Crawford conclude that the beam of electrons interacts with the plasma so strongly that the rf oscillations generated disperse the beam, in their experiments within about 1 cm for a beam of tens of electron volts energy, in a mercury discharge at 0.2 - 1.0 millitorr.

But we have still not explained *how* the primary electrons give up their energy to the glow electrons. The plasma waves have to be formed in the first place, and then they have to be persuaded to give up their energy. Bohm and Gross (1949, 1950) laid the foundations for solving these problems. Their papers show that if beams of sharply defined velocity or groups of particles with far above mean thermal speeds are present in a glow, such as the beam of electrons from the cathode sheath entering the glow, then there is a tendency towards instability so that small oscillations grow. They then go on to show how electrons in the glow can be trapped by a plasma wave, so that the trapped electron is forced to run with the wave, oscillating back and forth in the potential trough of the wave, with an average velocity equal to the wave velocity. This is the phenomenon of *electron trapping.*

Chen (1974) compares this situation with that of a surfer trying to catch an ocean wave. At first the surfboard merely bobs up and down and does not gain energy. The surfer then 'catches' the wave, is accelerated and gains energy, whilst the wave loses energy and is damped. In the same way, the plasma wave can trap electrons until it is completely damped.

An initial requirement for the surfer, and for the glow electrons, is that their velocity is close enough to the wave velocity for them to become trapped, and so only a fraction of the electrons will be affected. But there are many waves in the plasma other than those due to the primary electrons, and these propagating plasma oscillations can have a whole range of velocities, so that the entire distribution of glow electrons can be affected by waves. If the surfboarder in the analogy were moving faster than the wave, he could give energy to it; so electrons moving faster than the wave can become trapped and give energy to the wave.

The analysis of Bohm and Gross has been well substantiated subsequently. Chen describes some experiments which demonstrate the existence of both standing and travelling electron waves, again using probes.

There are many other wave phenomena to consider, such as Landau damping, wave-wave interactions, and ion waves, but such considerations are beyond the scope of this book and, quite frankly, beyond me at the moment. However, it does appear that the wave-electron interaction may be capable of explaining both the attenuation of the primary electrons when they enter the glow, in order to slow the primary electrons as observed and to 'heat' the plasma, and to account for the energy interchanges tending to push the plasma back towards a Maxwellian distribution. I wonder also how far one can extend the comparison with the klystron and argue that the plasma is like a distributed detector and external impedance, so that the power of the oscillations is amplified by the primary beam energy and then dissipated in the glow impedance.

The energies of the primary electrons in our cold cathode discharges are very much higher than in the hot filament discharges used by Langmuir, by Merrill and Webb, and by Cannara and Crawford, and so the attenuation of the energy of the primaries will take correspondingly longer. Apparently this process is not superefficient, because fast electrons are observed at the anode. Probably the reason is that it is much more difficult for a glow electron to become trapped in a higher energy electron beam because of the velocity mismatch. There are many other questions to be answered, such as why there isn't a uniform reduction of the energy of the primaries instead of some primaries apparently passing through the glow unchecked (or have they been retarded and then accelerated again?), and we still don't know the detail of the distribution of electron energies in the glow. Although there seems to be sufficient evidence that the glow is the main source of ionization, the reassuring numbers that we calculated for ion-electron pair generation by Maxwellian electrons in the glow would be worthless, and agreement with required rates fortuitous and illusory, if the distribution isn't Maxwellian. Other apparent agreement is also questioned. As we discussed earlier in the chapter, Brewer and Westhaver (1937) obtained excellent agreement between their measured values of negative glow lengths and the ranges of fast electrons obtained by Lehmann (1927), implying a close connection between the two. More recently, Woolsey et al. (1967) have used a magnetic lens arrangement to measure the energies of primary electrons in a helium glow and conclude that the range appears to be less than the length of the glow, contrary to Brewer and Westhaver's conclusion, being as little as two-thirds of the glow length in some cases. I have not been able to obtain a copy of Lehmann's paper yet, but I understand his results for range were obtained in an ionization chamber. In his case there would have been no plasma interaction, and the ranges obtained should therefore be longer than in a plasma using the same initial electron energy.

As a final dampening note, we should consider the probe measurements of Hirsch (1965). Pursuing some earlier observations by Gabor et al. (1955) of electron interactions with oscillations in electrode sheaths, Hirsch concludes that the apparent Maxwellian distribution of electrons, as measured by probes, is more a function of rf interactions in the probe sheath than of the electron energy distribution in the plasma, i.e. that Langmuir's Paradox is not based on reality!

The preceding discussion was intended to give some idea of the difficulties involved in plasma and discharge physics. We should heed the warning given by Cobine (1958) in his introduction, that no sources are infallible, that all proofs should be questioned, and that no discharge phenomena are so well understood that data can be applied precisely. The situation is not significantly different in 1979, at least not in sputtering and plasma etching discharges.

REFERENCES

F. W. Aston, Proc. Roy. Soc. **A84**, 526 (1911)

D. J. Ball, J. Appl. Phys. **43**, 7, 3047 (1972)

D. Barbiere, Phys. Rev. **84**, 653 (1951)

D. Bohm and E. P. Gross, Phys. Rev. **75**, 1851 (1949); Phys. Rev. **75**, 1864 (1949); Phys. Rev. **79**, 992 (1950)

A. Keith Brewer and J. W. Westhaver, J. Appl. Phys. **8**, 779 (1937)

S. C. Brown, in *Gaseous Electronics*, ed. J. W. McGowan and P. K. John, North-Holland, Amsterdam (1974)

H. Bruining, *Secondary Electron Emission*, Pergamon Press, London (1954)

A. B. Cannara and F. W. Crawford, J. Appl. Phys. **36**, 3132 (1965)

C. E. Carlston, G. D. Magnuson, P. Mahadevan, and D. E. Harrison Jr., Phys. Rev. **139A**, 729 (1965)

B. N. Chapman, unpublished results (1975)

F. F. Chen, *Introduction to Plasma Physics*, Plenum, New York and London (1974)

J. D. Cobine, *Gaseous Conductors*, Dover, New York (1958)

P. L. Copeland, Thesis, U. of Iowa (1931)

W. D. Davis and T. A. Vanderslice, Phys. Rev. **131**, 219 (1963)

A. J. Dekker, *Solid State Physics*, Macmillan, London (1963)

H. Dreicer, Phys. Rev. **117**, 343 (1960)

M. J. Druvestyn and F. M. Penning, Rev. Mod. Phys. **12**, 88 (1940)

E. W. Eckstein, J. W. Coburn, and Eric Kay, Int. Jnl. of Mass Spec. and Ion Phys. **17**, 129 (1975)

A. von Engel, *Ionized Gases*, Oxford Univ. Press (1965)

A. von Engel and M. Steenbeck, *Elektrische Gasentladungen*, Vols. 1 and 2, Springer, Berlin (1932-4)

K. Fataliev, G. Spivak, and E. Reikhrudel, Zh. eksp. teor. Fiz. **9**, 167 (1939)

G. Francis, in *Handbuch der Physik* **XXII**, ed. S. Flügge, Springer-Verlag, Berlin (1956)

D. Gabor, E. A. Ash, and E. D. Dracott, Nature London **176**, 196 (1955)

F. Howard Gillery, J. Vac. Sci. Tech. **15**, 2, 306 (1978)

A. Güntherschulze, Z. Physik **59**, 433 (1930)

O. Hachenberg and W. Brauer, Adv. in Electronics **11**, 413 (1959)

H. D. Hagstrum, Phys. Rev. **104**, 317 (1956a)

H. D. Hagstrum, Phys. Rev. **104**, 672 (1956b)

H. D. Hagstrum, Phys. Rev. **104**, 1516 (1956c)

H. D. Hagstrum, Phys. Rev. **119**, 940 (1960)

M. Healea and C. Houtermans, Phys. Rev. **58**, 608 (1940)

C. L. Hemenway, R. W. Henry, and M. Caulton, *Physical Electronics*, Wiley and Sons, New York and London (1967)

M. J. Higatsberger, H. L. Demorest, and A. O. Nier, J. Appl. Phys. **25**, 883 (1954)

A. G. Hill, W. W. Buechner, J. S. Clark, and J. B. Fisk, Phys. Rev. **55**, 463 (1939)

E. H. Hirsch, Inter. J. Electronics **19**, 537 (1965)

A. J. T. Holmes and J. R. Cozens, J. Phys. D Appl. Phys. **7**, 1723 (1974)

T. Holstein, Phys. Rev. **70**, 367 (1946)

J. E. Houston and J. E. Uhl, Sandia Research Report, SC-RR-71-0122 (1971)

J. H. Ingold, in *Gaseous Electronics*, Vol. 1, ed. M. N. Hirsh and H. J. Oskam, Academic Press, New York and London (1978)

J. B. Johnson and K. G. McKay, Phys. Rev. **91**, 582 (1953)

J. B. Johnson and K. G. McKay, Phys. Rev. **93**, 668 (1954)

C. Kenty, Phys. Rev. **44**, 891 (1933)

M. Knoll, F. Ollendorff, and R. Rompe, *Gasentladungstabellen*, Verlag Julius Springer, Berlin (1935)

I. Langmuir, Phys. Rev. **26**, 585 (1925)

J. F. Lehmann, Proc. Roy. Soc. **115**, 624 (1927)

R. G. Lye, Phys. Rev. **99**, 1647 (1955)

E. W. McDaniel, *Collision Phenomena in Ionized Gases*, Wiley, New York and London (1964)

K. G. McKay, Adv. in Electronics **1**, 65 (1948)

D. B. Medved, P. Mahadevan, and J. K. Layton, Phys. Rev. **129**, 2086 (1963)

H. J. Merrill and H. W. Webb, Phys. Rev. **55**, 1191 (1939)

E. Nasser, *Fundamentals of Gaseous Ionization and Plasma Electronics*, Wiley Interscience, New York and London (1971)

J. H. Parker, Phys. Rev. **93**, 1148 (1954)

V. Polin and S. D. Gvozdover, Phys. Z. Sowj Un. **13**, 47 (1938)

D. H. Pringle and W. E. J. Farvis, Phys. Rev. **96**, 536 (1954)

D. Rapp and P. Englander-Golden, J. Chem. Phys. **43**, 5, 1464 (1965)

V. K. Rohatgi, J. Appl. Phys. **28**, 951 (1957)

A. Rostagni, Ric. Scient. II/9, **1** (1938)

E. Rudberg, Proc. Roy. Soc. (London) **A127**, 111 (1930)

E. Rudberg, Phys. Rev. **4**, 764 (1934)

E. N. Sickafus, Phys. Rev. B **16**, 1436 (1977); Phys. Rev. B **16**, 1448 (1977)

R. F. Stebbins, Proc. Roy. Soc. **A241**, 270 (1957)

Y. Takeishi and H. D. Hagstrum, Phys. Rev. **137A**, 641 (1965)

J. Thornton, Pub. No. 5885, Litton Industries, Beverly Hills (1967)

T. C. Tisone and P. D. Cruzan, J. Vac. Sci. Tech. **12**, 1058 (1975)

L. Tonks and I. Langmuir, Phys. Rev. **33**, 195 (1929)

N. Wainfan, W. C. Walker, and G. L. Weissler, J. Appl. Phys. **24**, 1318 (1953)

G. Wehner, J. Appl. Phys. **21**, 62 (1950)

J. Woods, Proc. Phys. Soc. London **B67**, 843 (1956)

G. A. Woolsey, R. M. Reynolds, and L. P. Clarke, Phys. Letters **25A**, 656 (1967)

Chapter 5. RF Discharges

In current practice, glow discharge processes are almost always driven by high frequency power supplies, usually in the megahertz range. Such discharges are in some ways quite similar to, and in other ways very different from, the dc discharges that were discussed in the previous chapter. In such systems there is no real cathode or anode since the net flow of charge to either electrode is zero, unlike the dc discharge, and there is no uniquely defined floating potential, either.

WHY USE RF?

Charging of Insulator Surfaces

It often occurs, e.g. in sputter deposition or plasma etching (q.v.), that we wish to cover an electrode with an electrically insulating material. But if we place this insulator-covered electrode in an independently sustained dc discharge, the surface of the insulator will behave in the same way as the electrically isolated probe that was discussed in Chapter 3, "Sheath Formation at a Floating Substrate": the surface will charge up to floating potential, so that the fluxes of ions and electrons to the surface become equal, regardless of the potential applied to the electrode backing the insulator. These ions and electrons recombine on the surface, thus relieving the insulator of any need to conduct current, which it couldn't do anyway.

With the magnitude of plasma density ($\sim 10^{10}/cm^3$) found in the dc glow discharge processes under discussion, the sheath voltages developed at insulating or other electrically isolated surfaces are quite small, usually no more than 10 or 20 volts, and this may not be adequate for many purposes. We can see the problem more clearly, and also the solution, by following the various stages that result if one attempts to use an insulating target as the cathode in a dc discharge (Figure 5-1). In an equivalent circuit of this configuration, both the insulator and the discharge can be regarded as capacitors. By definition, capacitance is stored charge divided by the voltage across the capacitor plates ($C = Q/V$) and initially both capacitors will be uncharged with zero volts across them. Since Q is propor-

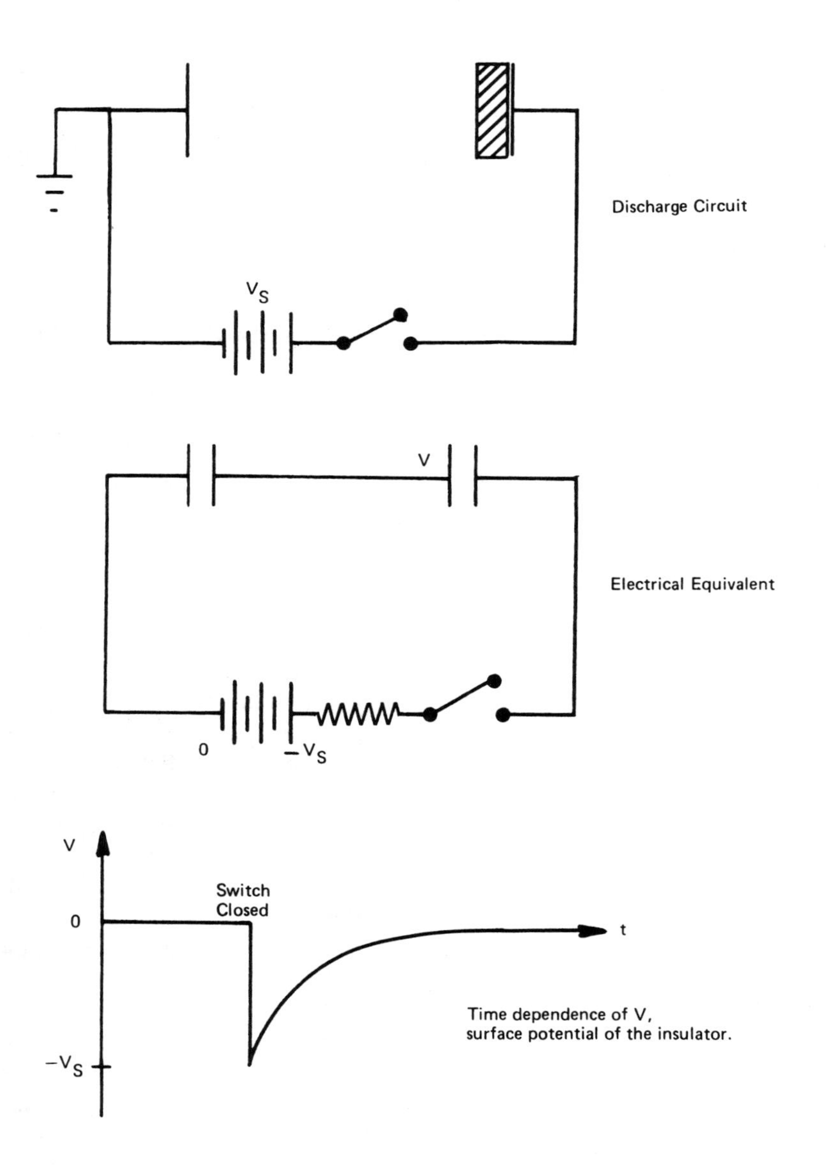

Figure 5-1. Surface charging of an insulating cathode

tional to V, and it takes time to change charge levels ($Q = \int i\, dt$), then the voltage across a capacitor cannot be changed instantaneously.

In our example (Figure 5-1), this means that both faces of the insulator will simultaneously drop to $-V_S$ volts when the switch is closed, V_S being the supply voltage. The glow discharge will be initiated and the negatively biased target will begin to be bombarded by positive ions. The insulator will start to charge positively (not because it collects the ions, but because it loses electrons as the ions are neutralized at its surface) and the potential V of the surface exposed to the discharge will rise towards zero (Figure 5-1). If the current to the target were proportional to V, then the potential rise would be exponential. Actually the current will not decrease proportionately as the sheath voltage decreases, so the form of the voltage rise will be more complex. But in either case, the discharge will be extinguished as soon as the insulator surface voltage drops below the discharge sustaining value.

The Use of AC Discharges

One proposal to deal with this problem was to use an ac discharge (Figure 5-2) so that the positive charge accumulated during one half-cycle can be neutralized by electron bombardment during the next half-cycle. Conventional mains frequency (50 Hz) was found to be not very effective because if the time during which the insulator charges up is much less than half the period of the ac supply, then most of the time the discharge will be off. Thus at low frequencies, there will be a series of short-lived discharges with the electrodes successively taking opposite polarities.

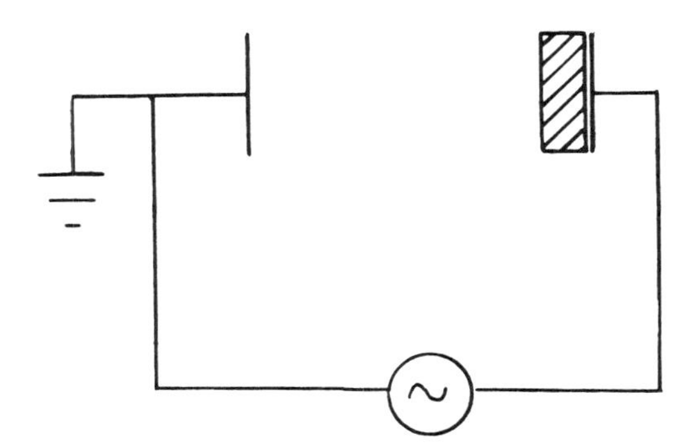

Figure 5-2. An ac glow discharge

Let's estimate the time it takes to charge up the insulator by considering the voltage rise across the capacitor in Figure 5-3. Although the current i to the target will actually decrease as the target charges up, it will be sufficient for an estimate to regard it as constant. Then the charge accumulated in t seconds will be Q = it, and so:

$$C = \frac{Q}{V} = \frac{it}{V}$$

and

$$t = \frac{CV}{i}$$

The capacitance of a piece of quartz 1/8" thick is about 1 pF/cm^2. Suppose that the applied voltage V is 1000 volts, and the ion current density is 1 mA/cm^2. (It's rather difficult to measure rf currents because of stray impedance effects at such high frequencies, and even if they could be measured accurately there would still be the problem of unravelling the ion current from the total rf current. However, by observing sputtering rates, we deduce that average rf ion currents are similar to dc sputtering currents, i.e. $\sim$ 1 mA/cm^2). These values give t $\sim$ 1 μS. This means that we can produce a discharge almost continuously at frequencies above about 1 MHz. Actually the insulator will not charge up so rapidly because the current will not be sustained at a constant value. In practice we can maintain a discharge quasi-continuously for frequencies above about 100 kHz. Wehner (1955) used a similar rationale in proposing rf discharges for sputtering purposes, and his proposals were successfully implemented some time later (Anderson et al. 1962).

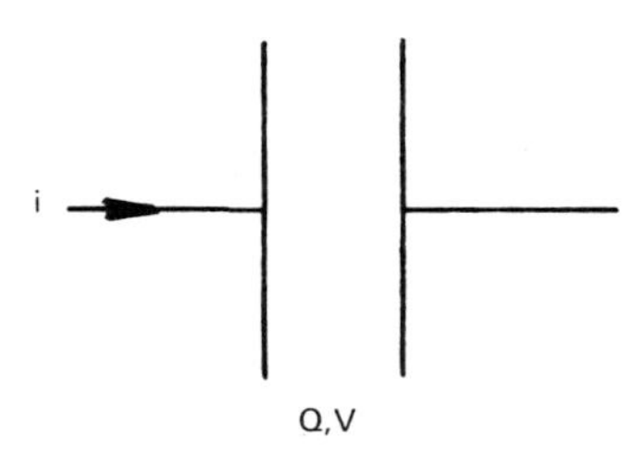

Figure 5-3. Charging of a capacitor

SELF-BIAS OF RF ELECTRODES

The simple argument just presented might suggest that ion bombardment of the insulator would occur for only half of each cycle at the most. Actually, nature is kinder than that and the much greater mobility of electrons in a discharge enables one to achieve almost continuous energetic ion bombardment if high enough frequencies are used. As we saw in Chapter 3, "Electron and Ion Temperatures", the acceleration f of a particle of mass m and charge e due to an electric field $\mathcal{E}$, is given by $f = \mathcal{E}\, e/m$, so that the lighter a particle is, the more it will accelerate due to a given force, and the greater the velocity it will acquire in a given time. But current is just the rate of flow of charge, and therefore depends on the velocity of the latter. So the lighter the charge is,

- the more current it will carry for a given electric field (the force), or
- the smaller field it will require to conduct a given current, or
- a combination of these.

In Chapter 3, we saw an effect of this behaviour: a biased probe immersed in a plasma can draw a large electron current but only a very much smaller ion current because of the much lower mean speed of the ions. This has a significant effect in rf discharges. Consider the glow discharge circuit shown in Figure 5-4, where C is the capacitance of an insulating target or is a blocking capacitor (the need for which will become apparent later) in the case of a conducting target. Let V_a be the (alternating) supply voltage, and let's see how the voltage V_b on

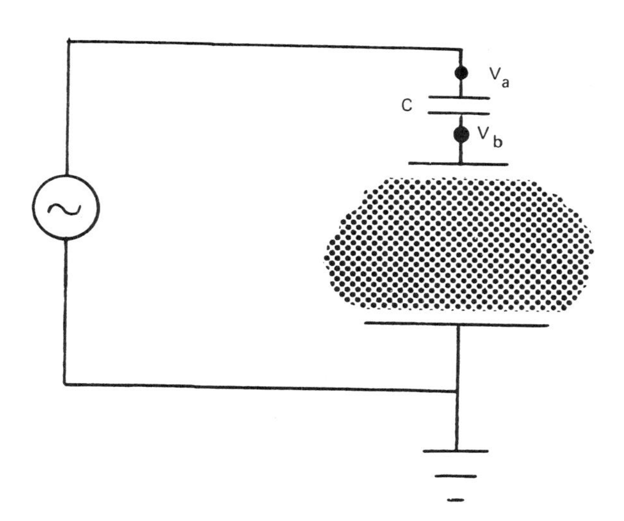

Figure 5-4. Schematic of a high frequency glow discharge circuit

the target surface varies. To simplify the argument, we'll assume that the plasma potential is close to zero (we'll see later when this assumption is justified; it wouldn't qualitatively change the argument anyway).

We begin by considering the application of a square wave power supply to the circuit (Chapman 1969). Let this have a peak-to-peak amplitude of 2 kV, so that V_a follows the form shown in Figure 5-5. The capacitor is initially uncharged, so that when V_a goes to -1 kV at time t = 0, then V_b takes the same value (the voltage across a capacitor cannot be instantaneously changed). The discharge is initiated, the target is bombarded by positive ions and the capacitor begins to charge positively (Figure 5-5) so that the target potential V_b rises towards zero. If the supply frequency is high enough, and in practice this turns out to be ~ 1 Mhz, then V_b will not have changed very much at the end of the half cycle, $\tau/2$; assume that it has risen to – 800V. At this instant V_a increases by 2 kV, and therefore so does V_b, in this case to + 1200 V. The positively charged target now draws a large electron current so that V_b decays towards zero much more rapidly than when subjected to ion bombardment. This is an example of the electron current being very much larger than the ion current for similar potential differences. Let's assume that V_b reaches + 100 V at the end of the first cycle (τ) and then as V_a switches, V_b will drop to - 1900 V and begin to rise. But since the ion current is small, V_b will not rise far before switching by + 2 kV again. After a few cycles, the voltage waveform will become repetitive with the main feature. apart from some distortion, that it has been dramatically displaced towards the negative, so that high energy ion bombardment of the target alternates with low energy electron bombardment. This is a manifestation of a much smaller potential being required for electrons than for ions, in order to conduct a given current.

The corresponding instantaneous current i will be something like that shown in Figure 5-5. I don't know quite what this waveform will look like; the principal requirement, though, is that the total charge flow per cycle sums to zero, so that the areas under the electron and ion portions of the current-time waveform must be equal. The square wave excitation was used only to illustrate the mechanism. The conventional sine wave excitation leads to the steady state waveforms shown in Figure 5-6. These voltage waveforms can be measured by attaching a high voltage probe to the electrode (the probe divides the voltage down to a safe and manageable level) and observing the waveform on a high frequency oscilloscope. V_b then has a sinusoidal waveform, again displaced to a negative value with a mean known as the *dc offset voltage*. The target has acquired a *self-bias*, which in this case will have a value equal to nearly half the applied rf peak-to-peak voltage. In contrast to the square wave excitation, V_b now is positive for only a very short fraction of each cycle, and ion bombardment of the target is almost continuous. As with the square wave, the charge flows during the positive and

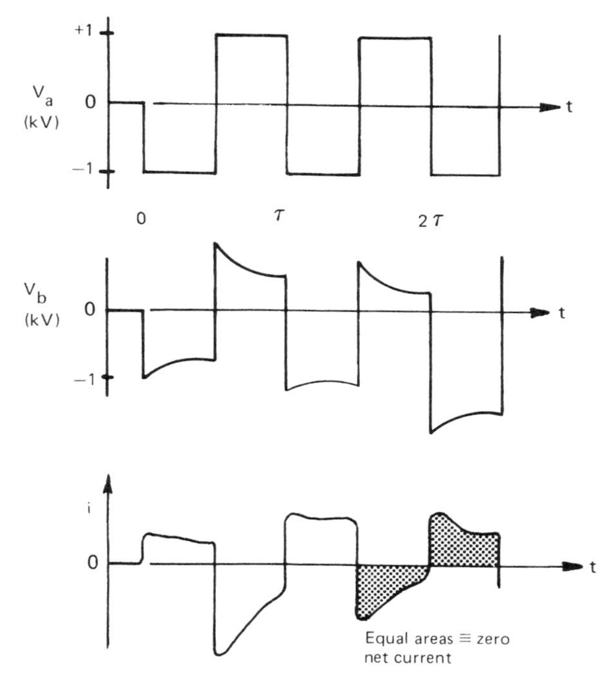

Figure 5-5. Voltage and target current waveforms when the circuit of Figure 5-4 is square wave excited

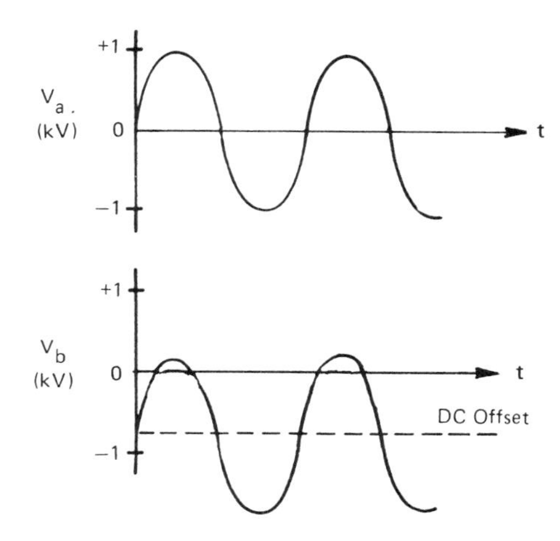

Figure 5-6. Voltage waveforms at generator (V_a) and target (V_b) in a conventional sinusoidally-excited rf discharge

negative portions of the V_b waveform must be equal and opposite; this time, the electrons demonstrate the combination of higher current and lower potential difference, than the ions.

An alternative approach to understanding the self-bias of an insulating electrode is given by Butler and Kino (1963) and is illustrated in Figure 5-7. From Chapter 3 we know that a probe at floating potential draws no net current. If a voltage is applied to the probe, the current drawn will be given by the probe characteristic. Figure 5-7a shows that when a probe is given an rf perturbation symmetrically about its initial potential, the asymmetry of the probe characteristic causes the probe to draw a net electron current. This charges the probe to a mean negative value with respect to floating potential, so as to draw a net zero current as shown in Figure 5-7b. In practical glow discharge processes, the rf perturbation will be very much larger than indicated in Figure 5-7. Additionally, Butler and Kino's experiments were made on a dc discharge; with an rf discharge the plasma potential has an rf perturbation which complicates the analysis.

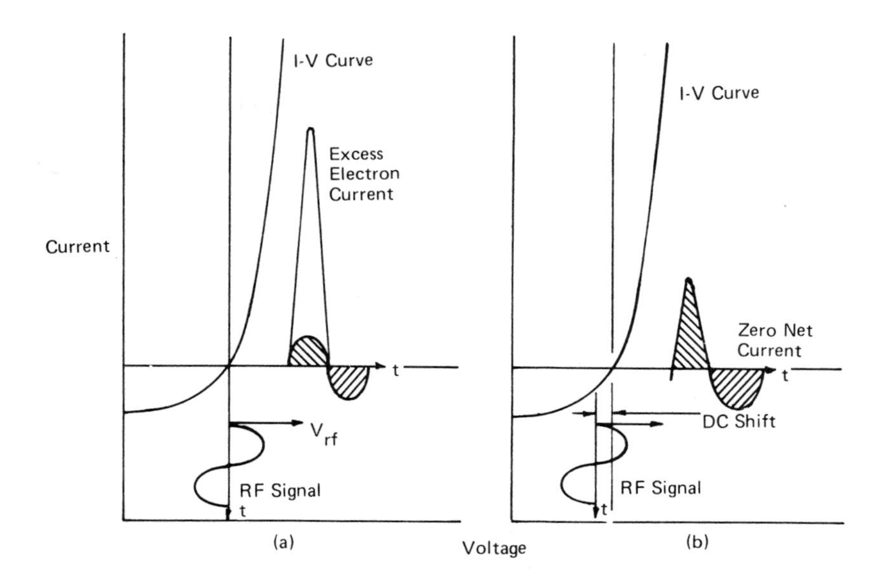

Figure 5-7. Self-biasing of a dielectric surface (Butler and Kino 1963)

THE EFFICIENCY OF RF DISCHARGES

As well as giving the ability to bombard insulating surfaces, it also transpires that the rf discharge is more efficient than its dc counterpart in promoting ionization and sustaining the discharge.

This can be shown by applying an ac source to a discharge. At low frequencies, this behaves like a double-ended dc discharge with similar limitations, particularly with regard to minimum operating pressure. But as the frequency increases, the minimum operating pressure begins to fall, reaching values of less than 1 mtorr at 13.56 MHz. There appears to be an additional mechanism occurring, or at least, an additional source of electron impact ionization.

Another manifestation of the same effect is that, for a given pressure, the impedance of a discharge decreases with increasing frequency, so that one can drive more current through the discharge with a given voltage. The system referred to in Chapter 4, "Architecture of the Discharge", with the V-I characteristics shown in Figure 4-1, has also been used as a dc sputtering system with an rf-induced substrate bias (Figure 5-8). The bias technique is described more fully in Chapter 6, and consists essentially of introducing another electrode with voltage (rf or dc) applied to it. Figure 5-8 shows how a comparatively small rf bias can increase the dc discharge current significantly. The *bias voltage* is conventionally taken as the dc offset voltage resulting from the applied rf.

Let us now ponder how this enhanced ionization might come about. Consider an electron oscillating along an x-axis in an ac field $\mathcal{E}$ of amplitude $\mathcal{E}_0$ and angular frequency ω:

$$\mathcal{E} = \mathcal{E}_0 \cos \omega t$$

Equation of electron motion is

$$m_e\ddot{x} = -e\mathcal{E}_0 \cos \omega t$$

and hence

$$\dot{x} = -\frac{e\mathcal{E}_0}{m_e\omega} \sin \omega t$$

and

$$x = \frac{e\mathcal{E}_0}{m_e\omega^2} \cos \omega t$$

The electron energy will be given by $\frac{1}{2} m_e\dot{x}^2$. Clearly the field strength and frequency are important in determining the electron motion. For the 13.56 MHz excitation frequency commonly used in sputtering, the electron amplitude and maximum energy are shown in Table 5-1 as a function of electric field strength. As we saw at the beginning of this chapter, the much greater mass of the ion will prevent it moving far in the rf field, or of acquiring much energy, compared with

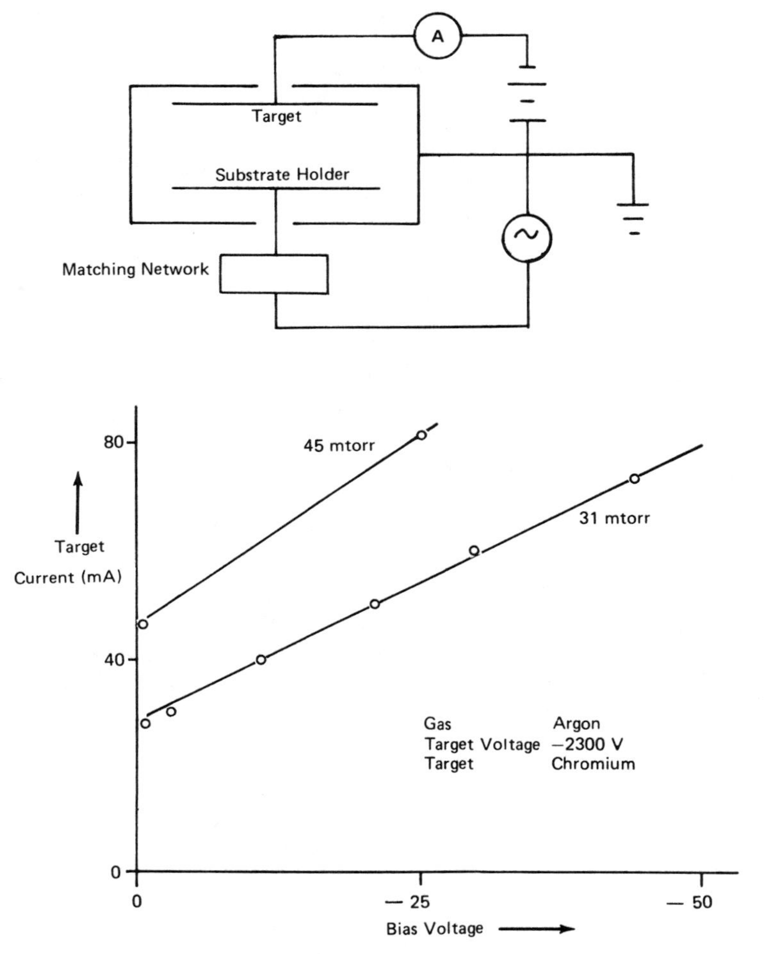

Figure 5-8. DC target current for a dc discharge in argon, vs. rf bias voltage on the substrate holder (Chapman 1975)

the electron. For an electron to acquire the 15.7 eV necessary to reach the ionization potential of argon, these values suggest that a minimum field strength somewhat above 10 V/cm would be required. However, from Figures 2-4 and 2-8 in Chapter 2, we know that the elastic collision cross-section of electrons around this energy is about 10^{-15} cm^2, whereas the ionization cross-section will be nearly two orders of magnitude smaller. This would suggest that much larger fields would be required. On the other hand, we know that the plasma will try to reduce any applied field in a time determined by the plasma frequency to be around 1 nS, as discussed in Chapter 3. However, as discussed in many articles on high frequency discharges, for example by MacDonald and Tetenbaum (1978), if an electron makes an elastic collision at an appropriate time with respect to the phase of the electric field, then its velocity and energy would continue to increase. Ideally, the electron would make an elastic collision with an argon atom, reversing its motion at exactly the moment the field changed direction. In this way, electrons could reach ionizing energies for quite weak electric fields.

Table 5-1

Amplitude and maximum kinetic energy for an electron oscillating in an rf field without collisions

Field Strength (V/cm)	Amplitude (cm)	Velocity (cm/sec)	Energy (eV)
0.1	0.02 cm.	$2.1\ 10^{6}$	$1.13\ 10^{-3}$
1	0.24 cm.	$2.1\ 10^{7}$	0.11
10	2.42 cm.	$2.1\ 10^{8}$	11.3
100	24.2 cm.	$2.1\ 10^{9}$	1130
*1000	242 cm.	$2.1\ 10^{10}$	$1.13\ 10^{5}$

*ignoring relativistic effects

This mechanism seems to be accepted as the dominant ionization source in microwave discharges. At such frequencies, of the order of a few GHz (10^9 Hz), the amplitude of electron motion becomes quite small even for large fields. In addition, a large applied field can exist for a longer period before the plasma screens it out.

Such high frequencies are used in glow discharge processes, for example in the Toshiba microwave plasma etching system discussed in Chapter 7. However, we are mostly concerned in this book with excitation frequencies around 13 MHz. Koenig and Maissel (1970) and Maissel (1970) invoke this same mechanism to explain ionization in rf sputtering. But Holland et al. (1974), making similar calculations to those shown above, conclude that the secondary electrons which are emitted from the walls and target and are accelerated across the positive ion sheath into the plasma, act as an additional supply of electrons. In this case, electron collisions with the wall take the place of elastic collisions in the microwave discharge. The explanation of Holland et al. is essentially that of the phenomenon of *multipacting*, which is discussed by MacDonald and Tetenbaum, and relies on secondary electron emission from the walls. If the electric field reverses at the right time, this can lead to efficient ionization; it can be a resonance phenomenon. MacDonald and Tetenbaum report the work of Hatch and Williams (1958); apparently the lower limit of the product of the frequency f and the electrode spacing d, for multipacting to occur, is 70 MHz cm. Many rf sputtering systems would be on this bottom end, but according to the predicted values, resonance would not occur until fd had a value of several hundred, for an rf peak to peak voltage of 1000 V.

Jackson (1970) also has discussed the conditions for rf sputtering discharge initiation and maintenance. He refers to the multipacting phenomenon, and also to the *electric field amplification* ideas of Vacquie et al. (1968) which also are based on secondary electron emission from the walls. Keller and Pennebaker (1979) reject the possibility of ionization by fast electrons from the target, and conclude that ionization is instead due to the large rf currents flowing through the glow or to electrons *surf riding* on the oscillating edge of the sheath. Apparently (Keller 1978) this surf riding effect, which must be distinguished from the analogy used to describe electron trapping in Chapter 4, relies on a result we derived in Chapter 1, "Energy Transfer in Binary Collisions": the rf field causes the electrode sheaths to grow and decay, modulating the sheath voltage and sheath length. So the sheath edge has an effective velocity v_W. An electron in the glow coming under the influence of the repulsive field at the edge of the moving sheath, would regard the encounter as a 'collision' with a massive particle. If the electron velocity perpendicular to the sheath edge is v_{el}, then by thinking in terms of relative velocity and using the result that a light particle striking a very heavy particle speeds away at twice the impact velocity, then we find that the velocity of the electron after impact is $v_{el} + 2v_W$ back into the glow; i.e. the electron has picked up energy from the oscillating sheath.

This result brings up another point. In Chapter 4, we saw that electrons tend to be trapped in the glow by the positive ion sheaths at the electrodes. In rf discharges, the effect is likely to be much larger. Although there are portions of the rf cycle when electrons can freely escape to the boundaries – in fact must do so to ensure zero net current – during most of the cycle the barrier to escape will be much larger than in dc discharges. This results from the larger plasma potentials generally extant in rf systems. Electron 'reflection' will occur at both target and counterelectrode sheaths, and will presumably be enhanced by Keller and Pennebaker's 'surf riding' effect at both sites.

So we have various possibilities for glow space ionization in rf discharges. But without clear experimental evidence, which is notably lacking for rf sputtering and plasma etching discharges, the detail of the process is not all clear. Unfortunately, although there have been many experiments on dc discharges and microwave discharges, the sputtering and etching discharges have received comparatively little attention. This is a frequent source of difficulty in trying to understand some of the strange effects that occur in 'our' discharges.

RF SHEATHS – COLLISIONS AND MODULATION

For sputtering purposes, in order to avoid gas phase scattering, rf systems are usually operated at the lower end of their operating range, from 1 mtorr up to about 40 mtorr. The dark space thickness is much less dependent on pressure than in the dc case, and is usually around 1 cm down to a pressure of a few mtorr, when it starts to increase. The comments in Chapter 4 about collision phenomena in dc sheaths being fairly independent of pressure apparently does not apply to rf sheaths.

Let's make some educated guesses about what happens in an rf sheath. Assume an argon pressure of 10 mtorr and a dark space length L of 1 cm. The probability of ionization for a secondary electron ejected from the target will be less than $nqL = (3.54\ 10^{14})\ (2.6\ 10^{-16})(1) = 0.09$, where we have used the maximum value of the ionization cross-section. However, the charge exchange cross-section is about a factor of 10 higher, giving a charge exchange probability of 1. So even at 10 mtorr, the sheath will not be collision-free, although there will not be any significant amount of ionization occurring there. So ionization is confined to the glow.

Although the lower frequency of collisions in the sheath will lead to less attenuation of the ion energy, the energy distribution of the ions will be considerably affected by the rf modulation of the sheath voltage and sheath thickness. A 500 eV argon ion has a velocity of $4.9\ 10^{6}$ cm/sec, which means that it would take 200 nS to travel 1 cm. But the period of a 13.56 MHz oscillation is 74 nS, and therefore the ion will undergo several oscillations on its way across the sheath.

There is a paper by Tsui (1968) which is frequently referred to. It considers the time-dependent ion and electron motions in an rf sheath, first to calculate the time-independent (dc offset) component of the sheath voltage, and then to calculate the energy distribution of ions at the cathode. Unfortunately the assumptions on which the analysis is based do not seem very realistic. It assumes that there are no collisions in the sheath over the range 2 – 20 mtorr, whereas there will be some charge exchange collisions, and that the electric field increases linearly across the sheath; this would seem to be even less acceptable at these dark space distance than in the Davis and Vanderslice (1963) analysis, although the high frequency dynamic sheath modulation makes this uncertain. The model also assumes that the ion velocity on entering the sheath is $n_i \overline{c_i}/4$, although this is somewhat offset by the further assumption that $T_i \simeq T_e$. Finally it appears to treat the sheath thickness as a constant, and the plasma potential as constant and zero.

Coburn and Kay (1972) have made measurements of the energy distributions of various ions incident on a grounded electrode in an rf discharge. Their apparatus, which is conceptually the same as that used by Davis and Vanderslice, will be discussed further in Chapter 6. The sheath voltage at a grounded electrode should be numerically equal to the plasma potential. Therefore the energy of singly charged ions on the ground plane will be equivalent to the plasma potential, modulated by the rf component of the plasma potential, and attenuated by charge exchange collisions. Figure 5-9 shows the energy distributions of several ions at the ground plane, obtained by Coburn and Kay for a situation where the plasma potential was 100 V. Coburn and Kay quote several references to other relevant work and compare their results with a theoretical expression for the energy spread ΔE of the ion distribution, which predicts that $\Delta E \propto M^{-1/2}$, where M is the mass of the reterant ion. The basic rationale is that a fast light charged particle transits the sheath rapidly and responds to the instantaneous potential, whereas a heavy particle transits the sheath during several cycles and is aware only of the average potentials. This behavior is substantiated by the results of Figure 5-9, where the energy width of Eu^+ (molecular weight 151) is very narrow, whereas that of H_3^+ is very large. Measuring the distributions of a range of ionic species, they obtained reasonably good agreement with $\Delta E \propto M^{-1/2}$, although there were some problems near the origin.

For small sheath voltages, the thickness of the sheath is also quite small, thus minimizing charge exchange collisions. To obtain accurate values of V_p, Coburn and Kay used the energy distribution of Ar_2^+; compared with Ar^+, this has higher mass and hence lower ΔE, and a smaller charge exchange cross-section. Theory and experiment seem to be in comparatively good accord in this case.

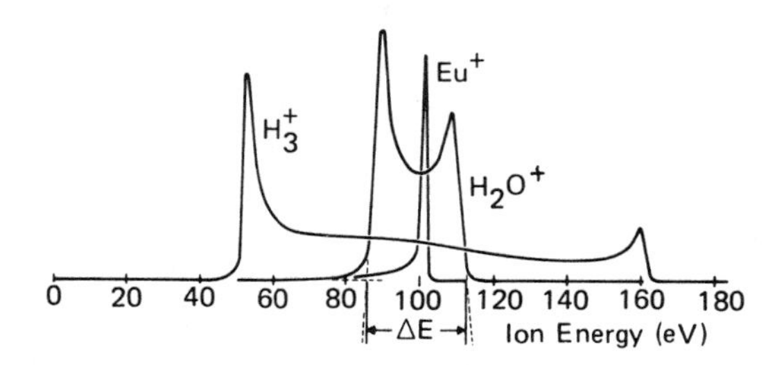

Figure 5-9. Energy distribution of H_3^+, H_2O^+, and Eu^+ at the substrate plans in a confined discharge. 13.56 MHz rf power = 100 W, argon pressure = 75 mTorr, target = 5-cm diam. (Coburn and Kay 1972)

MATCHING NETWORKS

It's common practice to use a matching network between the rf generator and the glow discharge (Figure 5-10). The purpose of this network is to increase the power dissipation in the discharge, and to protect the generator.

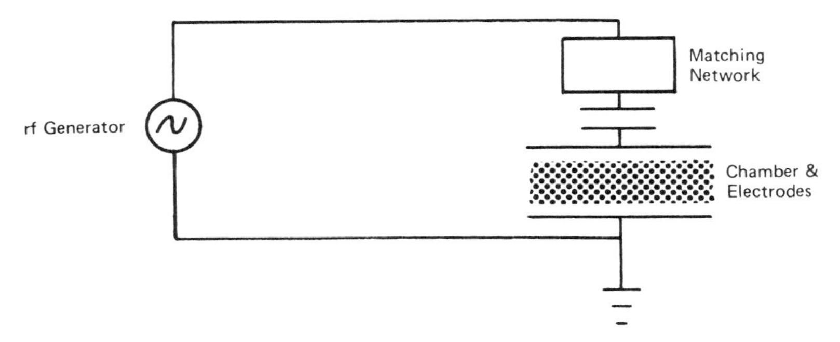

Figure 5-10. RF sputtering circuit with matching network

We can understand this better by considering the dc counterpart (Figure 5-11) in which a cell of emf E and internal resistance r is supplying power to an external load of variable resistance R. The current in the circuit is $E/(r + R)$ and therefore the power P dissipated in R is given by:

$$P = \frac{E^2 R}{(r + R)^2}$$

The power P will vary with the value of R. The maximum value of P is obtained by differentiation:

$$\frac{dP}{dR} = \frac{E^2 (r + R)^2 - 2(r + R)E^2 R}{(r + R)^2}$$

This differential has a zero value for $R = r$, which is for maximum P. Therefore to dissipate maximum power in a load, we match the resistance of the load to the resistance of the power supply.

There is an ac counterpart to this *dc maximum power theorem*. If the ac power supply in Figure 5-12 has an *internal* (also called *output*) *impedance* of $a + jb$ ohms, then a similar calculus argument suggests that the impedance Z of the load must be equal to $a - jb$ ohms (i.e. the conjugate of the generator output impedance) for maximum power transfer. The conjugate impedance is required so that the total load (internal plus external) is purely resistive. The equality of the resistive parts of source and load follows from the dc proof.

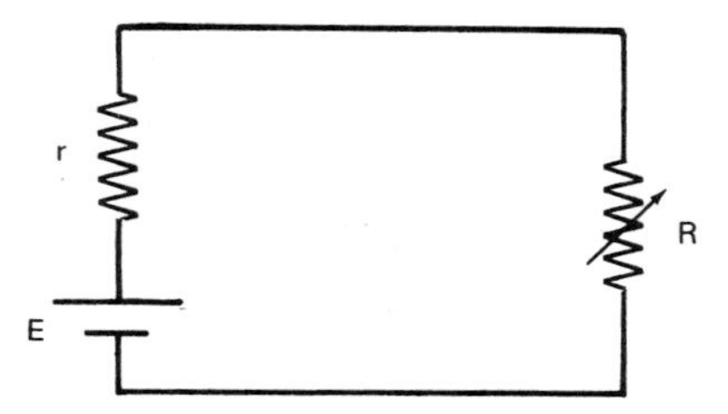

Figure 5-11. DC circuit with load R

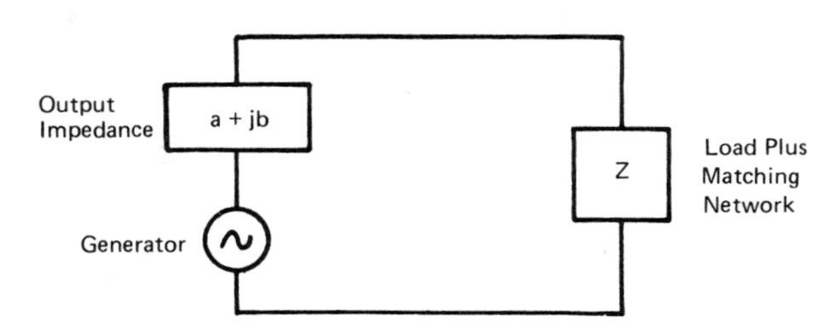

Figure 5-12. RF circuit with load impedance Z

In practice, to avoid large rf currents flowing round a circuit, generators are designed to have purely resistive outputs with a value that is usually 50 ohms. An rf discharge normally has a numerically larger and partly capacitive impedance which we cannot adjust without compromising the discharge process. We therefore simulate a load equal to the generator output impedance by combining the discharge load with a variable matching network load. The two loads will be reactive, and the matching network is therefore placed close to the discharge chamber so as to avoid power losses due to the large reactive currents flowing between these two components. A typical matching network configuration is shown in Figure 5-13.

Logan et al. (1969) have measured the impedance of an rf sputtering discharge in argon, and have employed the values obtained to design a matching network. Their technique was to use a matching network to tune their discharge to a generator of 50 Ω output impedance, and then to replace the generator with a 50 Ω load and measure the input impedance of the matching network (as seen from the discharge end) for the same dial settings. The value obtained should be the conjugate of the discharge impedance. This also follows from the maximum power theorem.

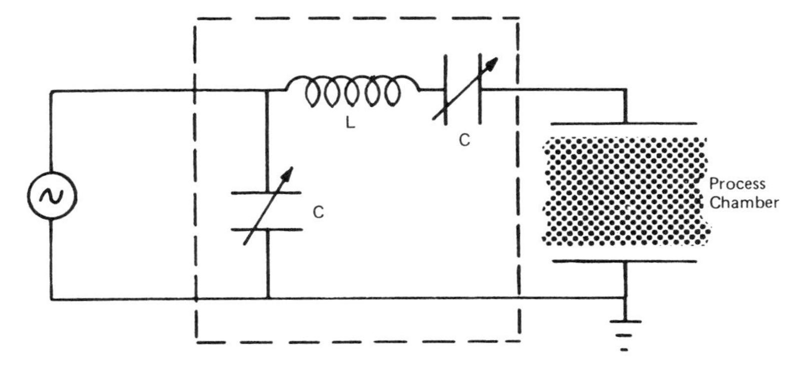

Figure 5-13. A typical rf matching network

Logan et al. found that the glow impedance was always capacitive, with little apparent dependence on power level (although there was some conflicting evidence on this point). The capacitance values (of a parallel capacitor-resistor representation) ranged from 0.08 pF/cm^2 at 5 mtorr to 0.12 pF/cm^2 at 20 mtorr, but increased at both pressures up to a value of 0.25 pF/cm^2 with an axial magnetic field of 150 gauss; 0.1 pF is equivalent to 1.2 $10^5\,\Omega$ at 13.56 MHz. The parallel conductance ranged from 1.4 micromhos/cm^2 (i.e. the reciprocal of 7 $10^5\,\Omega$ for each cm^2, presumably of the target in this case) at 5 millitorr up to 1.8 μmhos/cm^2 at 20 mtorr, and then increased with magnetic field up to values of around 4 μmhos/cm^2 at 150 gauss.

It is a condition of this analysis that the power losses in the matching network should be small, and unfortunately this is unlikely to have been the case. However, a design exercise based on the impedance values obtained gave good agreement between calculated and empirically found values of the matching network components.

WHY 13.56 MHz?

Many rf glow discharge processes operate at 13.56 MHz. There is nothing 'magic' about this number as far as the glow discharge is concerned. It just happens to be a frequency allotted by international communications authorities at which one can radiate a certain amount of energy without interfering with communications. Unfortunately this isn't such a great help for rf glow discharge systems because the glow discharge has so many nonlinear effects that it generates many harmonics of the fundamental frequency, and the radiation requirements on the harmonic frequencies are far more stringent. The sixth harmonic falls in the VHF broadcast band, and the seventh and eighth fall in aircraft communication bands. At the Allen Clark Research Centre in Northampton, the converted 'hot dog' machine with which we powered an early rf sputtering system incapacitated the internal radio paging system. So it goes.

There seem to be two schools of thought regarding operating frequencies. One school restricts itself to the magic 13.56 MHz or to multiples at 27.12 MHz and 40.68 MHz, with the oscillator crystal-controlled at this frequency. The smaller school prefers to use frequencies that can be chosen to optimize performance; usually the load is made part of the oscillator circuit so that the operating frequency may change during a process (Jackson 1970, Vossen and O'Neill 1975, McDowell 1969). These systems tend to have greater stability as they remain 'tuned' over a broader range of operating parameters, i.e. they have a lower Q, and are less susceptible to damage when running untuned; however they do require more careful coupling and shielding. Their stability advantage is also somewhat offset now by the availability of improving automatic tuning systems.

VOLTAGE DISTRIBUTION IN RF SYSTEMS

In the first part of this chapter, in illustrating the action of an rf discharge, we made the assumption that the plasma potential was close to zero. This is relatively true for most sputter deposition systems, but quite untrue for other systems, such as a high pressure parallel plate plasma etcher.

The 'classical' treatment of voltage division in an rf discharge is by Koenig and Maissel (1970) and Koenig (1972). They considered the relationship between the unequal areas A_1 and A_2 of two electrodes, and the sheath voltages and thicknesses V_1 and V_2, D_1 and D_2 respectively, developed at the electrodes due to an rf discharge. If we make the assumption that the glow is equipotential, then the configuration of Figure 5-14 would suggest that $V_1 = V_2$ since these are the only dc voltages in the system and must therefore be equal and opposite. However, a blocking capacitor (Figure 5-15) changes that situation and is commonly used in glow discharge systems just to create asymmetry. Koenig and Maissel treat this situation by making the following assumptions:

1. That positive ions of mass m_i come from the glow space and traverse the dark spaces without making collisions, and with a space charge limited flux j_i:

$$j_i = \frac{KV^{3/2}}{m_i^{1/2} D^2}$$

   where K is a constant.

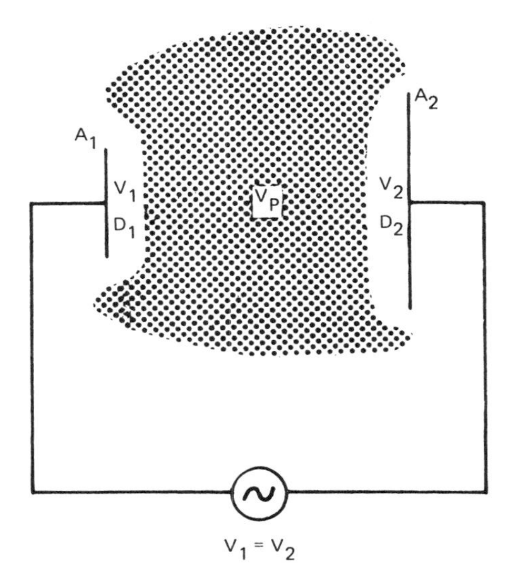

Figure 5-14. Voltage distribution – without blocking capacitor

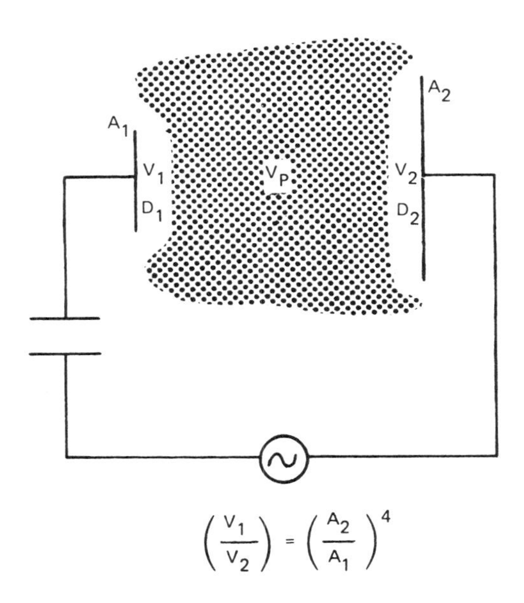

$$\left(\frac{V_1}{V_2}\right) = \left(\frac{A_2}{A_1}\right)^4$$

Figure 5-15. Voltage distribution – with blocking capacitor

2. That the current density of the positive ions is uniform and is equal at both electrodes. Combining this with 1.,

$$\frac{V_1^{3/2}}{D_1^2} = \frac{V_2^{3/2}}{D_2^2}$$

3. That the capacitance across a dark space is proportional to the electrode area and inversely proportional to the dark space thickness:

$$C \propto \frac{A}{D}$$

4. That the rf voltage is capacitively divided between the two sheaths:

$$\frac{V_1}{V_2} = \frac{C_2}{C_1}$$

combining 3. and 4. gives:

$$\frac{V_1}{V_2} = \frac{A_2}{D_2}\frac{D_1}{A_1}$$

and substituting this into 2.,

$$\frac{V_1^{3/2}}{V_2^{3/2}} = \frac{D_1^2}{D_2^2} = \left(\frac{A_1}{A_2}\frac{V_1}{V_2}\right)^2$$

$$\therefore \frac{V_1}{V_2} = \left(\frac{A_2}{A_1}\right)^4$$

This result suggests several things:

- The larger voltage sheath appears at the smaller electrode
- The fourth power dependence exaggerates the effect of geometrically asymmetric systems.
- A substantial sheath voltage can be set up at an electrode even when it is grounded. Either of the electrodes could be grounded in the preceding example, although normally the chamber is part of the grounded electrode and helps make that electrode the larger in a system.

Generalization of the Koenig Model

The area ratio model for sheath voltage distribution was developed for rf sputtering systems operating at a few mtorr. Three problems occur in its application, particularly at higher pressures:

a. The glow density can vary considerably throughout the volume of a system and sometimes will be confined essentially only between the electrodes. So there is a difference between the total geometric areas of the electrodes, and the areas as they appear to the plasma.

b. The assumption that the ion current density is equal at both electrodes is questionable.

c. The version of the Child-Langmuir used in the derivation is for a collisionless sheath where the ions free-fall (see Chapter 4). This is generally untrue in rf sheaths except at very low pressures ~ 1 mtorr or perhaps a little higher if change transfer does not impede ion motion. At higher pressures, one should use a mobility-limited or ionization-limited version of the space charge equation. For example, the former would suggest a voltage division varying as the cube of the area ratio if other assumptions remained unchanged. However, the existence of collisions in the sheath implies power absorption there, so that the sheaths can no longer be regarded as purely capacitive, and the equivalent circuit also has to change.

Experimental Test of the Voltage Distribution Model

Coburn and Kay (1972) have measured the relationships between target and substrate sheath voltages in a rf diode system operating at 13.56 MHz, as a function of area ratio. The plasma potential V_p was measured by the energy analysis technique mentioned in "RF Sheaths – Collisions and Modulation".

The area ratio was changed by using a series of pyrex confining cylinders of equal height and a range of radii. (Figure 5-16). A parameter R was defined as the ratio of the carbon target area to the total area of all other surfaces in contact with the plasma. This is not strictly the situation addressed by Koenig and Maissel, since the pyrex cylinder is insulating and therefore not part of the second (grounded) electrode. Coburn and Kay plotted the results they obtained for the substrate sheath voltage against the target sheath voltage $V_p - V_t$, and obtained a family of straight lines for different values of R (Figure 5-17).

The Koenig model predicts that the ratio of substrate to target voltage should be constant for given area ratio, and this appears to be well substantiated by the linear results of Figure 5-17, although the lines appear to have a common intercept which is not at the origin. The Koenig model also predicts that the sheath voltage ratio should be proportional to the fourth power of the inverse area ratio. In the context of Figure 5-17, that would translate into the slope (sheath voltage ratio) being proportional to R (the inverse area ratio) to the fourth power. In fact, the values obtained by Coburn and Kay suggest that the power dependence should be much closer to unity rather than the fourth power (Table 5-2). The reason for this disagreement is not at all clear. Although the range of area ratios tested was quite small, and could not be considered an exhaustive test, the magnitude of disagreement is striking. The plasma potentials obtained are much larger than predicted; for example $V_p - V_t = 300$ V and $R = 0.289$ should give $V_p = 2.1$ V compared with an actual value of 101 V. These values are for the smallest containing cylinder, when the plasma should have been most aware of the wall and probably most uniform.

The overall conclusion seems to be that although the target and substrate sheath voltages are linearly dependent, and their ratio dependent on the inverse area ratio of their electrodes, both as predicted, there is some doubt about the power of the dependence.

Application to Sputtering and Reactive Ion Etching Systems

In a sputter deposition or etching system (Chapter 6), we want sputtering to be confined to specific electrodes. We do not want wall materials to be sputtered since these will act as contamination sources in a sputter deposited film. We therefore prefer to keep wall sheaths down to about 20 volts, which is a typical threshold for sputtering. This is achieved by making the target area much smaller than the wall area, with the resultant voltage division we saw previously.

Reactive ion etching systems (Chapter 7) are almost identical to sputtering systems, and the same rationale applies. This virtually guarantees electrode material being sputtered and backscattered onto the substrates, so compatible materials must be used.

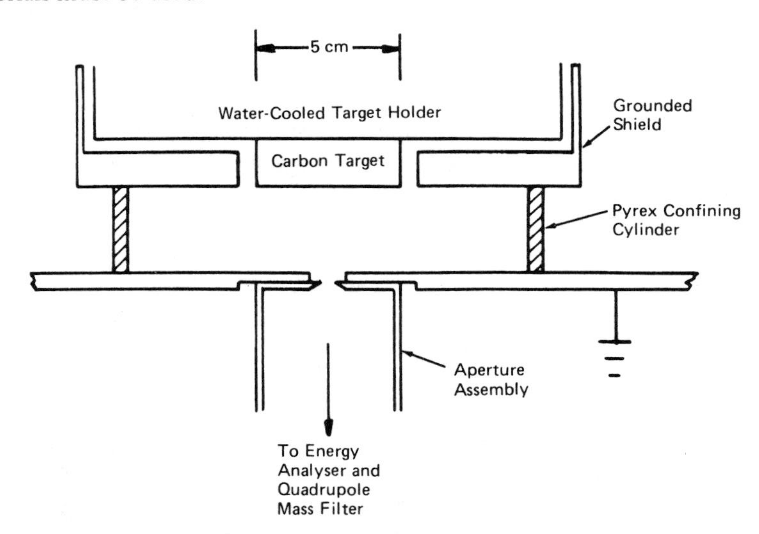

Figure 5-16. Chamber configuration for the area ratio experiments of Coburn and Kay (1972)

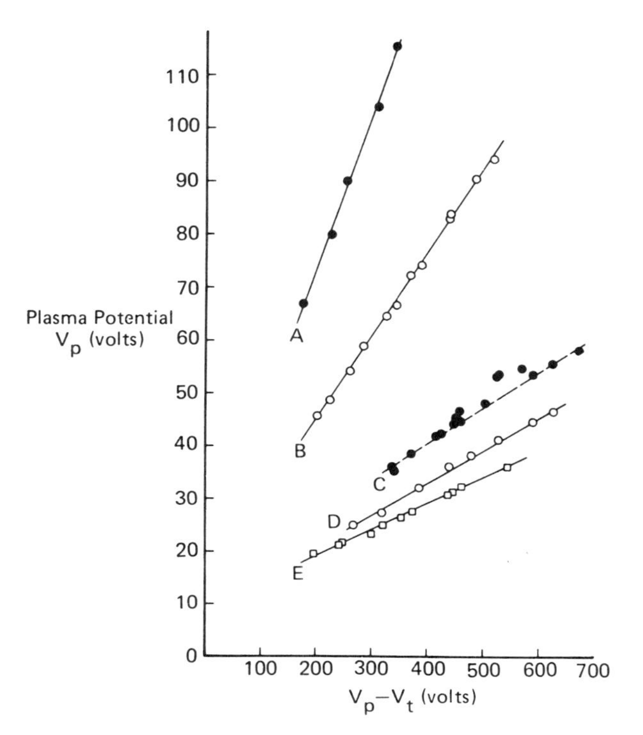

Figure 5-17. Plasma potential V_p vs. the dc voltage across the target-plasma sheath $(V_p - V_t)$ for the five confining cylinders; argon pressure = 50 mtorr; interelectrode spacing = 1.88 cm. Curve A: R = 0.289, slope = 0.298; B: R = 0.208, slope 0.160; C: R = 0.142, slope ~ 0.07; D: R = 0.114, slope = 0.06; E: R = 0.092, slope = 0.05 (Coburn and Kay 1972)

Table 5-2

The power dependence y of the sheath voltage ratio on the inverse electrode area ratio, i.e. $V_1/V_2 = (A_2/A_1)^y$, as suggested by the results of Coburn and Kay (1972) shown in Figure 5-17.

Line	A	B	C	D	E
Slope	0.298	0.160	~0.07	0.063	0.052
Area Ratio	0.289	0.208	0.142	0.114	0.092
Power Dependence	0.98	1.17	~1.4	1.3	1.2

There is a tendency with these systems to make the target (sputtering) or wafer electrode (reactive ion etching) larger and larger so as to increase the system capacity. This reduces the area ratio of the electrodes so that the wall sheath voltage increases and sputtering inevitably results (Bresnock 1979) – contaminating the wafer surfaces. Microcircuits are extremely sensitive to certain materials, such as fast-diffusing metals – Cu, Fe, and Ni; quite tiny amounts can ruin devices.

Application to Planar Diode Reactors

In Chapter 7 on plasma etching, I have described a type of plasma reactor, often known as a *planar diode reactor*, which is used for etching and deposition. This type of reactor typically has two large electrodes of equal area, one grounded, inside a chamber of not-much-larger internal diameter; the chamber may be grounded or insulating (Figure 5-18). This configuration approaches the other extreme to the asymmetric sputtering system.

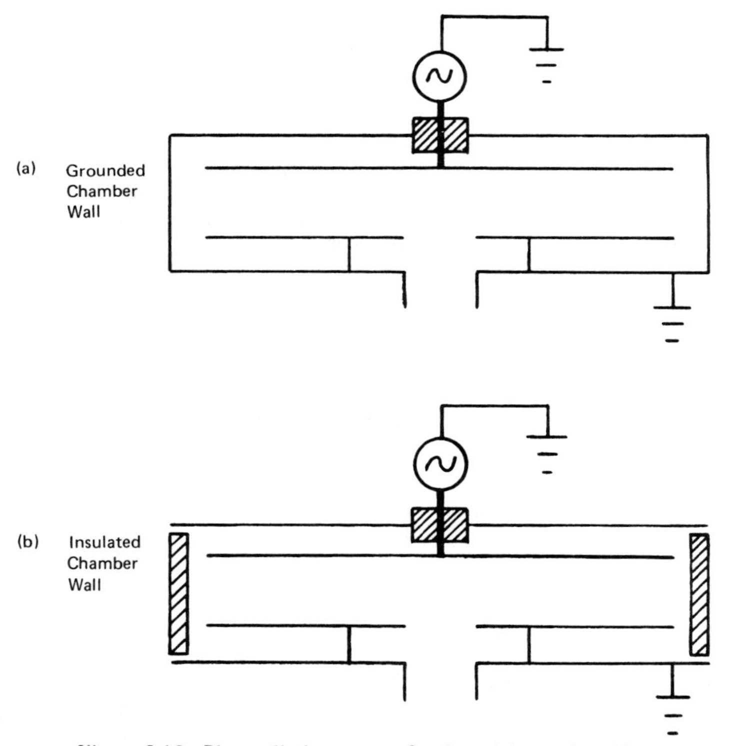

Figure 5-18. Planar diode reactors for deposition and etching

In these planar diodes, the wafer is placed on the grounded electrode. This appears to have caused some confusion and has mistakenly led some people to believe that only low energy ion bombardment can occur there. The Koenig model described earlier makes it clear that sheath voltages, and hence ion energies, can be quite substantial at a grounded wafer in such systems. The higher pressures (typically 500 mtorr) used might seem to ensure low ion energies due to the large number of collisions per unit length of travel, even though the sheath voltage is substantial. However, the sheaths at the electrodes are extremely thin at such high pressures, so that the total number of collisions attenuating the ion energies might not be so different from the low pressure case. Indeed we need quite substantial ion energies to explain directional plasma etching (q.v.) in such systems.

There are considerable advantages to being able to place wafers on a grounded electrode. However, we must recognize that it is easy to sputter material from both electrodes and from the wall, if too much power is applied. Vossen (1979) has identified aluminium contamination in a planar diode.

SYMMETRICAL SYSTEMS

It is interesting to look at this type of system in rather more detail, to get some idea of the sheath voltages involved. Consider the truly symmetrical system in Figure 5-19. The system retains its electrical symmetry regardless of the position of the matching network and regardless of the position of the system ground – both of which the discharge would be unaware of. Assume that the bottom electrode is well grounded (and this is a non-trivial matter because connecting wires can easily have a considerable impedance at radio frequencies). Let the (time dependent) voltage on the top electrode be V_{rf} (Figure 5-19). Let the two sheath voltages, which will be equal in magnitude (the symmetry demands that) although not in phase, be V_s and V_s' as shown, measured from their respective electrodes; each will have a dc component and an rf component, and the two rf components will be 180° out of phase as the sheaths alternate. So V_{rf} will not have a dc component, as expected for a symmetrical system, and can be represented by

$$V_{rf} = V_{rfo} \sin \omega t$$

as shown in Figure 5-20. If we assume that the glow is equipotential and of value V_p, then we also have

$$V_s = V_p - O$$

$$V_s' = V_p - V_{rf}$$

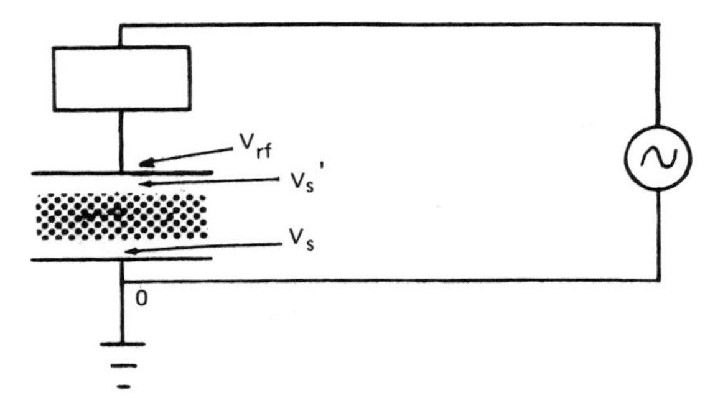

Figure 5-19. Schematic of a symmetrical rf system. $|V_s| = |V_s'|$

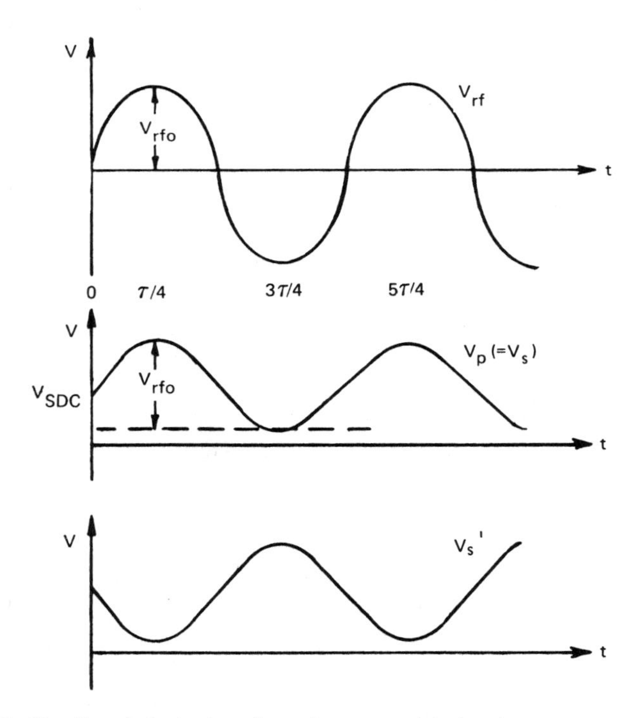

Figure 5-20. Sheath and electrode voltages in a symmetrical system
(a) Electrode voltage $V_{rfo} \sin \omega t$
(b) Sheath voltage V_s = plasma potential $V_p = V_{SDC} + ½V_{rfo} \sin \omega t$
(c) Sheath voltages $V_s' = V_{SDC} + ½V_{rfo} \sin (\omega t + \pi)$

The symmetry of the situation requires that V_p must exceed the potential V_{rfo} of the upper electrode at $\tau/4$ (τ = period) by exactly as much as it exceeds the potential of the lower electrode (OV) at $3\tau/4$. So the amplitude of V_p, and hence of V_s and V_s', must be exactly half of that of the applied rf in this symmetrical system, as shown in Figure 5-20, with a positive dc offset of V_{SDC}. This offset will be approximately equal to $V_{rfo}/2$; its precise magnitude will be determined by the electron current required during the portions of the cycle when V_p is close to the potential of either electrode.

The waveform of V_p is uncertain; a sinusoidal function has been assumed in Figure 5-20.

From the symmetry of the system, it is clear that the sheath voltage is just as large at the grounded electrode as at the ungrounded electrode. This is a very important point, and would apply just as well to a symmetrical system with insulated electrodes. We therefore have a clear distinction between the dc and rf cases. In the dc case, an insulator would charge up to floating potential (see Chapter 4) so that the sheath voltage was just a few volts. In the rf case, the sheath voltage at the insulator can be very substantial. It doesn't matter whether the plasma potential V_p is substantially dc and the potential of an electrode backing the insulator is rf powered, as in the example ("Self-Bias of RF Electrodes") earlier in the chapter; or whether the backing electrode is grounded and the plasma has a substantial modulation, as in the present example.

To illustrate the point further, we could build an assymmetric system for sputtering purposes, rather like a conventional system except that the insulating target would be grounded instead of the walls (Figure 5-21); provided that the target area is smaller than the wall area, the larger sheath voltage will still be developed at the target, regardless of the position of the electrical ground.

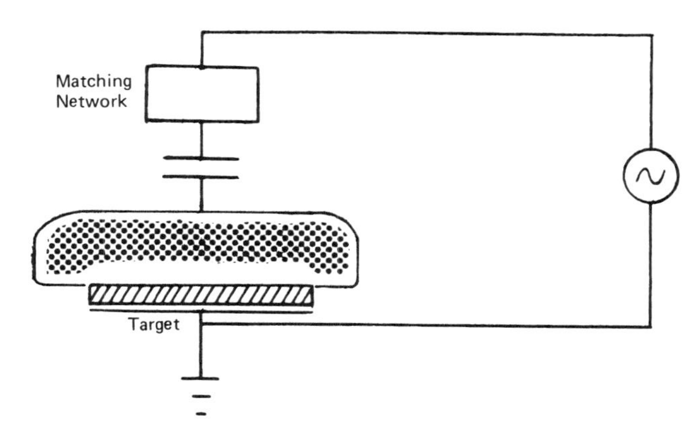

Figure 5-21. RF sputtering system with grounded target

Hence my comment at the beginning of the chapter that there is no uniquely defined floating potential in an rf system. Although it is generally accepted that an insulator backed by a conducting rf powered electrode will not charge up to floating potential (otherwise sputtering of insulators would not work!), it is not generally realized that an insulator backed by a grounded conducting electrode is in qualitatively the same situation if the plasma potential has an rf modulation. The magnitude of the modulation is important, of course. It may well be that the asymmetry of a conventional sputtering system results in a small enough plasma potential modulation that raising the plasma potential without changing the modulation does not increase the ion bombardment energy on an electrically isolated substrate, as observed by Coburn and Kay (1972), but this is generally untrue and particularly untrue for the symmetrical system.

Let's return to the symmetrical system and re-examine the assumption that all waveforms were sinusoidal (Figure 5-20). In this case the mean values of V_p, V_s, and V_s' would be equal to $V_{rfo}/2$ plus the minimum amount by which V_p exceeds zero (at $t = 3\tau/4$ in Figure 5-20). This minimum amount could be negative, but is unlikely to be significantly different from zero. So the plasma potential will be closely equal to $V_{rfo}/2$, or one quarter the rf peak-to-peak voltage.

But will the plasma potential waveform be sinusoidal? Let's consider, in Figure 5-22a, what will happen to the plasma potential V_p in a low frequency ac discharge, which is just like a double ended dc discharge. Initially ($t = 0$) there is no discharge and $V_p = 0$. As V_{rf} increases, it will reach, at t_1, a large enough value to initiate the discharge and V_p will rise to become a little more positive than V_{rf}. This continues until t_2 when V_{rf} is no longer large enough to sustain the plasma, and V_p drops to zero. At low enough frequencies, the times for establishment and extinction of the discharge will be negligibly small, so that V_p will change comparatively instantaneously.

The situation is essentially the same through the next half-cycle; the discharge exists between t_3 and t_4, during which time V_p remains just above OV, i.e. a little more positive than the counterelectrode. Note that this waveform for V_p satisfies the equal amplitude requirements of V_s and V_s' (Figure 5-19).

Two effects become significant at higher frequencies. The time for the discharge to decay can no longer be ignored. Decay occurs by diffusion of ions and electrons to the wall, recombining there, and by recombination in the gas phase. Such decay is studied in *afterglow* measurements; i.e. the discharge driving force is suddenly removed, and the resultant decay of plasma density is observed. Diffusion effects lead to an exponential decay of density; recombination is proportional to the square of density and so is more significant at higher densities where it may sometimes dominate. Decay times can be quite significant and are probably of the order of 1 msec for a sputtering discharge.

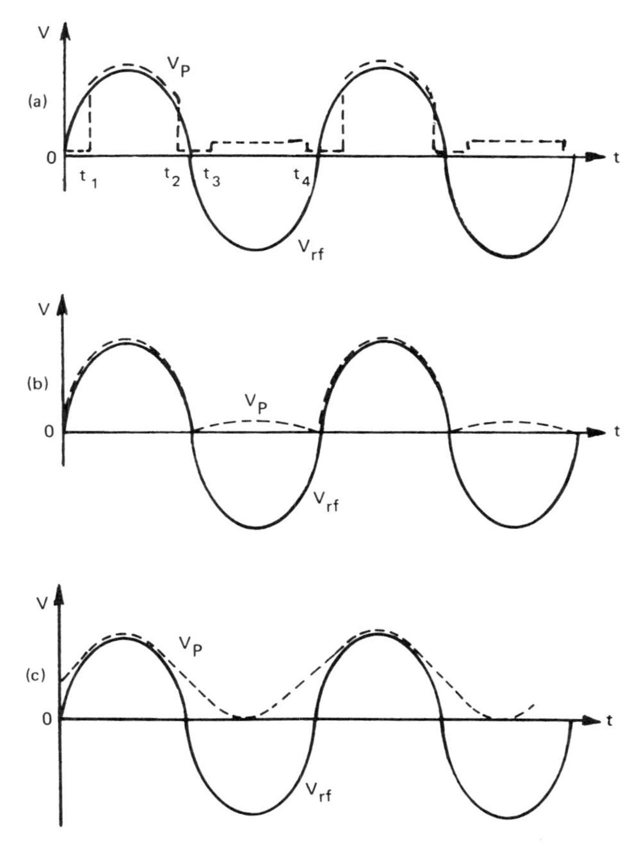

Figure 5-22. Voltage-time dependence in rf systems
(a) Low frequency discharge
(b) High frequency discharge?
(c) Assuming sinusoidal dependence

The second effect is the enhanced ionization that occurs in high frequency discharges, an effect we have already discussed.

The implication of the combination of these two effects is that, at high frequencies, there will be no initiation or destruction times as in Figure 5-22a; the plasma will be continuous even though its density will be modulated.

It is not clear to me what the high frequency waveform of V_p will be. Certainly the equivalence of V_S and V_S' must be maintained, but this can be achieved with many waveforms. I'm not sure whether the waveform should look like Figure 5-22b or like 5-22c. Perhaps someone else can help. It matters because

Figure 5-22b would lead to a mean plasma potential V_p of about ¼ the rf peak-to-peak, whereas Figure 5-22c would suggest about 1/8 peak-to-peak. One would expect that this could be resolved by measuring V_p, but only the mean is currently accessible and measurements by Vossen (1979) on a quasi-symmetrical system suggest mean values for V_p higher than either of the above. The above predictions would be wrong by the extent to which V_p exceeds ground potential at $3\tau/4$, and we may be wrong to ignore this. Vossen's measurements were made by conventional probe measurements. This technique must be re-evaluated in dealing with systems with large modulations of V_p; the nonlinear effects on ion and electron flow make it dangerous to assume that modulation of V_p will be averaged out, and it seems that the capacitance to ground of the probe will have a sizeable effect on the results.

ASYMMETRIC SYSTEMS AND MEASUREMENT OF PLASMA POTENTIAL

The discussion above relates directly to a method of measuring plasma potential proposed by Christensen and Brunot (1973). If we measure the voltage V_{rf} in a symmetric system, then it will be symmetrical about zero, and the mean target voltage V_T will be zero. As the system becomes asymmetric, then V_{rf} also becomes asymmetric and V_T becomes negative, as we saw in the earlier example ("Self-Bias of RF Electrodes"). The resultant waveforms for V_{rf} which would be observed on the ungrounded electrode are shown in Figure 5-23. Christensen and Brunot use the electrical equivalent circuit of Koenig and Maissel (1970) to conclude that the plasma potential varies sinusoidally as shown in Figure 5-23. It is implicit that the minimum by which the plasma potential V_p exceeds the electrode potential is negligible at certain points ($\tau/4$, $3\tau/4$) during the cycle, and that when insulating targets are used, the voltage drop across them is also negligible. The latter assumption is questionable.

If the mean target voltage is V_T, then Figure 5-23 predicts that the mean plasma potential will be given by:

$$\overline{V_p} = \frac{V_{rfo} + V_T}{2}$$

This gives an alternative to probes for the measurement of plasma potential and requires only an oscilloscope to observe the target waveform, or, as Christensen and Brunot propose, electronic techniques to automatically measure V_{rfo} and V_T, and hence calculate V_p; they report good agreement between this technique and conventional probe measurements.

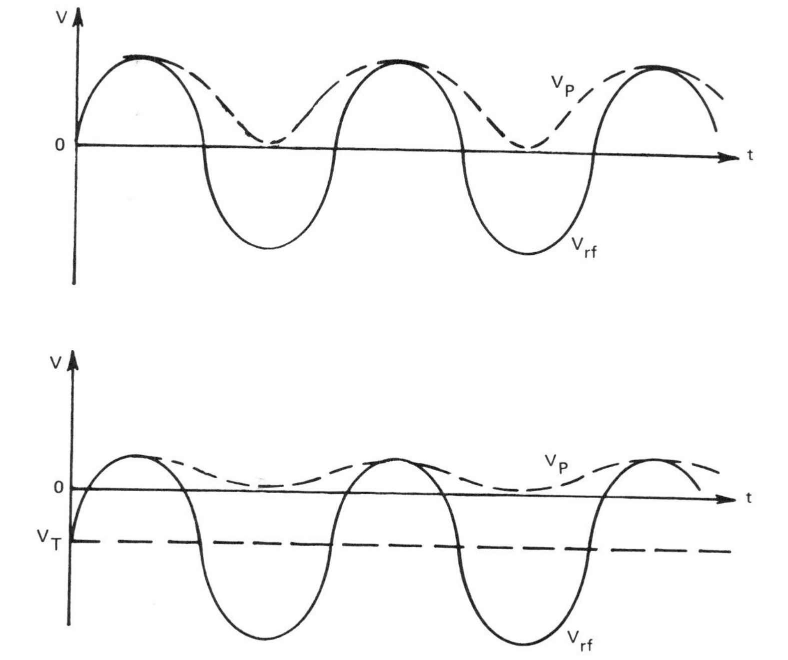

Figure 5-23. Voltage V_{rf} across rf system
(a) Symmetrical system, $V_T = O$
(b) Asymmetric system, V_T is negative (Christensen and Brunot 1973)

EQUIVALENT CIRCUITS OF RF DISCHARGES

It is often convenient to represent an rf discharge system by an electrical equivalent circuit. This has been done by Koenig and Maissel (1970), and by many others. Although the details of the various proposals differ, the basic features are as shown in Figure 5-24a.

The elements Z_t, Z_s and Z_w are the impedances of the sheaths at the target, substrate and walls, respectively. The detail of these sheath impedances is shown in Figure 5-24b. The sheath is primarily capacitive due to the positive ion space charge in the sheath and the compensating negative surface charge on the adjacent surface (target or deposited film). The diode allows for the asymmetry of the ion and electron currents. The resistors r and R are often omitted; r limits the electron current and will be small, wheareas R allows for power dissipation due to collisions, and will be very large at very low pressures, decreasing (more power dissipation) at higher pressures. The final element Z_g represents the relevant portion of the distributed glow impedance; this is sometimes simplified to a resistance (to represent power losses due to inelastic collisions) and sometimes even omitted.

If the target is insulating, then the capacitances of the target and the films on the substrate and walls must be included, as must the capacitances of the substrate and wall if they are not conducting. The target-ground shield capacitance

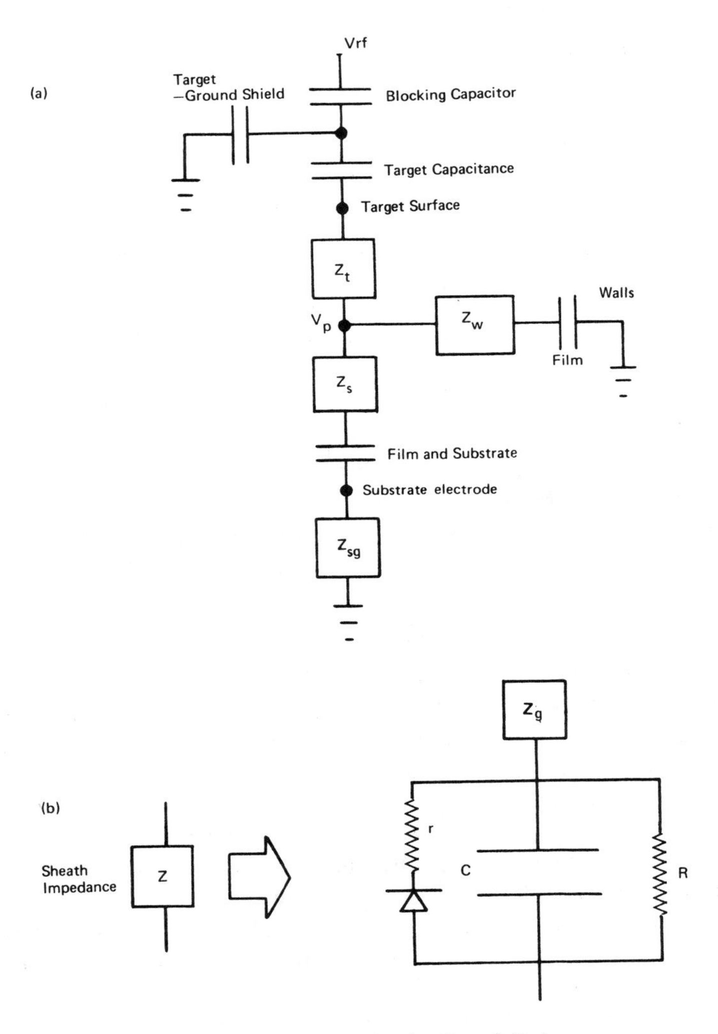

Figure 5-24. Equivalent circuit of an rf discharge

has an effect on the reactive current flowing through the matching network and therefore has an effect on the power losses in the network.

It is not a trivial matter to electrically ground an object in an rf system; even a straight piece of conducting wire will have some inductance L which can lead to sizeable voltage drops along the wire because of the large reactive currents involved. With increasing frequency, the reactance ωL of the inductance will increase whilst its resistance will also increase because of the reduced *skin depth* to which the current is confined. In the equivalent circuit, this impedance is represented by Z_{sg}.

Sometimes one wishes to minimize Z_{sg}, and then special precautions are taken. However, one can also positively control the dc offset voltage on the substrate holder by controlling Z_{sg}, for example to control the bias voltage in a sputtering system. This is done in the *tuned anode* system of Logan (1970). Figure 5-25 shows the dc substrate potentials developed with inductive and capacitive values of Z_{sg}. The larger effect with an axial magnetic field is interpreted as due to the constraining effect of the magnetic field which increases the impedance Z_w to the wall and forces more current through Z_{sg}. The success of this approach is a further confirmation of the earlier contention that there is no uniquely defined floating potential on an insulator in the presence of a modulated plasma potential, and that the 'floating' potential is determined by the impedance to ground.

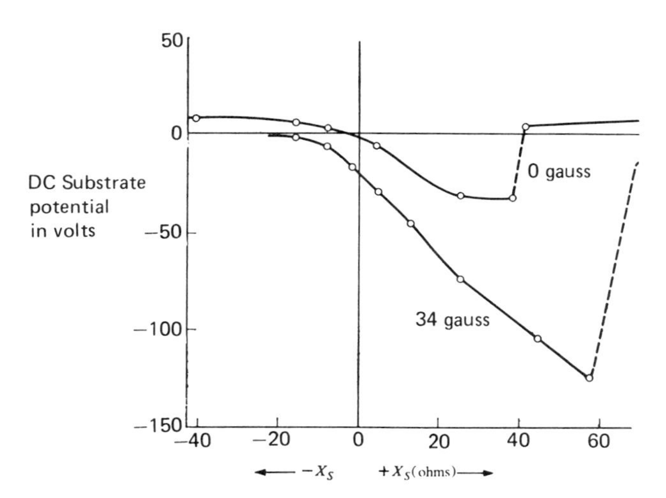

Figure 5-25. Substrate bias dependence on substrate tuning impedance. Frequency 13.56 Mhz, power density 3.9 W/cm^2, argon pressure 15 mtorr (Logan 1970)

PLASMOIDS

There are certain subjects that rarely are discussed in glow discharge articles, but are common knowledge amongst folks who play regularly with these systems. Such subjects include things such as initiating a discharge with a transient burst of pressure or voltage, things that seem too obvious to mention (but only in retrospect!). Another common observation is of *plasmoids*, and these seem not to be discussed because of lack of comprehension, though I might be wrong. A plasmoid is generally described as a volume of more intense glow than its surroundings. Some plasmoids are easy to explain, including those inside hollow cylindrical regions such as entry ports in a chamber, where the enhanced glow is due to a *hollow cathode effect*: the normal ejection of electrons by the sheath field around the inside of a cylinder causes an enhanced electron density along the axis and in the glow. But more mysterious are those plasmoids that occur in the body of the glow. They sometimes occur in groups symmetrically arranged about the axis of the discharge at about the target radius, or sometimes as a single plasmoid similarly located, and are either stationary or rotate in either direction, with continuous motion or regular discrete jumps.

The only reference to these plasmoids in the sputtering literature that I know of, is by Lamont and DeLeone (1970). They compare the form of the plasmoids to the spokes (1, 2, 6 and 8) of a wagon wheel, either stationary or rotating. They report that plasmoids were observed and named as early as 1930, and briefly mention some mechanisms and give some further references. I haven't studied these and can't elucidate the proposed mechanisms. Of practical import, Lamont and DeLeone propose that the plasmoids increase the current density at the periphery of a sputtering target and thus have an effect on the thickness uniformity across a substrate.

The plasmoids are believed to be due to some type of plasma instability; plasmas are very rich in instabilities and a goodly amount of plasma literature is concerned with them. Apparently these instabilities can be reproducible, though, and rf processes can achieve reliable and reproducible output when plasmoids are present. My own experience is that the incidence of plasmoids in sputtering and plasma etching systems is higher in California than anywhere else, but I am not sure how to interpret that observation yet!

REFERENCES

G. S. Anderson, W. N. Mayer, and G. K. Wehner, J. Appl. Phys. **33**, 2991 (1962)

F. Bresnock, Proc. IEEE Intern. Conf. on Plasma Science, Montreal (1979)

H. S. Butler and G. S. Kino, Phys. Fluids **6**, 1346 (1963)

B. N. Chapman, Proc. MRC Conference on Sputtering, Brighton (1969)

B. N. Chapman, previously unpublished results (1975)

O. Christensen and M. Brunot, Le Vide, Les Couches Minces **165**, 37 (1973)

J. W. Coburn and Eric Kay, J. Appl. Phys. **43**, 4965 (1972)

W. D. Davis and T. A. Vanderslice, Phys. Rev. **131**, 219 (1963)

A. J. Hatch and H. B. Williams, Phys. Rev. **112**, 681 (1958)

L. Holland, W. Steckelmacher, and J. Yarwood, eds., *Vacuum Manual*, Spon, London (1974)

G. N. Jackson, Thin Solid Films **5**, 209 (1970)

J. M. Keller, private communication (1978)

J. M. Keller and W. B. Pennebaker, IBM J. Res. Develop. **23**, 3 (1979)

H. R. Koenig and L. I. Maissel, IBM J. Res. Develop. **14**, 168 (1970)

H. R. Koenig, U.S. Patent 3 661 761 (1972)

L. T. Lamont Jr. and J. J. DeLeone Jr., J. Vac. Sci. Tech. **7**, 155 (1970)

J. S. Logan, N. M. Mazza, and P. D. Davidse, J. Vac. Sci. Tech. **6**, 120 (1969)

J. S. Logan, IBM J. Res. Develop. **14**, 172 (1970)

A. D. MacDonald and S. J. Tetenbaum, in *Gaseous Electronics*, Vol. 1, ed. Merle N. Hirsh and H. J. Oskam, Academic Press, New York and London (1978)

L. I. Maissel, in *Handbook of Thin Film Technology*, ed. L. I . Maissel and R. Glang, McGraw Hill (1970)

R. B. McDowell, Solid State Tech., p. 23, February (1969)

W. B. Pennebaker, IBM J. Res. Develop. **23**, 16 (1979)

R. T. C. Tsui, Phys. Rev. **168**, 107 (1968)

S. Vacquie, J. Bacri and G. Serrot, Comptes Rendu **266**, 387 (1968)

J. L. Vossen and J. J. O'Neill Jr., J. Vac. Sci. Tech. **12**, 1052 (1975)

J. L. Vossen, J. Electrochem. Soc. **126**, 319 (1979)

G. K. Wehner, Adv. in Electronics and Electron Physics **VII**, 253 (1955)

Chapter 6. Sputtering

"WHAT IS ALL THIS SPUTTERING NONSENSE ANYWAY?"

A few years ago, a medical conference and a sputtering conference were taking place simultaneously at Imperial College. The conferees were as always demonstrating the well-known scientific phenomenon that conference systems tend towards a condition of being in the bar, where a well-oiled medic accosted a group of the sputterers and demanded to know "What is all this sputtering nonsense, anyway?". "Well", replied one of the sputterers, "we're in a branch of the medical profession too, old chap – in speech therapy actually. Sputtering's like stuttering, you know, except our chaps say p . . . p . . . p . . . p . . . instead of t . . . t . . . t . . . t . . .". The medic warmly thanked his newly-discovered professional colleague and hurried back to enthusiastically convey the freshly-gleaned information to his cronies.

The medic might have been a bit closer, though not very much, if he had looked in the dictionary. It seems that the word 'sputter' appeared in the English language (The Shorter OED, 1959) as early as 1598 and is adapted from the imitative words 'sputteren' in Dutch and 'sputterje' in West Frisian. 'To spit out in small particles and with a characteristic explosive sound', says the dictionary; 'to utter hastily and with the emission of small particles of saliva, to ejaculate in confused, indistinct or uncontrolled manner, especially from anger or excitement – His tongue was too large for his mouth; he stuttered and sputtered (1878)'.

Compared with the above, you may be disappointed with the type of sputtering I'm going to describe. I must confess that I have never heard the sound of sputtering, although the sound of rotary pumps will ring in my ears forever. On the other hand, my type of sputtering is rather colourful!

INTERACTIONS OF IONS WITH SURFACES

Let us consider what happens when an ion appraches the surface of a solid (of the same or different material); the solid is usually called the *target*. One or all of the following phenomena may occur (Figure 6-1):

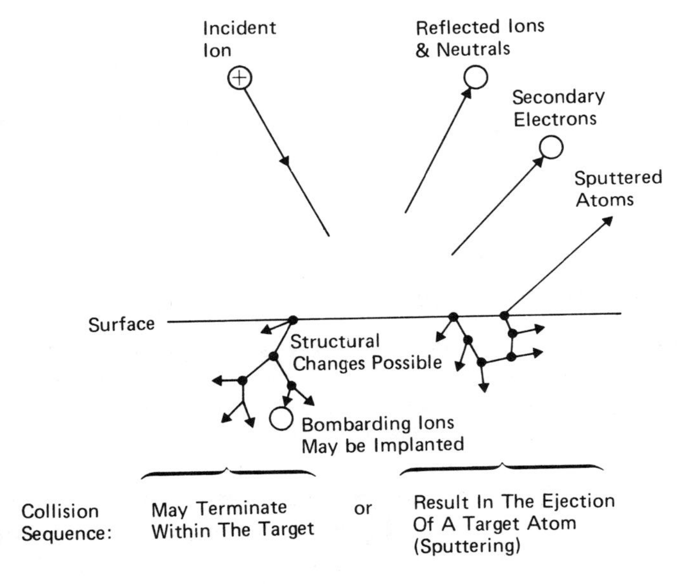

Figure 6-1. Interactions of ions with surfaces

- The ion may be reflected, probably being neutralized in the process. This reflection is the basis of an analytical technique known as *Ion Scattering Spectroscopy*, which enables us to characterize the surface layers of the material, and also tells us a lot about the fundamental ion-surface interaction.
- The impact of the ion may cause the target to eject an electron, usually referred to as a *secondary electron* (Chapter 4, "Secondary Electron Emission").
- The ion may become buried in the target. This is the phenomenon of *ion implantation*, which is already used extensively in integrated circuit technology for selectively doping silicon wafers with precisely controlled amounts and depth profiles of specific impurities, and is likely to find many other applications such as surface treatment of steels.
- The ion impact may also be responsible for some structural rearrangements in the target material. 'Rearrangements' may vary from simple vacancies (missing atoms) and interstitials (atoms out of position) to more gross lattice defects such as changes of stoichiometry (i.e. relative proportions) in alloy or compound targets, or to changes in electrical charge levels and

distributions, and are usually collectively referred to as *radiation damage*, which is a subject of great importance, particularly with relation to nuclear energy. Radiation damage can often be removed by annealing (heat treatment) but it is not always unwanted, and perhaps the alternative name of *altered surface layers*, used mostly to describe the stoichiometry changes, is more apt.

- The ion impact may set up a series of collisions between atoms of the target, possibly leading to the ejection of one of these atoms. This ejection process is known as *sputtering*.

The Mechanisms of Sputtering

In the energy range most relevant to sputter deposition, the interaction between the impinging ion and the target atoms, and the subsequent interactions amongst the latter, can be treated as a series of binary collisions. The sputtering process is very often compared to the break in a game of atomic billiards (Figure 6-2) in which the cue ball (the bombarding ion) strikes the neatly arranged pack (the atomic array of the target), scattering balls (target atoms) in all directions, including some back towards the player, i.e. out of the target surface. In the real process, the interatomic potential function (the variation of interatomic repulsion or attraction with separation distance) is rather different from the hard sphere billiard ball case, but nevertheless the billiard model is not too unrealistic.

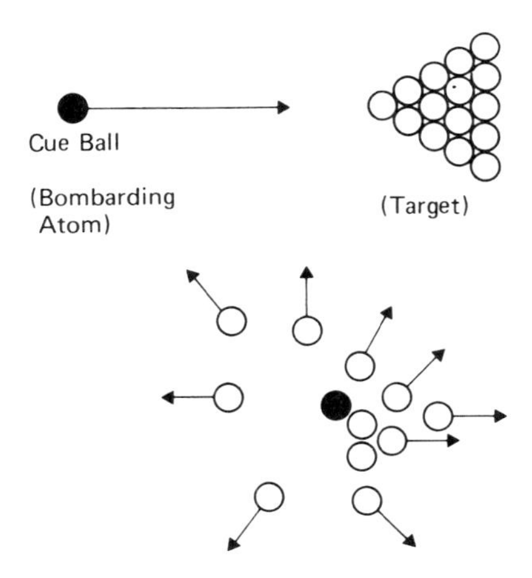

Figure 6-2. Sputtering – the atomic billiards game

It is implied in our description of the basic interaction that the incident particle could be either an ion or a neutral atom. Ions are normally used since they can easily be accelerated by an electric field, whereas neutrals pose a problem in this respect. Furthermore, the ions are likely to be neutralized anyway by the Auger emission of an electron from the target as the ion approaches, so that the impacting species are actually mostly neutral.

The series of collisions in the target, generated by the primary collision at the surface, is known as a *collision cascade* (Figure 6-1). It will largely be a matter of luck whether this cascade leads to the sputter ejection of an atom from the surface (which will require at least two collisions) or whether the cascade heads off into the interior of the target, gradually dissipating the energy of the primary impact, ultimately to lattice vibrations, i.e. heat. It's not surprising then that sputter ejection is rather inefficient, with typically 1% of the incident energy reappearing as the energy of the sputtered atoms.

The collision phenomena occurring in a target, often referred to as *target kinetics*, are a fascinating and important subject for study. They relate not only to sputter deposition and etching, but also to ion implantation and radiation damage. Life is rather short, however, and there is neither room in this book nor am I adequately informed to pursue the topic much further. But in this chapter we shall be looking at the applications of sputtering rather than the collision phenomena leading to sputtering, and fortunately we need consider only certain aspects of the process in order to do this. In the next section we shall look briefly and (unfortunately) superficially at some of the quantitative relationships involved in sputtering target kinetics.

Sputtering Target Kinetics

A generalized treatment of target collision phenomena would have to consider the detailed interatomic potential function, but fortunately the interactions in a sputtering target are sufficiently short range that we need consider interactions only between immediate neighbours (including the incident ion). A binary collision is characterized by the energy transfer function which we met earlier (Chapter 1, "Energy Transfer in Binary Collisions"):

$$\frac{4m_i m_t}{(m_i+m_t)^2}$$

where m_i and m_t are the masses of the colliding atoms. The sputtering process is the result of a series of such collisions. A detailed consideration and experimentation show that the binary model is a useful representation of the interactions under sputtering conditions.

A useful parameter that we shall encounter frequently is the *sputtering yield* S, defined as the number of target atoms (or molecules) ejected per incident ion. From our model above, we would expect the sputtering yield to depend on the masses of the incident ion and the target atom, m_i and m_t respectively, and on the energy E of the incident ion. However, consider sputtering as the overall process of transferring energy from the incident ion to the sputtered atoms. Then, since the sputtered atoms can come only from the surface layers of the target, it is not just a question of transferring energy to the target atoms, but also that this energy should be transferred mostly to the surface layers. We would therefore expect the sputtering yield S to be proportional to the energy deposited in a thin layer near the surface, and this is determined by the *nuclear stopping power* s (E); for low bombardment energies up to about 1 keV, an expression due to Sigmund (1969), which not suprisingly involves the energy transfer function, is

$$s(E) = \frac{m_i m_t}{(m_i + m_t)^2} E \times \text{constant}$$

and this is used to predict the following form for the sputtering yield S:

$$S = \frac{3\alpha}{4\pi^2} \frac{4 m_i m_t}{(m_i + m_t)^2} \frac{E}{U_o}$$

Here U_o is the surface binding energy of the material being sputtered, and α is a monotonic increasing function of m_t/m_i which has values of 0.17 for m_t/m_i = 0.1, increasing up to 1.4 for m_t/m_i = 10.

This expression for S predicts that the yield will increase linearly with E. In practice, this seems to be satisfied up to above 1 keV, above which S becomes relatively constant; Figure 6.3a is typical. It appears that the higher input energy is being distributed through a larger volume, so that the energy transmitted to the surface layers remains virtually constant. At very high energies, S even decreases as ion implantation becomes dominant (Figure 6-3b).

So our original expression for S is apparently valid only up to about 1 keV, and this is due to various assumptions about the atomic interactions. Above 1 keV, a modified interaction yields

$$S = 3.56\alpha \frac{Z_i Z_t}{(Z_i^{2/3} + Z_t^{2/3})} \frac{m_i}{(m_i + m_t)} \frac{s_n(E)}{U_o}$$

where $s_n(E)$ is a reduced stopping power and is a function of a reduced energy based on the actual energy, masses and atomic numbers Z_i and Z_t of the atoms involved. The interested reader is referred to Winters (1976) for further details.

The success of these theoretical models can be demonstrated by comparing experimental and theoretical results, with good agreement resulting in most cases. This is illustrated in Figure 6-4 for the case of argon on copper, which compares the yield predicted by the equation above, with experimental values.

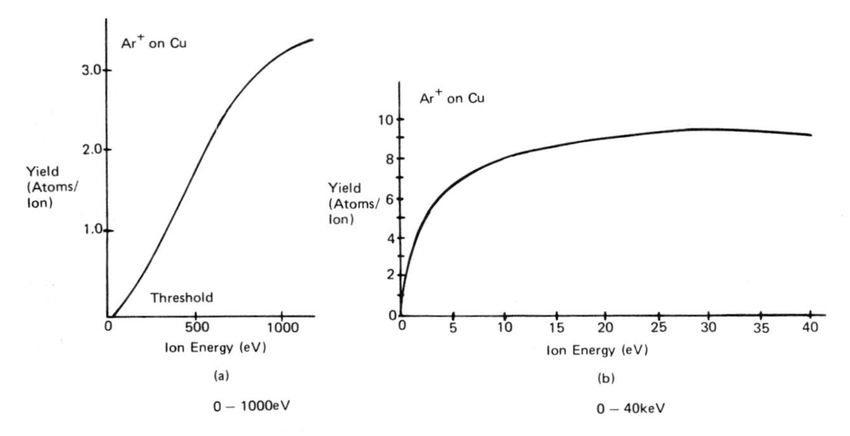

Figure 6-3. The variation of sputtering yield, for argon ions on copper, as a function of the ion bombardment energy (Carter and Colligon 1968)

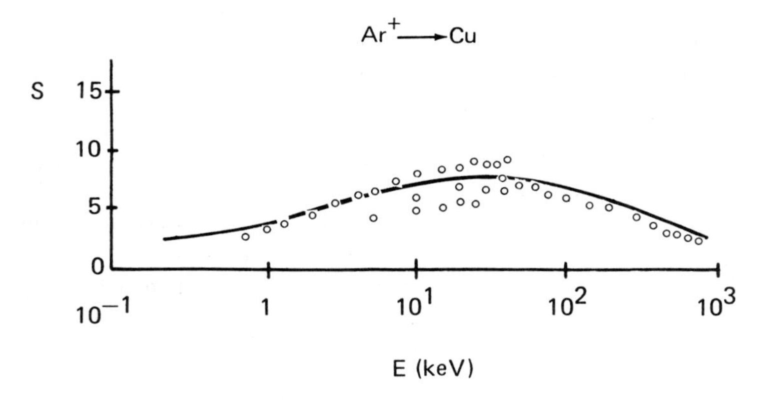

Figure 6-4. Theoretical (solid line) and experimental values for the energy dependence of the sputtering yield of copper in argon. Data from various authors; see Winters (1976)

It is apparent from examining the two sputtering yield expressions that we should no longer expect to find a maximum yield when $m_i = m_t$ as suggested by the energy transfer function alone. This prediction seems to be borne out in practice: Figure 6-5 shows the results obtained by Almen and Bruce (1961) for the sputtering of copper by inert gas ions over a wide range of ion energies.

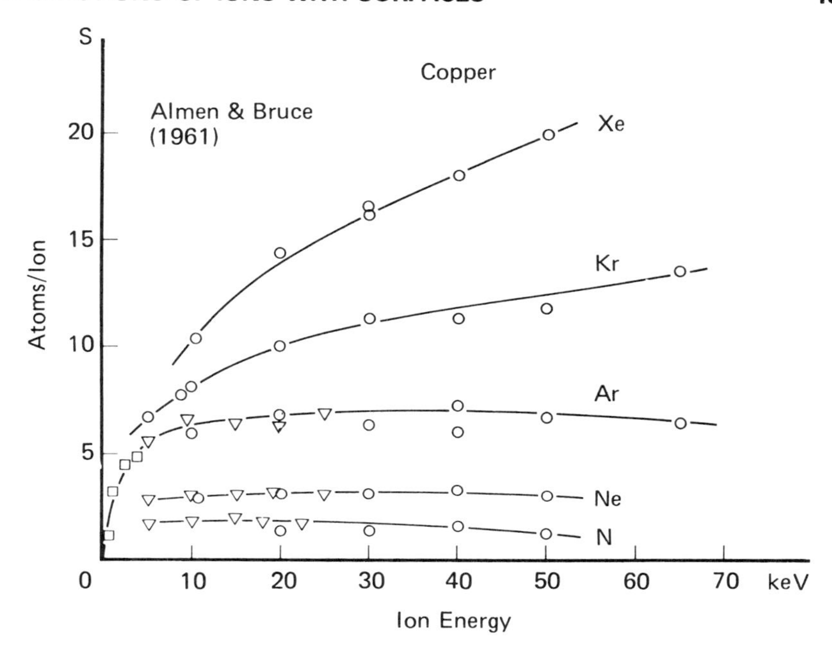

Figure 6-5. Sputtering yields of the noble gases on copper, as a function of energy (Almen and Bruce 1961)

These results show that the sputtering yield S obtained with xenon (atomic weight 131) is the largest of all these gases at all energies, even though the masses of krypton (83) and argon (40) are much closer to that of copper (64).

The high energy (>1 keV) expression for S is reasonably successful at predicting the observed mass dependences but the low energy yield expression shows quite the wrong mass dependence. And indeed, the low energy model becomes quite incorrect for energies below 100 eV, which is an energy range which we shall discuss further in connection with bias sputtering.

The sputtering model which we used above shows only the general features of the process. A more accurate model is being developed and the interested reader should refer to the review articles by Wehner and Anderson (1971), Townsend and Kelly (1976), Winters (1976), and Kelly (1978). A realistic model is rather complex: for example, one has the pragmatic consideration that a sputtering target eventually becomes a mixture of the original target and the bombarding element embedded in or otherwise combined with it, so that one is no longer dealing with the sputtering of the original target. Nevertheless, we shall see that the simple ideas discussed above will be quite valuable when we consider later how the sputtering process is put into practice.

In the appendix to this chapter, we have shown more sputtering yield data, concentrating on typical sputtering ion energies and the usual sputtering gas, argon.

Summary of the Overall Process

Although we have looked a little at what's going on in the target, that knowledge isn't really essential to an understanding of sputter deposition. What we do need to know, however, is what the overall results of these target processes are (Figure 6-6):

- A target atom may be sputter ejected.
- The incident ion will either become implanted or be reflected, probably as a neutral and probably with a large loss of energy.
- The ion impact and the resulting collision cascade will cause an amount of structural reordering in the surface layers.
- A secondary electron may be ejected.

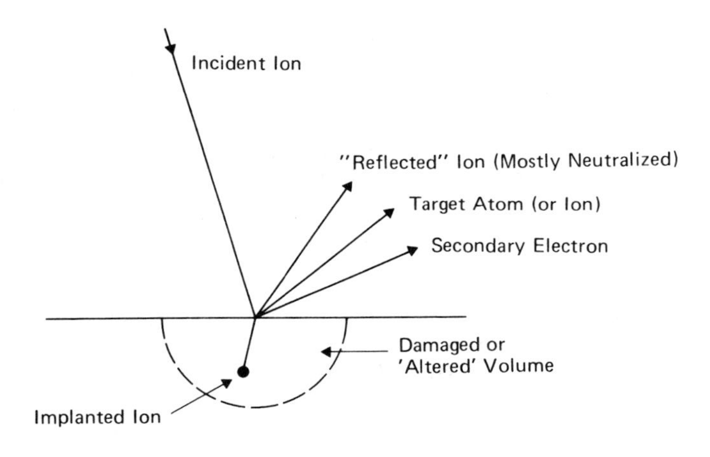

Figure 6-6. Summary of the target processes

These are just *possible* consequences of the impact; the possibilities will not generally occur in any specific ratio, nor even necessarily occur at all. We shall see shortly how each of these aspects of the sputtering process are manifested and/or utilized.

APPLICATIONS OF SPUTTERING

Sputter Etching

As we have already seen, the sputtering process essentially involves knocking an atom out of the surface of a target. By repeating this process over and over again, we can evidently *sputter etch* the target. We may wish to do this over the whole of the target surface, for example to clean it or to make it thinner, or selectively to generate a topographic pattern on the surface. We shall return to this application later in the chapter.

Sputter Deposition

The ejected atom can also be used. After ejection it can, under the right circumstances, move through space until it strikes and condenses on the surface of a receiver, which is known as a *substrate*. By repeating the process over and over, we can build up a coating of several or many atomic or molecular layers of target material on the substrate. The coating, which is generally less than about 1 μm, is called a *thin film*, and the process is known as *sputter deposition*; this process is currently the main application of sputtering.

Although I shall be frequently referring to the properties of thin films prepared by sputter deposition, there is no room here for a general treatment of thin film science, although the initial formation and growth stages are discussed below (in "Thin Film Formation"). The reader who has no background in this area is advised to read one of the several reviews on the subject (Berry, Hall and Harris 1968, Lewis and Campbell 1967, Chapman and Jordan 1968, Neugebauer 1970, Leaver and Chapman 1971; the last is very elementary).

Limitations of Sputtering

There are some devotees of sputtering who would claim that anything can be sputtered. There are others who would claim that if a material can't be sputtered then that material isn't worth bothering with!

These views are rather extreme, though it's certainly true that sputtering is a widely applicable and versatile process. How then do we decide whether to sputter? When it comes to specific applications and specific materials, there are so many factors to be considered that it would be misleading of me to generalize. However, ignoring my own advice, there are a few general restrictions. The capital expenditure for sputtering equipment is higher than for virtually all other coating processes. As a coating or etching process, it's too slow for some applications (for a review of the many other surface coating techniques available, see Chapman and Anderson 1974). Deposition rates are typically 50 – 500Å per minute, and etch rates around 500 – 10,000Å per minute, although there are higher rate versions of the process available.

As we shall see, sputtering is carried out in partial vacuum and there are some materials which are incompatible with good vacuum practice, either due to their inherent properties or due to their products under the various types of bombardment involved in the process. Organic solids are frequently unable to withstand this bombardment, and completely degrade; the temperature rise alone is often adequate to produce this effect. On the other hand, it is remarkable that polytetrafluoroethylene, which has a long chain molecule and cannot possibly be sputtered intact, and which can easily be reduced to a carbonaceous mess at quite low temperatures, can nevertheless be sputtered to produce quite useful films with properties similar to those of the target.

The presence of the vacuum pumps also means that one frequently encounters problems when using target materials containing volatile components. The resulting films are often deficient in the volatile component, although this can sometimes be rectified.

To summarize, sputtering is a rather versatile process. It's applicable widely but naturally has some shortcomings and drawbacks. One is advised to read the published literature on specific applications and specific materials before becoming too enthusiastic.

We now begin to look at some of the details of the process by considering a dc sputtering system. Actually, most sputtering systems are rf driven, but it is helpful to first look at the dc counterpart.

A CONVENTIONAL DC SPUTTERING SYSTEM

In this section we shall meet briefly some ideas which are discussed in more detail later. This probably means that this will make more sense after you've read later sections. On the other hand, I think you should read this first so that you see the overall picture before we plunge into the detail.

How do we deposit thin films by sputtering? It's very informative to look at a conventional dc sputtering system (Figure 6-7) to see the elements used to turn the sputtering phenomenon discussed in the last section into a practical deposition process.

The material we wish to sputter is made into a *sputtering target* which becomes the cathode of an electrical circuit, and has a high negative voltage V (dc) applied to it. The target is nearly always solid, although powders and even liquids are sometimes used. The substrate which we wish to coat is placed on an electrically grounded anode a few inches away. These electrodes are housed in a chamber which is evacuated. Argon gas is introduced into the chamber to some specified pressure. The action of the electric field is to accelerate electrons

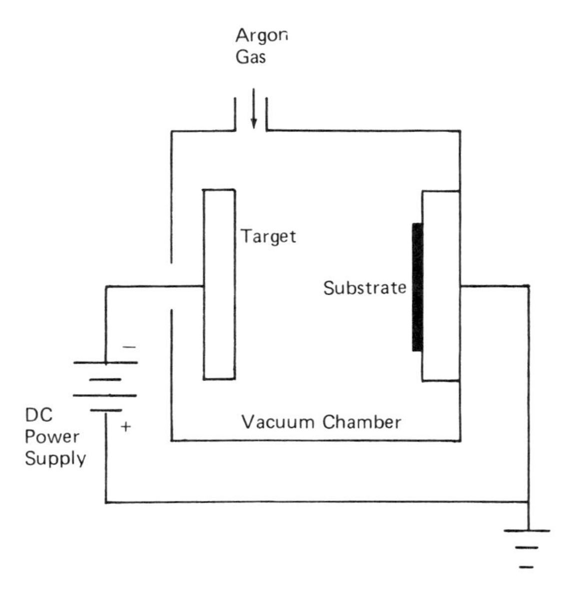

Figure 6-7. Schematic of a dc sputtering system

which in turn collide with argon atoms, breaking some of them up into argon ions and more electrons to produce the glow discharge that we discussed in Chapter 4. The charged particles thus produced are accelerated by the field, the electrons tending towards the anode (causing more ionization on the way) and the ions towards the cathode, so that a current I flows.

When the ions strike the cathode, they may sputter some of the target atoms off. They may also liberate secondary electrons from the target and it is these secondary electrons which are responsible for maintaining the electron supply and sustaining the glow discharge (Chapter 4, "Maintenance of the Discharge"). The sputtered atoms from the target fly off in random directions, and some of them land on the substrate (on the anode), condense there, and form a thin film.

The voltage V which is required to drive the current I through the system is a function of the system pressure p. The rate of thin film formation on the substrate will depend on the amount of sputtering at the target; this in turn will depend on the ion flux at the target and so (linearly) on the current. However, the amount of sputtering also depends on the sputtering yield S and so on the ion energy and hence on V, which determines the sheath voltage at the target. The choice of sputtering pressure p and the implied choice of the V-I characteristic, are thus rather important and so we'll now look at the criteria used in selecting operating parameters and in deciding on the nature of the sputtering gas.

Choosing the Sputtering Gas

As we saw earlier, it doesn't really matter whether we use neutral atoms or ions as far as the actual sputtering process is concerned. However, it is easy to accelerate ions to the energies required by using an electric field and much more difficult to accelerate neutrals; so we normally use ions.

It's much easier to ionize atoms when they are in a gaseous state, and naturally the latter state is easier to achieve with materials already in a gaseous form at room temperature. So we use gases for our ion source. A glow discharge happens to be a particularly convenient method of producing a significant flux of gas ions.

We usually don't want the ions to react with the target or growing film, which therefore requires noble gas ions. We shall see later that some sputtering ions will become incorporated into the growing film and become trapped by the depositing film atoms. Since we are usually concerned with the purity of the film, we'd prefer these incorporated ions to be as innocuous as possible; this is another reason for using noble gas ions, with their closed shell electronic structures and (almost always) chemical inactivity. We saw earlier that the heaviest inert gas will give the highest sputtering yield (Figure 6-5). Radon ($Z = 86$) is the heaviest, but it is also radioactive. Xenon (54) and krypton (36) are next in line, but argon (18) is almost always used in sputter deposition because it is easily available and cheaper, and the sputtering yield is only a factor of about two down on xenon at sputter deposition energies. The cost of the sputtering gas is not usually a significant factor, though, because a cylinder may last a year in a small sputtering system. Most gas suppliers have several grades of argon; their standard grade is usually intended for applications such as argon arc welding, and is not pure enough for sputtering.

Choosing the Pressure Range

A vacuum system enables us to control the operating pressure inside the sputtering system. Operating pressure limitations are imposed by the requirements of both the glow discharge and of film deposition.

The glow discharge sets a lower pressure limit. The discharge is sustained by electrons making ionizing collisions in the gas. The number of ionizing collisions will decrease with decreasing gas density, and hence gas pressure, so that the discharge current (for constant voltage) will also decrease. This is shown for a typical case in Figure 6-8. Below about 30 mtorr, the current (and hence ion flux at the target) and sputtering rate in a dc discharge become quite small.

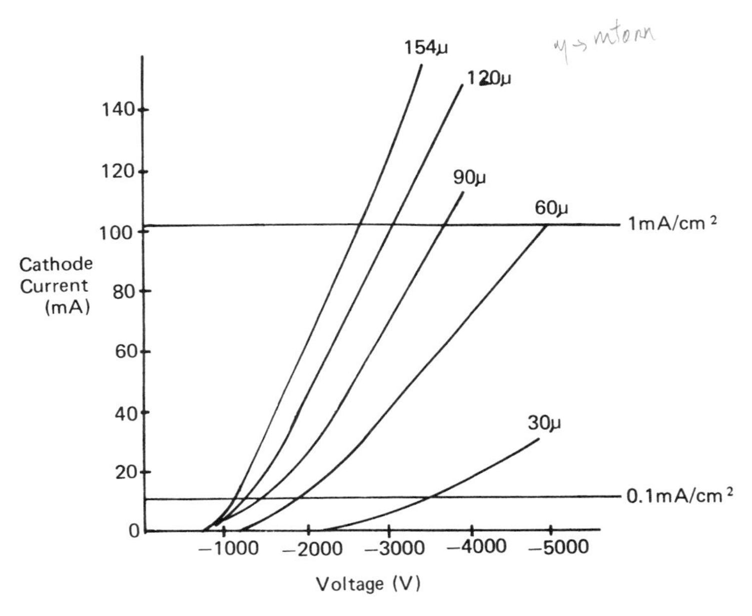

Figure 6-8. Typical I-V curves at different argon pressures using a 4½ inch nickel cathode – area 103 cm² (Kay 1969)

A different problem arises at the other end of the pressure range. In the same way that electrons undergo collision, material sputtered from the target may collide with gas atoms on its way to the substrate, at a rate which will increase with increasing pressure; we discussed this in Chapter 1, "Probability of Collision". The result of the collision is to deflect the sputtered atom, sometimes back towards its parent, and hence decrease the deposition rate. This is demonstrated in Figure 6-9, which shows how the apparent sputtering yield of a nickel target (obtained by measuring its weight loss) decreases with increasing pressure due to sputtered nickel being backscattered in the gas phase and redeposited on the target. The mean free path (see Chapter 1) of the nickel is about 1 mm at 120 mtorr, so this result is expected. With increasing pressure, deposition becomes less a line-of-sight process and more a diffusion process. The scattering process becomes serious above about 100 mtorr, so that taking both of our limitations into account, an overall operating range of about 30 – 120 mtorr is usual for dc sputter deposition. This isn't a very wide operating range, but at least the two limits haven't overlapped! Later we'll see modifications of the basic sputtering process which permit lower operating pressures.

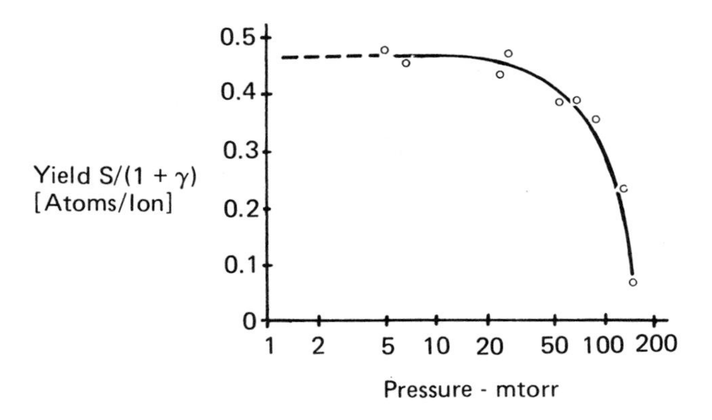

Figure 6-9. Variation of the apparent sputtering yield of nickel vs argon gas pressure; 150 eV ion energy (Laegreid and Wehner 1961)

Choosing Electrical Conditions for the Glow Discharge

We have an operating range of 30 – 120 millitorr, but what voltage and current should we use? To some extent that is Hobson's choice since, for each operating pressure, target material and sputtering gas, there is a specific voltage – current relationship. The V-I relationships at several pressures, for a nickel target sputtered in argon, are shown in Figure 6-8. Actually the target material isn't that important even though the secondary electron coefficient γ does vary from material to material. Table 6-1 shows the current densities obtained from various target materials under similar conditions, and the densities are very much alike, at least for this restricted case of conducting targets.

The sputtering rate of the target is determined by the flux of ions and energetic neutrals, and the sputtering yield. An ion current of 1 mA/cm^2 corresponds to a flux of 6 10^{15} singly charged ions per cm^2 per second. However, the target current also includes a secondary electron component, and the target bombardment flux is due to energetic neutrals as well as ions.

Figure 6-8 suggests that, by applying a high enough voltage, we could have any target current we want. But maybe that's the wrong approach. What ion

Table 6-1

Target current density as a function of target material and discharge gas. This data was acquired over a period of years and is presented to show that the target current density is relatively insensitive to target material and additive gases. Target voltage = -2000 V, gas pressure 70 mtorr, and interelectrode spacing ~4 cm. From Winters et al. (1977)

Target Material	Current Density (mA/cm^2)	Gas
Si	1.35	CF_4
Si	1.25	CF_4 + 6½% O_2
Si	1.20	CF_4 + 10½% O_2
Si	1.05	CF_4 + 16% O_2
Si	1.15	CF_4 + 22½% O_2
Si	1.30	CF_4 + 34% O_2
Si	1.35	CF_4 + 42% O_2
Si	0.45	Ar
Mn_5Ge_3	0.55	Ar
GdFe	0.69	Ar
CoNi	0.49	Ar
Ni	0.46	Ar
CoNi	0.42	Ar
Ti	0.43	Ar
V	0.43	Ar
Ag	0.49	Ar
C	0.41	Ar
Ta	0.63	Ar
Ni	0.56	Ar + 10% N_2
Au	0.39	Ar + 10% N_2
Au	0.51	Ar + 10% N_2
W	0.51	Ar + 10% N_2
W	0.49	N_2
Ni	0.44	N_2

energies should we use for sputtering? We saw earlier that the sputtering yield rises monotonically with ion energy up to several tens of keV, where it begins to decrease. This latter region, where we're getting less for more, is obviously not an energy bargain basement but what other criteria are important? A prime matter is that of safety. Electrically we'd like low voltages and because X-rays can be produced by fast ions and electrons, we ought to keep energies and voltages below 10 kV. Within this limit, many other criteria become important and so the material being deposited and the requirements made of it must be considered. The broad range, though, is usually determined by the fact that we will be restricted (if only economically) to a sputtering power supply of limited output. By varying the operating pressure of the sputtering system, we can change the V-I relationship of the discharge (within certain limits) whilst maintaining the same power input VI. So rather than consider only the sputtering yield S, we need to look also at the yield per energy input, S/E; this is shown in Figure 6-10 for the case of xenon on copper, and in Figure 6-11 for argon on tungsten over a narrower energy range. These suggest a most efficient ion energy of a few hundred volts. Other materials would give similar results.

This dependence of S/E on E is not too surprising. Our sputtering kinetics model ("Sputtering Target Kinetics") suggested that S should increase linearly with E up to about 1 keV. If so, then S/E should be constant, and this seems to be the case except below about 100 eV, where the model begins to break down. Above 1 keV or so, the sputtering yield is fairly constant, so that S/E falls with increasing energy.

Several points, which are discussed in Chapter 4, have to be stressed. A V-I discharge relationship does not mean that the flux of sputtering particles at the cathode is equivalent to I, since some of the cathode current is carried by electrons, and some of the sputtering particles are neutral. It also doesn't mean that the sputtering particles will all have an energy of V electron volts; instead they will have a wide energy distribution with a maximum of V because they too (like the electrons) collide with gas atoms and usually slow down in the process.

Another way of saying all of the above is that not all of the VI power input goes into the target. Of course there is no general reason why we should try to get the maximum amount of sputtering per unit energy input, since there may be other more important criteria governing our choice of V-I such as a particular operating pressure or a low operating voltage.

The sputtering yield per unit energy input (S/E) data tell us that we shouldn't expect to sputter very rapidly below about 100 V and that the process is becoming very inefficient above 10 kV. In practive, a lower limit of about 500 V is used to achieve adequate current density, and an upper limit above 5 kV is rarely found necessary. Even with a sheath voltage of 500 V, the ion bombardment energy is generally much lower because of charge exchange collisions in the sheath.

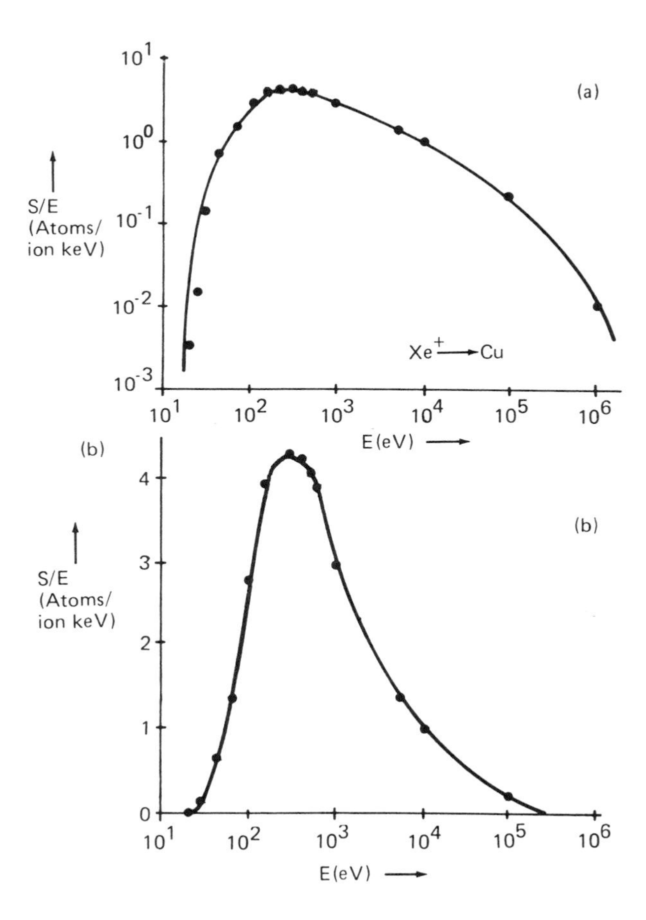

Figure 6-10. Variation of the sputtering yield per unit energy input of xenon on copper vs ion energy. Data from: Dupp and Scharman (1966), Almen and Bruce (1961), Guseva (1960), Wehner (1962), Stuart and Wehner (1962)

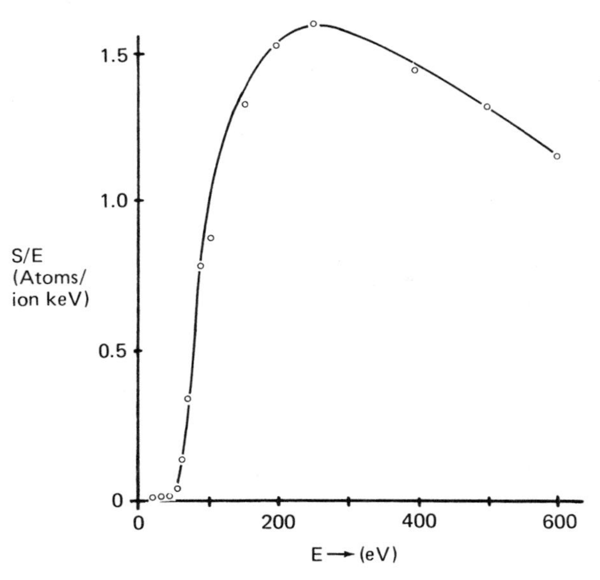

Figure 6-11. Variation of the sputtering yield per unit energy input for argon on tungsten vs ion energy. Data from: Stuart and Wehner (1962), Wehner (1962)

Summary

We have seen in outline how sputter deposition can be carried out in a dc system using a glow discharge as an ion source. The ions are usually of a noble gas which is introduced into the system and maintained at a pressure in the 30-120 militorr range. The voltage required to maintain a glow discharge having a current density of 0.1 - 2.0 mA/cm^2 is usually in the range 500 – 5000V.

In the next sections we shall look more closely at sputtering discharges so that we can better understand and control the sputter depositon process.

SPUTTER ETCHING AND DEPOSITION OF INSULATORS

In the previous discussion of a dc sputtering system, it was assumed that the target was conducting, or at least reasonably so. One wouldn't want to drop more than about 100 V across the target, so with typical numbers of 1 mA/cm^2 for the target current density and 0.5 cm for the target thickness, this sets a limit of $\rho = 2\ 10^5\ \Omega$cm, which is quite resistive. So some nominal insulators could be sputtered, but it isn't usually a very good solution because of problems such as stress developed in the target by the resistive heating there.

Instead, two common ways of depositing insulating thin films are by *rf sputtering* and *reactive sputtering*.

RF Sputtering

The technique of rf sputtering uses an alternating voltage power supply at rf frequencies around 10 MHz, so that the sputtering target is alternately bombarded by ions and then electrons so as to avoid charge build-up. The rf discharge was discussed in some detail in Chapter 5, where we also looked at the choice of operating frequency, crystal controlled and self-excited oscillators, and matching networks. It also seemed that the rf discharge made more efficient use of the electron impact ionization, so that operating pressures could be practically extended down to 1 mtorr. Because the detail of the sheath mechanism is slightly different in the rf case, the sheath is still about 1 cm thick at 10 mtorr, and so the lower pressure leads to less ion energy attenuation due to charge exchange. The lower operating pressure also reduces the amount of scattering of material sputtered from the target (cf. Figure 6-9).

There are more subtle advantages: in dc sputtering systems, *arcing* is sometimes a problem, due to patches of dirt (with higher secondary electron coefficient), pockets of outgassing (higher pressure, higher j locally), or asperities (higher $\mathcal{E}$). These *unipolar arcs* (Maskrey and Dugdale 1966) can be quite troublesome, leading to the necessity of *conditioning* a sputtering target before general usage, by slowly increasing the applied power and sputtering (or evaporating) away the arc-forming defect.

These arcs are less likely to form in rf discharges because the field is maintained in one direction for less than one cycle, and reduces to zero twice in each cycle, making it more difficult for the arc to be sustained. Even so, target conditioning is still necessary.

With its various advantages, rf discharges are almost always used for sputtering purposes. The two principal exceptions are magnetron sputtering systems and ion beam sputtering systems; these are discussed later.

Reactive Sputtering

Reactive sputtering avoids the problem of target charging instead of solving it. A conducting elemental target is dc sputtered, and the sputtered material is combined chemically with a component from the gas phase, e.g. oxygen. In Chapter 7, we shall see how electron impact dissociation can turn the discharge into a chemically active environment. To illustrate the technique, silicon nitride can be produced by sputtering silicon in a dc (or rf) discharge containing nitrogen. Even when a compound target is rf sputtered, reactive sputtering is often used to restore the stoichiometry of the film. We shall return to this topic in "Deposition of Multicomponent Materials".

PRACTICAL ASPECTS OF SPUTTERING SYSTEMS

In the previous section, we dealt with some of the process conditions required for sputtering. Of course, there are many other practical considerations in making a sputtering system. This book is primarily about the glow discharge aspects of the process, and so such details are discussed only briefly.

Ground Shields

Looking into a sputtering system, one is immediately aware of several features (Figure 6-12). The target is surrounded by a *dark space shield*, also known as a *ground shield*. The purpose of this is to restrict ion bombardment and sputtering

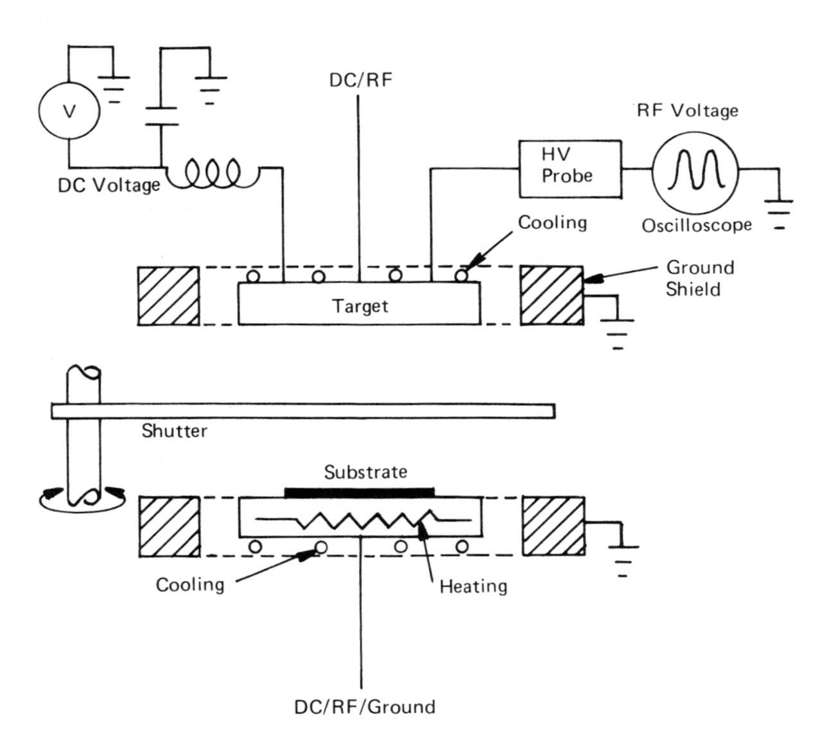

Figure 6-12. Schematic of a sputtering system showing ground shields, shutter, electrode cooling and heating, rf and dc voltage measurement

to the target only. Otherwise, the target backing plate, mounting clips (if any) and mechanical supports would also be sputtered and cause the film to be contaminated. In order to prevent ion bombardment of the protected regions, the space between the target and the ground shield must be less than the thickness of the dark space. From the descriptions of the glow discharge maintenance mechanisms in Chapters 4 and 5, we can see that this criterion is such as to prevent the establishment of a self-sustained discharge in the space between the target and the shield. Occasionally one finds that sharp points or patches of dirt cause local discharges or arcs, particularly with dc discharges, and these must be eliminated. Some possible ground shield arrangements are shown in Figure 6-13.

Since the thickness of the dark space decreases with pressure, the size of the gap between the target and shield sets an upper pressure limit for operating the system. In principle, the gap could be extremely small, but in practice this is limited by spurious discharges, and in the case of rf discharges, by increasing capacitive target-to-ground coupling as the gap is decreased. The dark space thickness also decreases with frequency, so systems operating above 13.56 MHz need to have correspondingly closer ground shields.

Many systems have the capability of applying electrical power to the substrate for the technique known as *bias sputtering* (q.v.). In this case, a ground shield is required around that electrode too.

Shutters

Figure 6-12 shows a shutter that can be rotated into place between the electrodes. This has its use during a *presputtering* period when the first few atomic layers of the target are removed by sputtering in order to clean it. During the time that the system is open to air to load or unload it, the target is liable to become contaminated by atmospheric pollution, by handling, or by chemical combination with the atmosphere to form an oxide or other compound. If this were left in place, the first period of sputtering would transfer the contamination to the substrate. This is avoided by interposing the shutter during that initial period, in order to prevent deposition on the substrate. For the same reasons, it is often extremely helpful to clean the substrate in the same way, and to do this immediately prior to deposition so that the surface does not become contaminated. So the ability to power the substrate is useful for this pre-sputter period as well as for applying bias during sputter deposition.

One has to be careful in the use of shutters. The operating pressure range of sputtering is such that the sputtered material makes many collisions in the gas phase, and so it can diffuse around shutters to an extent. This is in contrast to the use of shutters in high vacuum ($< 10^{-5}$ torr) evaporation where shutters block everything in the line of sight of the source.

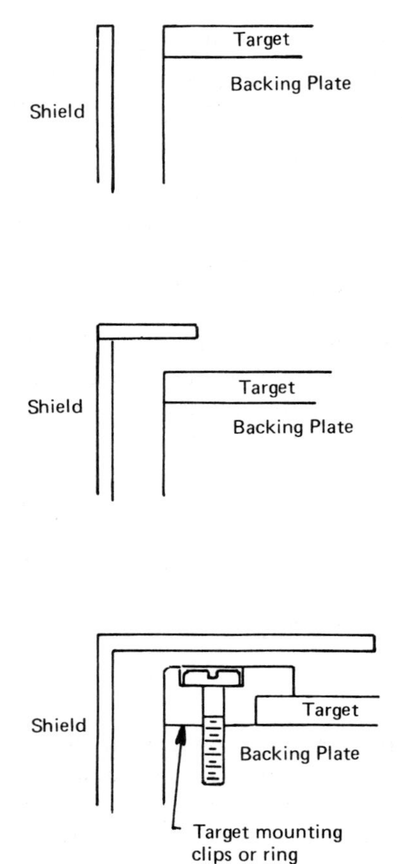

Figure 6-13. Some possible ground shield arrangements

Target Cooling

Returning to Figure 6-12 again, we see that the target is cooled. Sputtering is a very inefficient process, and most of the power input to the system appears finally as target heating. Such heating can become excessive – local temperatures of 400° C have been reported – and can lead to damage of the bonding between the target and the backing electrode, of the target itself, of associated vacuum O-rings, etc. This is usually avoided by cooling the target with water or another suitable liquid. On the other hand, such cooling is a complexity and it can be avoided if the power input to the target is not too great. We have successfully run several systems without cooling; indeed, this probably contributed to their reliability. There are many of us who have suffered from water leaks and their consequences! Cooling systems seem to have a propensity for blocking up and then failing at their weakest point, usually a connection. As usual, nature likes to show her influence; aluminium seems to be attacked by recirculating water systems and algae seem to thrive in rf environments, both being causes of blockage. As a last resort against water leaks, Widmer (1978) has proposed a rather 'sweet' solution, using an electromechanical arrangement with a sugar 'fuse'!

Substrate Temperature Control

The temperature of the substrate surface is an important and yet difficult parameter to control. We shall see in the section on "Thin Film Formation" that the substrate temperature influences the formation stages of a thin film and its final structure, and in "Bias Techniques" how the temperature influences gas incorporation in the film. It is not terribly difficult to incorporate water cooling or even liquid nitrogen cooling into the substrate platform, although if the substrate platform is biased, one has to ensure that the liquid flow and its associated pipework do not cause a partial electrical short of the target by excessive resistive or capacitive coupling to ground.

Heating of the substrate platform can be achieved by circulating a hot liquid or by electrical resistance heating. If electrical isolation is required, the resistance heater can be decoupled with a suitable isolation transformer. The temperature is usually measured and controlled with a thermocouple feeding a power controller, and in this case the electrical isolation problem is a little more difficult since we need to preserve the thermoelectric emf of the couple whilst removing both the rf and dc offset components. The former may be removed by rf chokes; the latter is normally removed by using a thin insulator of mica or ceramic between the couple junction and the substrate platform or dummy substrate. In a dc sputter etch system, we have been able to successfully run the whole assembly of heater, thermocouple, and temperature controller at the dc target voltage by using a single isolation transformer at the power source, and this arrangement eliminated the need for electrical isolation (and consequent thermal isolation) between the thermocouple and substrate (Chapman et al. 1973).

There is an inherent problem in substrate temperature control in thin film deposition, whether by sputtering or evaporation. What we are really interested in is the temperature of the *surface* of the substrate rather than of its bulk, and it is difficult to measure even the latter, because of the thermal barriers inevitably present, e.g. due to electrical isolation between the substrate and a thermocouple pressed down onto it, or due to thermal isolation between a substrate and the substrate platform. In this latter case, the problem arises because the substrate will generally make only three point contact to the substrate holder; it is sometimes possible to fill the intervening space with a suitable heat-conducting liquid or solid such as gallium, in order to heat-sink the substrate to the holder, but this is usually inconvenient and would be unacceptable in manufacturing processes.

The problem is exacerbated by the power input to the substrate from the glow discharge, which is liable to make the surface temperature greater than that of its bulk. The use of a thin film thermocouple evaporated onto the surface of the substrate has been proposed as a solution to this difficulty, but this is not very convenient. An alternative is to use an infra red thermometer that measures the infra red radiation emitted by the substrate; one needs to know the transmission characteristics of the window through which observation is made, but often this can be done empirically.

So it seems that measurement of the absolute substrate surface temperature is quite a difficult matter, although practically it is possible to reproduce the same conditions from run to run.

Electrode Voltage Measurement

Although many rf systems are controlled by the power input to the matching network and process chamber, some people prefer to measure and control with the target voltage since this eliminates the uncertain power losses in the matching network. It is more usual to measure the dc offset voltage, although the rf peak-to-peak is sometimes used. The dc voltage is normally obtained by filtering out the rf components with an LC circuit, as in Figure 6-12. The rf voltage waveform can be observed by using a high voltage probe (which is essentially a resistive network voltage divider) to reduce the rf signal to a suitable size for display on an oscilloscope. The probe is essentially a resistive or capacitive network voltage divider; if the ac and dc components of the waveform are required, the resistive type should be used. If only the rf peak-to-peak magnitude is required, a clamping circuit can be used, preferably immediately after the probe to avoid long leads carrying rf.

John Vossen has pointed out that when insulating targets are used, the dc offset voltage depends on leakage around the target edge to the backing plate; the unreliability of this can be avoided by using the peak-to-peak voltage. It is not clear to me why this leakage does not change the sputtering rate, if it really does change the target surface potential.

Whatever parameter is measured, care is required, both in safety and in interpretation. There are large rf currents flowing in the external circuitry, and the inductance of even a straight piece of wire can become significant at radio frequencies. These combine to cause significant drops along current-carrying connecting cables, particularly that between the matching network and the target. One can observe these voltage changes with the probe . . . carefully. To obviate this problem, the probe should be attached to the back of the electrode.

All of the same considerations apply to the measurement of substrate voltage, which is almost always done in a bias sputtering system by measuring the dc offset of the applied rf.

At both electrodes, the sheath voltage is determined by the difference between plasma potential and electrode potential, as discussed in previous chapters. Christensen and Brunot (1973) have proposed a method of monitoring the sheath voltage continuously (their technique is discussed in Chapter 5) but this method has not been generally adopted. Target voltage is instead used to *reproduce* conditions rather than give absolute magnitudes.

SPUTTERING AS A DEPOSITION PROCESS

The current main application of sputtering is for the deposition of thin films.

Thin Film Formation

In sputter deposition, as with the other standard vacuum deposition process of evaporation, material arrives at the substrate mostly in an atomic or molecular form (Figure 6-14). The atom diffuses around the substrate with a motion determined by its binding energy to the substrate and is influenced by the nature as well as the temperature of the substrate. Energetically, the surface of the substrate is like an egg carton, with each of the depressions constituting a temporary resting point or *adsorption site* for the depositing and diffusing atoms. At each 'hop', the atom will either jump over the barrier into an adjacent site, or might even hop right out of the egg carton – i.e. re-evaporate. After a certain time, the atom will either evaporate from the surface or will join with another diffusing single atom to form a doublet, which is less mobile but more stable than the single atom. I like to think of two men (please insert your favourite ethnic group) tied together inside a giant egg carton, and trying to jump over the barriers into the adjacent depressions. The chance of them being well enough co-ordinated to jump together is extremely slim, so their mobility is severely limited, as is the chance of their 're-evaporation'.

The chance of forming the atomic pair will depend on the single atom density and hence on the arrival or deposition rate. In time, the doublets will be joined by other single atoms to form triplets, quadruplets and so on. This is the *nucleation* stage of thin film growth, leading to the formation of quasi-stable *islands*,

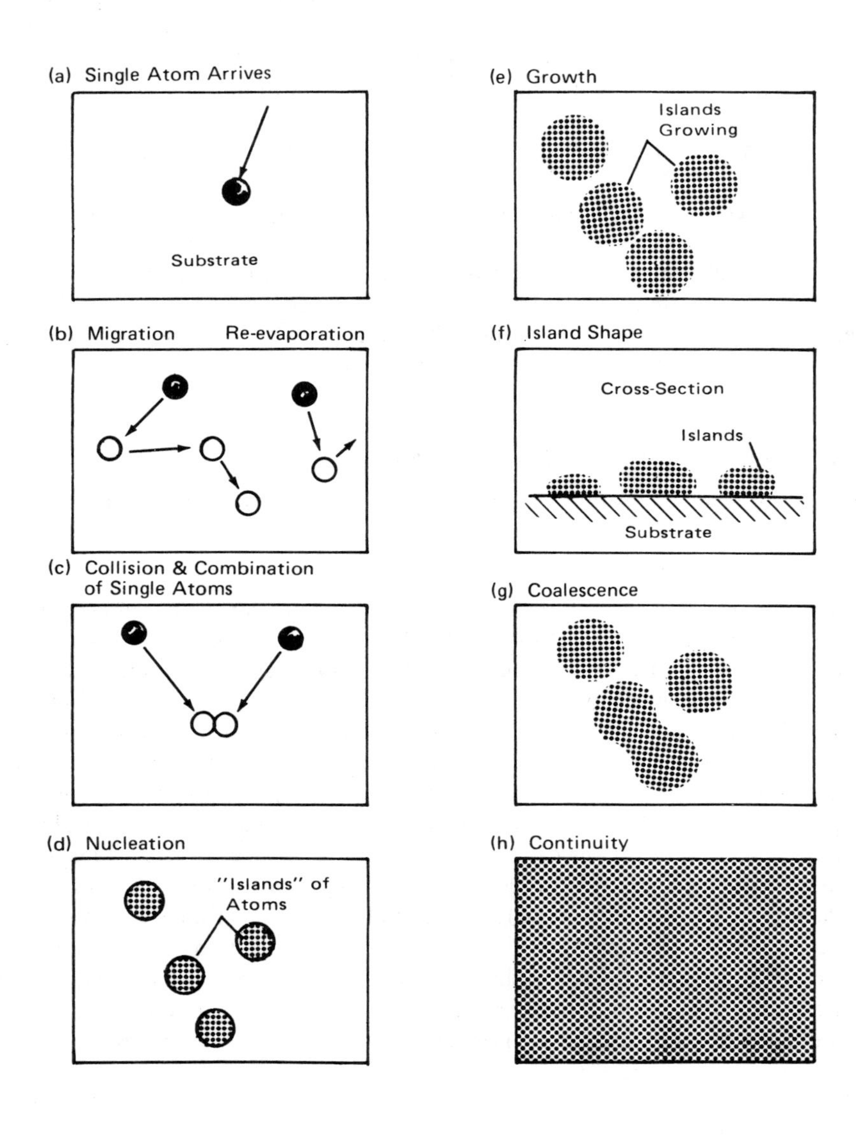

Figure 6-14. Formation of a thin film (Leaver and Chapman 1970)

each containing tens or hundreds of atoms and typically having densities of $10^{10}/cm^2$. During the next, *island growth* stage, the islands grow in size rather than in number. Eventually they grow large enough to touch; this is the *agglomeration* or *coalescence* stage. From observations in the transmission electron microscope, it appears that the islands often display liquid-like behaviour during coalescence, and there are often crystallographic reorientations as a result of competition between the structures of the coalescing islands. Coalescence proceeds until the film reaches *continuity*, but this may not occur in some cases until the film is several hundred Ångstroms in average thickness. During the coalescence stages, the film therefore typically consists of hills and valleys.

During the island stage, each island is usually single crystal or contains just a few crystals. On a polycrystalline substrate, the orientation of each island will be random, so that the resulting film is polycrystalline. On a single crystal substrate, the island orientations may be determined by the substrate structure so that growth and coalescence leads to a single crystal film. This is the phenomenon of *epitaxy* (Bauer and Poppa 1972).

If surface atoms are mobile, they have a greater opportunity of finding low energy positions, consistent with crystal growth, in the growing film. Mobility is enhanced by increased substrate temperature. But since it also takes time to find an energetically favourable lattice position, crystal growth is also encouraged by low deposition rates. Hence, on single crystal substrates, for each deposition rate there will be a temperature, the *epitaxial temperature*, above which single crystal films can be grown.

It is more likely that polycrystalline films on polycrystalline substrates will be required. During the island stage, each island will contain one or a few crystallites. The same mechanisms obtain as in single crystal growth, so that high substrate temperature and low deposition rate lead to large grains, low density of crystal defects, and large film thickness for continuity. The reverse (low temperature and high rate) associations are also generally true.

All of the relationships above were found for the comparatively simple case of deposition by vacuum evaporation. The structure of the growing film was found to be extremely sensitive to deposition conditions. Electron bombardment either prior to or during deposition was found to encourage film continuity and reduce epitaxial temperatures (Stirland 1966). Ion bombardment (Wehner 1962) and increased arrival energy of the depositing atoms (Chapman and Campbell 1969) also reduced the epitaxial temperature.

Life on the Substrate

The purpose of the preceding discussion was to illustrate the very sensitive dependence of thin film structure on growth conditions, even for deposition by evaporation. By contrast, the sputtering environment is extremely complex and has many variables.

Figure 6-15 shows a substrate on which we wish to deposit a sputtered film. We have seen that the nature and temperature of the substrate are important in determining the nature of the film. During thin film growth, the substrate and growing film will be subjected to many types of bombardment (Figure 6-16) which will now be described.

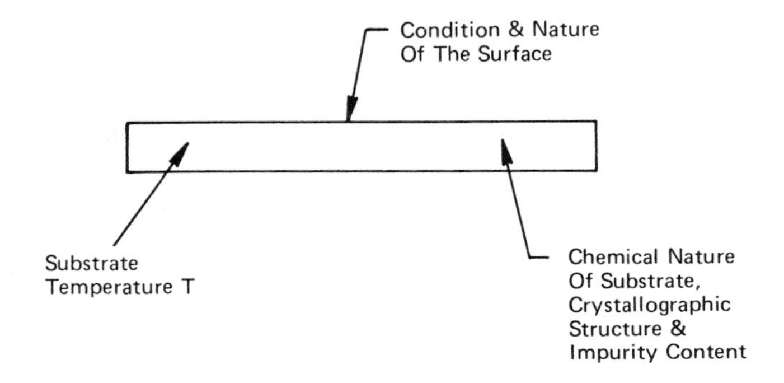

Figure 6-15. The influence of the substrate on thin film structure

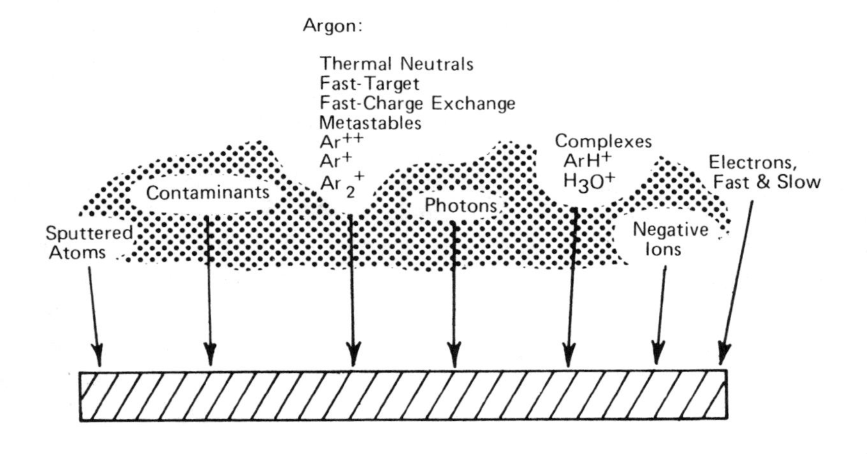

Figure 6-16. Particles bombarding the substrate in sputter deposition

Sputtered Atoms and Contaminants

Let's refer again to the example given in Chapter 1, "Monolayer Formation Time". A typical sputter deposition rate is one monolayer per second, i.e. $\sim 10^{15}$ atoms/cm^2 second or 200 Å/minute for a 'typical' atom of about 3Å diameter. A contaminant gas having a partial pressure of 10^{-6} torr will contribute a numerically equal flux at the substrate. Such contamination will be particularly effective if it is chemically active. The example in Figure 6-17 is from deposition by evaporation, at a higher rate (1200 Å/min) than the sputtering example, but the point is the same; the aluminium film begins to oxidize at an oxygen pressure $\sim 10^{-7}$ torr. We must remember that a contaminant partial pressure of 10^{-6} torr in a total sputtering pressure of 20 mtorr amounts to a contamination level of only 50 ppm! We are unlikely to achieve such a low level, and the contaminant flux will increase proportionately with its partial pressure. If the contamination results from an internal source such as outgassing from a heated substrate, then its partial pressure (which results from an equilibrium between the rates of introduction and pumping) can be minimized by maximizing the pumping rate and hence gas flow rate. (See Chapter 1, "Conductance"). But this expedient is ineffective if the contamination is introduced with the gas, indicating a vital need for pure sputtering gas.

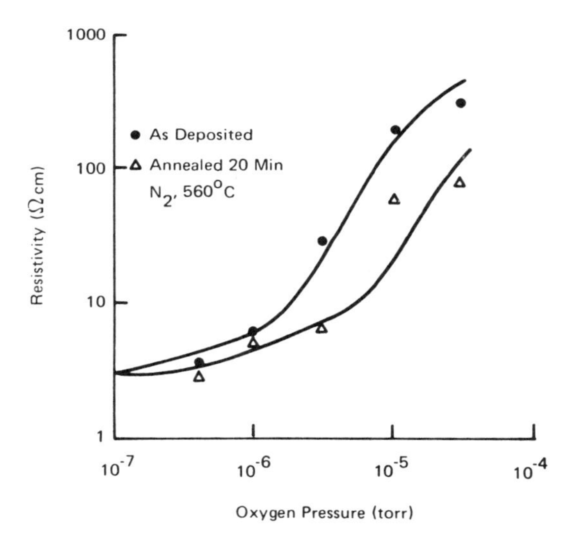

Figure 6-17. Room temperature resistivity vs oxygen pressure during evaporation of 5000Å aluminium films at 200° C and 20Å/sec (d'Heurle et al. 1968)

Sputtering Gas Atoms — Fast and Slow

Compared with the fluxes of sputtered atoms and contaminants at the substrate, the flux of argon (or other sputtering gas) is truly enormous. At 20 millitorr, the argon flux would be about 10^4 times greater than the arrival rate of sputtered material. It would not be surprising, therefore, if argon were trapped in the growing film. Indeed, trapped argon is observed in sputtered films (Winter and Kay 1967), although not for the obvious reason. Winters and Kay evaporated a nickel film under similar conditions of argon pressure and deposition rate as for the sputtered films, but found that the argon content in the evaporated films was very much lower ($< 1\%$).

They determined the content of their nickel films by vaporizing the film and measuring the resultant gas evolution with a mass spectrometer. The argon content was measured as a function of substrate temperature (Figure 6-18) and total argon pressure (Figure 6-19). The temperature dependence is as expected; argon is likely only to be physisorbed, so is less likely to be initially adsorbed and more likely to be subsequently desorbed, with increasing substrate temperature.

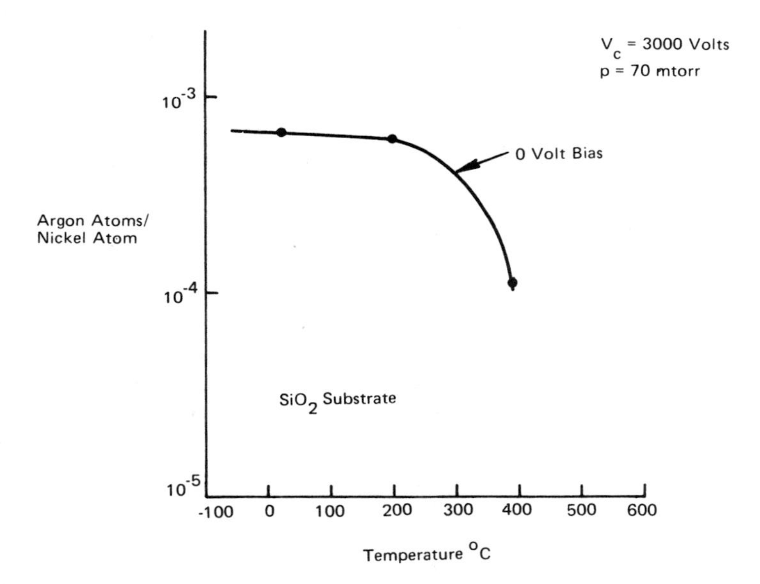

Figure 6-18. Argon content in sputtered nickel films as a function of deposition temperature (Winters and Kay 1967)

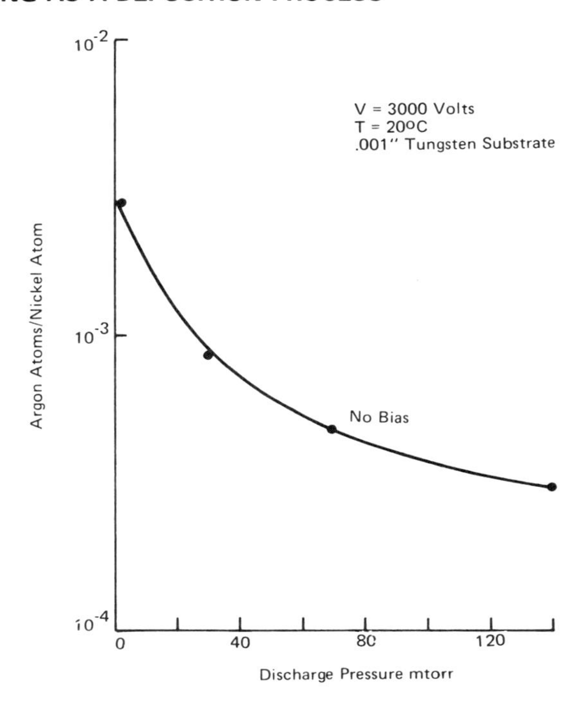

Figure 6-19. Argon content in sputtered nickel films as a function of argon discharge pressure (Winters and Kay 1967)

The pressure dependence (Figure 6-19) is rather interesting. The effect is ascribed to the small flux of high energy argon neutrals striking the substrate, rather than the large flux of thermal neutrals. Energetic argon ions striking the target are neutralized in the process and rebound as energetic neutrals. These energetic neutrals, arriving at the substrate, are likely to be embedded in the growing film; Comas and Wolicki (1970) have demonstrated how argon ions are entrapped in silicon, and fast neutrals are likely to behave in a similar way. Travelling across the sputtering chamber, the energy of these fast neutrals is attenuated by gas phase collisions and so their incorporation into the growing film (per unit film atom) decreases with increasing pressure.

One would expect fast neutrals to have a smaller collision cross-section than thermal neutrals. The existence of fast electrons and negative ions from the target similarly traversing the sputtering system without collision and bombarding the substrate has been clearly demonstrated, as is discussed later. In the Winters and Kay work there is apparently an effect of the fast neutrals even at a pressure of 100 mtorr. Since there will be about one fast neutral leaving the target for each sputtered target atom, the measured argon content of $\sim 10^{-4}$ argon atoms per nickel atom at 100 mtorr (Figure 6-19) implies that at least this proportion reach the substrate with enough energy left to be adsorbed. Based on collision probabilities (Chapter 1), this is a little surprising, but one can't argue with the results. It is presumably a manifestation of the lower collision cross-section and pronounced forward scattering that one expects for higher energy particles (see Appendix 4).

Excited Neutrals

To return to Figure 6-16, we have so far been considering bombardment by ground state neutrals. A further source of bombardment is due to excited neutrals, of which metastables of the sputtering gas would be most abundant. These can presumably lose their potential energy at the growing film and hence influence its growth, although Kaminsky (1965) indicates that these metastables should be resonance ionized and Auger neutralized before they reach the substrate.

Positive Ions

In addition to these neutrals, there will be bombardment by charged particles. Argon ions will be the most abundant positive ions, with a flux of the order of $n\bar{c}/4$. The figure of 20 $\mu A/cm^2$ from our example in Chapter 3 corresponds to a flux equivalent to a few tenths of a monolayer per second. There will also be ions of the sputtered material, produced both by electron impact ionization and by the Penning process of collision with metastables (Chapter 2, "Metastable Collisions"). These ions, and any others, will be accelerated across the sheath at the substrate. Under normal conditions, the sheath will be quite thin and there will be little attenuation of the ions due to collisions in the sheath.

Valuable information about ion bombardment at the substrate has been acquired via the work of John Coburn (1970 et seq.), who has been able to analyze the energy and mass of ions striking the substrate. His work is described in more detail in the section on bias sputtering. For now, we note that he was able to identify contaminant ions, the less abundant argon ion species such as Ar^{++} and $Ar_2{}^+$, and also complex ions such as ArH^+ and short-lived ions such as H_3O^+. (We discussed in Chapter 2 how ions can change their chemical identity as their electronic shell structure changes). So we have to add these ions to our list in Figure 6-16.

Negative Ions

Negative ions of the target material may form also. The space charge sheath established at the substrate will tend to repel and slow down these ions, but they will still reach the substrate if they are energetic enough (which will be so if they're formed at the target or in the target sheath). Negative ions at the substrate were detected by Koenig and Maissel (1970) in their work on sputtered quartz. Hanak and Pellicane (1976) have shown how fast negative ions from the target can sputter etch the substrate, and their findings have been confirmed more recently by the experimental work of Cuomo et al. (1978) and the theoretical work of Robinson (1979). Presumably, negative ions can also be formed from gas phase contaminants, although they would be energetic enough to reach the substrate only if they were formed in the target sheath.

Electrons

A major source of charged particle bombardment at the substrate is due to electrons. With a conducting substrate, the average current density will be about 1 mA/cm^2, which is equivalent to 6.25 10^{15} electrons/cm^2 second or a few electrons for each depositing atom. The majority of these electrons will be thermal electrons from the glow where they have energies of a few electron volts (Chapter 3), although only the more energetic of them will be able to surmount the sheath at the substrate. An insulating substrate on the anode in a dc discharge will charge up to floating potential and will receive a much smaller electron flux, equal to the ion flux.

In addition to these slow electrons, there will be bombardment by fast electrons. These electrons are emitted from the target by ion and other impact, are accelerated across the target sheath, and then travel across the sputtering system without making collisions, as described in Chapter 4. So they have energies equivalent to the sheath voltage. These electrons have been detected by Koenig and Maissel (1970); their results in an rf system, obtained by retarding potential measurements, are shown in Figure 6-20. Ball (1972), again using a retarding potential technique but in a dc system, obtained the results shown in Figure 6-21; these results suggest that a large fraction of the electrons striking the substrate have almost the full target sheath voltage, but it is not clear that an allowance was made for the transmission of the analyzer.

The energy spectrum of electrons bombarding the anode has also been measured by Leopoldo Guimarães (Chapman et al. 1974) using the apparatus shown in Figure 6-22. Typical results for the collector current as a fraction of the total current, using a copper target, are shown in Figure 6-23. For reasons involving the resolution and mode of use of the analyzer, these curves do not show electron flux against energy but rather power input to the substrate due to charged

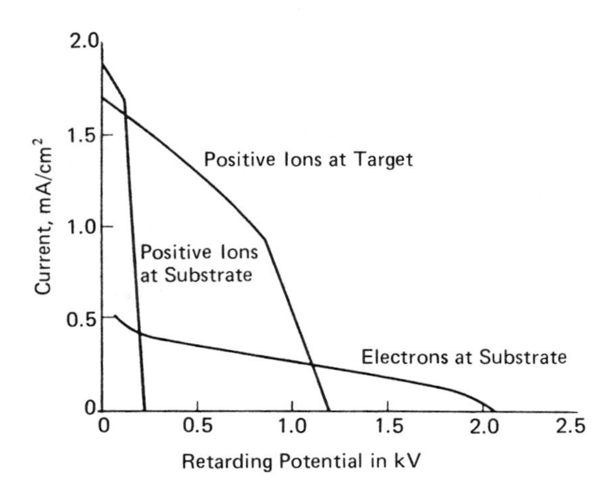

Figure 6-20. Electron and positive ion currents at substrate and target in an rf sputtering system (Koenig and Maissel 1970). Pressure ~ 5 mtorr. Maximum ion energy ~ equivalent dc sheath energy.

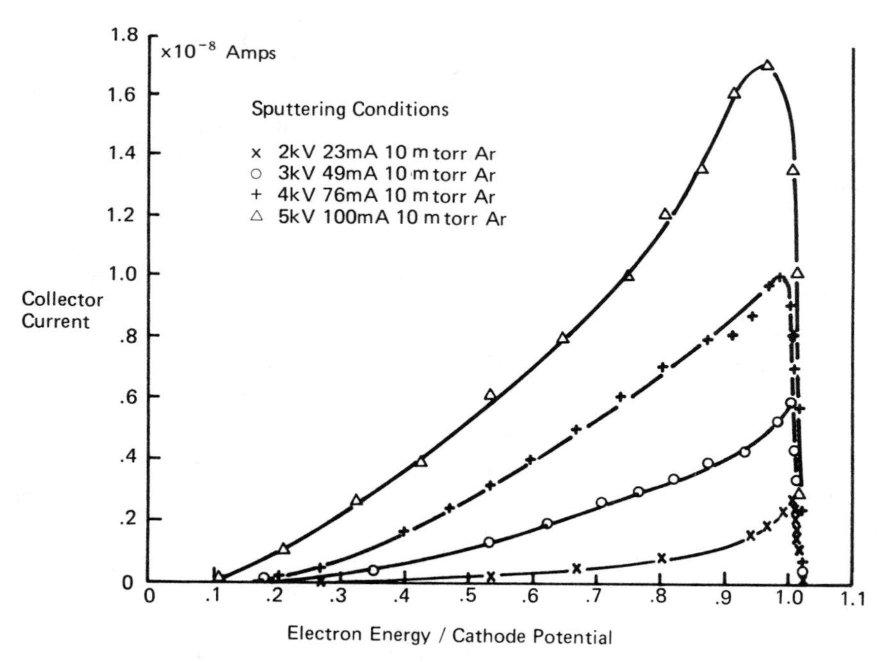

Figure 6-21. Energy spectra of secondary electrons arriving at the anode of a dc sputtering discharge (Ball 1972)

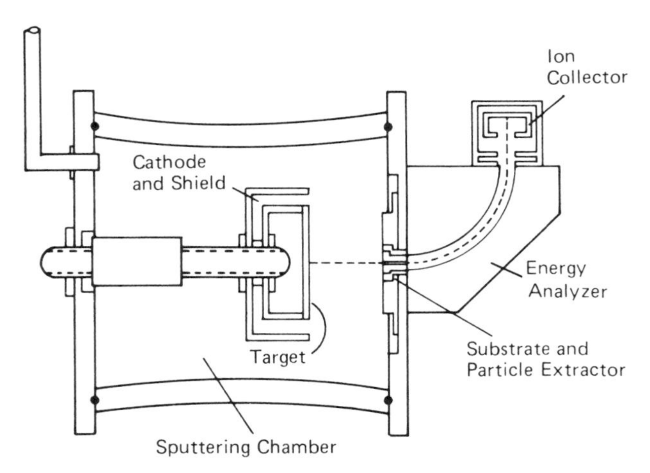

Figure 6-22. Energy analysis system for electrons on the substrate in dc sputtering (Chapman et al. 1974)

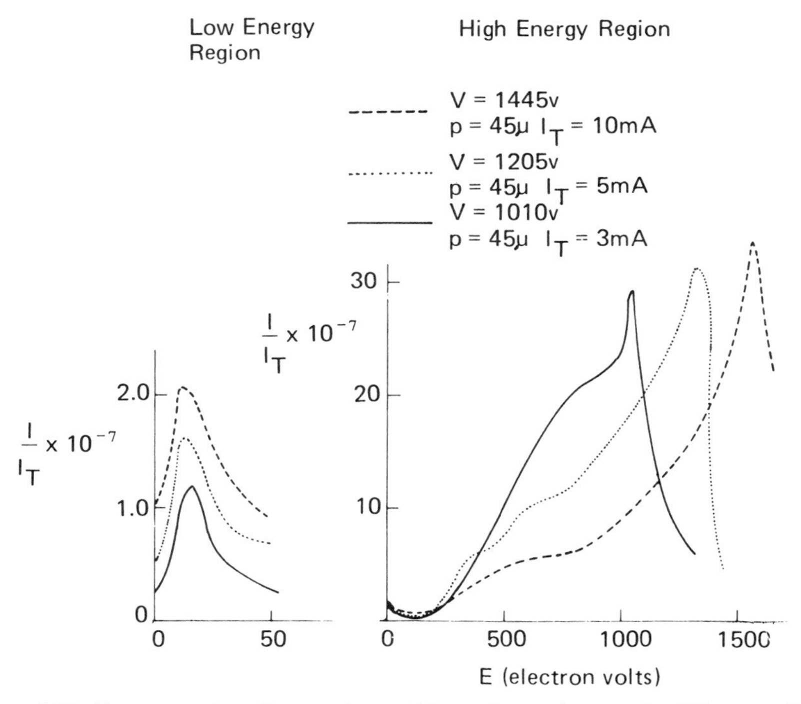

Figure 6-23. Energy spectrum for negative particles at the anode; note the different ordinate scales (Chapman et al. 1974)

particle bombardment versus particle energy, or, more precisely, dP(E)/dE vs. E, where dP(E) is the power carried to the electrode by particles with energies between E and E + dE. The flux distribution can be unravelled by noting that dP(E) = E dN(E). The peak of the flux distribution occurs at quite low energy as expected, but there are a significant number of negative particles, which appear to be secondary electrons emitted from the target, that travel from the target to the substrate without making collisions, and hence travelling along the essentially straight field lines; we discussed these fast electrons in Chapter 4. This collision-free electron travel is presumably a manifestation of the total collision cross-section for electron scattering in argon (Figure 2-27) becoming quite small for electron energies above about 100 eV. However, although small in number, these electrons are responsible for almost all of the power input into the substrate. (The ion-electron recombination energy can also be considerable, amounting to $4.5 mW/cm^2$ for an ion current component of 0.3 mA).

In the same series of experiments, a composite sputtering target of copper and glass (Figure 6-24) was used in an rf sputtering system. The glass had a much larger secondary electron coefficient than the copper. The electron emission pattern was directly observed by coating a glass substrate with a fluorescent material and placing this on the counterelectrode of the sputtering system, several inches from the target. It was then possible to directly observe changes in the

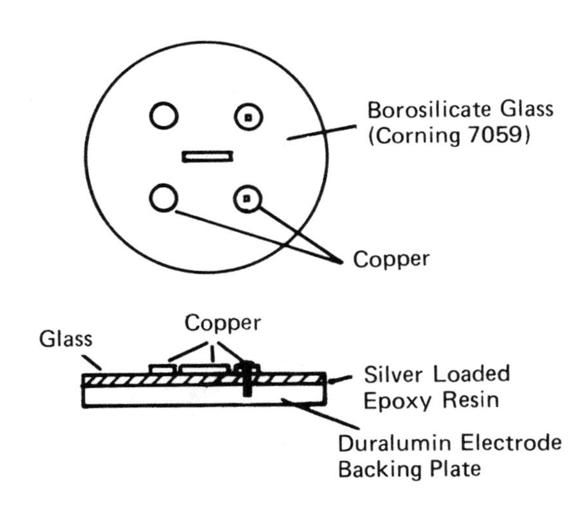

Figure 6-24. Composite sputtering target for secondary electron experiments (Chapman et al. 1974)

electron bombardment pattern as the sputtering conditions were varied. The threshold for glowing was at several hundred volts, so that only fast electrons were detected by the screen. The screen glowed very brightly opposite the glass sections of the copper/glass target. Figure 6-25a is a photograph of the fluorescent screen taken at an angle of about 45° through the vacuum chamber wall. The pattern on the screen could easily be deflected with a weak magnetic field (Figure 6-25b), showing the particles to be electrons rather than much heavier (and hence more difficult to deflect) ions.

These fast electrons can have a major influence on the structure and properties of the growing film on the substrate. The large energy input causes a good deal of substrate heating, and there are more subtle effects due to the electron interaction with the surface, as discussed in "Thin Film Formation". These electrons have been observed to discourage as well as enhance thin film growth (Chapman et al. 1974).

Photons

The final type of bombardment that the substrate experiences is due to photons. Photons can be produced by ion or electron bombardment on any surface, and the photon can be as energetic as the ion or electron producing it, which therefore means a thousand electron volts or more in a sputtering system. Such energies put these photons in the soft x-ray class. Lower energy photons will also result from relaxation of excited atoms in the glow.

We have already discussed how photon bombardment can cause electron emission from a surface, and I would be surprised if these photons did not affect the growth of a film, as does every other type of energy input to the substrate. However, there appears to have been very little work on photon effects in thin film growth.

Radiation Damage: Creation and Removal

There is a recent paper by DiMaria et al. (1979) on neutral charge traps produced in silicon dioxide films, actually in a reactive ion etching system which is nevertheless very much like a sputtering system. By measuring the centroid of the damage, they concluded that this damage was due to soft x-rays rather than charged particle bombardments. This conclusion is consistent with the additional observation that similar damage could be produced with the substrate anywhere in the system; photons, being uncharged, bombard all surfaces within the system. DiMaria et al. also found more gross damage at the surface of their samples, which they concluded was due to energetic ion, neutral, and electron bombardment. Their results are consistent with the earlier work by Hickmott (1969), who studied radiation damage in rf sputtered SiO_2 films.

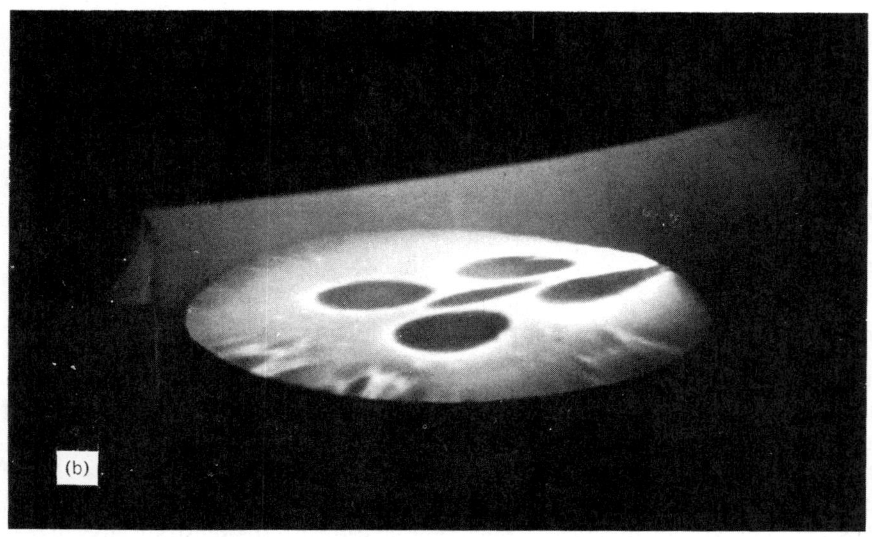

Figure 6-25. Fluorescence patterns on the counterelectrode screen
(a) without magnetic field
(b) with magnetic field
(Chapman et al. 1974)

There is a general observation that metallization of semiconductor devices in conventional sputtering systems leads to radiation damage of various types (neutral traps, interface states, etc.). The magnitude of the damage increases with increasing sputtering target voltage. Probably a similar photon bombardment mechanism is responsible for the damage. In order to minimize the damage, magnetron sputtering devices using lower target voltages are currently used for metallization; it seems to be necessary to reduce the target voltage below 500 V. We shall be discussing magnetron sputtering systems later in the chapter. Inconsistent with this sputtering experience, DiMaria et al. have not found any significant energy dependence of damage, at least over the range 300 V - 800 V in their CF_4 reactive ion etching system.

The general phenomenon of radiation damage in thermally grown SiO_2 due to the various microcircuit fabrication processes of sputtering, plasma etching, ion beam implantation, and e-beam lithography, has been reviewed by Gdula (1977). A more recent application of rf discharges is the *rf annealing* process, a method of removing radiation damage (Ma and Ma, 1977), or at least a way of removing charge states. The process consists essentially of exposing wafers to a low pressure rf discharge of fairly low power, inside a barrel type plasma etching system. The gas composition does not appear to be critical. The annealing mechanism is postulated to be a plasma assisted electrical-thermal effect.

Bias Techniques

With so many particles bombarding the film, and with the sensitivity of the nucleation and growth processes to this bombardment, one would expect to be able to influence the properties of the film by changing the flux and energy of incident particles. It is difficult to directly modify the behaviour of the neutral particles, but the charged particles can be controlled by changing the local electric field, and this is the basis of the technique of *bias sputtering*.

Voltage Distribution in Bias Systems

Figure 4-4 showed the distribution of voltage in a dc diode system when a voltage of -2000 V was applied to the target. The plasma potential V_P was at +10 V in that example. Suppose that we now electrically isolate the substrate platform from ground and apply a negative potential of 50 V to it (Figure 6-26). Electrical ground at 0 V is still present as the conducting parts of the chamber walls and baseplate. The plasma potential likes to remain positive with respect to everything in the chamber and, to a first approximation, will be unaffected by the negatively biased electrode, so that V_P remains at +10 V (Figure 6-27).

If the applied bias is V_B volts, then a sheath of potential difference $V_P - V_B$, which would be 60 V in the example, will be established in front of the substrate

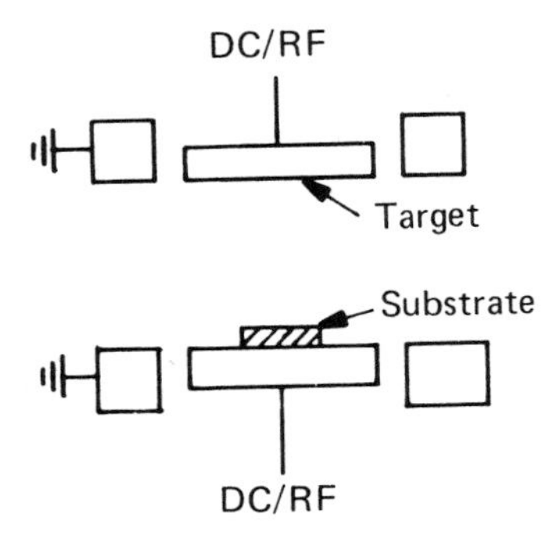

Figure 6-26. Schematic of a bias sputtering system

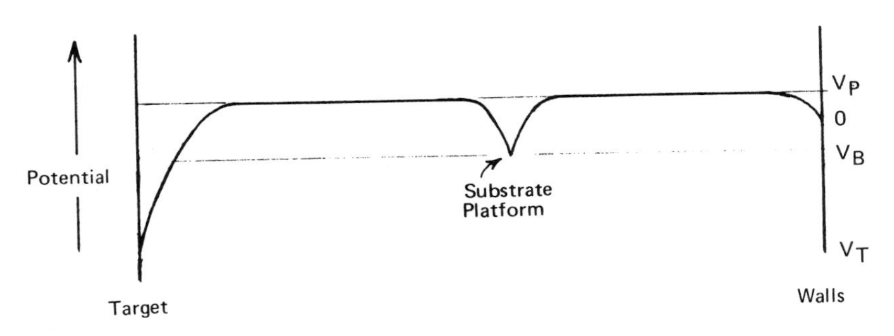

Figure 6-27. Potential distribution in a bias sputtering system

with polarity such as to accelerate positive ions onto the substrate. In effect, the substrate platform has become a secondary sputtering target. By this means, the flux and energy of all charged particles bombarding the substrate (Figure 6-16) can be modified. The energy of the ions striking the substrate would be equal to the energy with which they enter the sheath (which is usually small) plus the energy difference across the sheath, provided that there were no inelastic collisions en route. For small bias voltages, the sheath at the substrate will be thin enough that it will be collision free. For increasing bias voltage, the sheath thickness grows so that collisions may become important and attenuate the ion energy (Chapter 4, "Charge Exchange in the Sheath"). This means that the average energy of particles striking the substrate will increase less rapidly than the applied bias voltage. So a change in film properties as a result of bias sputtering may demonstrate the effects of a much smaller range of ion energies than is apparent.

If we tried to apply a positive bias potential, then we would encounter the results already discussed in Chapter 4: the plasma potential would rise so as to try to maintain the sheath in front of the bias platform, and the functions of the original anode and the bias table would interchange. This has been observed experimentally by Orla Christensen (1975) in a paper concerned specifically with voltage distributions in bias sputtering. He points out that a positive bias can cause sputtering of the (original) anode and consequent contamination.

Coburn and Kay (1972) have measured positive bias potentials slightly (< 15 V) greater than the plasma potential, which implies that the sheath polarity has reversed. Although this result is not surprising for the very small probes that were used, and which would be limited by their small current-collecting surface areas, much the same result was found for very large positively biased probes. This unexplained result may have been due to the limited fast electron flux on the large probes due to their geometric shapes.

Bias – dc or rf?

An insulating substrate placed in a dc discharge will charge up to floating potential, and placing that substrate on a dc biased platform will not substantially change the situation until the substrate is coated with a conducting film, at which time the bias becomes effective. This situation can be changed by using rf bias, with the same rationale as for the sputtering of insulating targets. There is the same latitude in using dc or rf to power the main target, so that insulating films can be bias sputtered from insulating targets. The sputtering systems used become double ended. Figure 6-28 shows a system used for dc sputtering with rf bias. Systems with rf on both electrodes are probably more common.

It is usual, in both rf and dc bias sputtering systems, to measure the dc bias level with respect to ground. Although a more relevant parameter would be the sheath voltage – the plasma potential minus the bias potential – this is not easy to measure directly because of the relative inaccessibility (on a routine basis) of the plasma potential. The bias voltage is however a useful parameter to ensure reproducibility of the system.

It is probably true to say that, without the additional flexibility of bias sputtering, sputter deposition would be used to a much lesser extent than it is; the bias sputtering technique enables one to control so many film properties.

Control of Film Properties

Figure 6-29 shows how the resistivity of sputtered gold films can be controlled with the use of bias, reaching the bulk resistivity value for a bias of about -40 volts (this figure refers to the dc offset of the rf waveform – see Chapter 5).

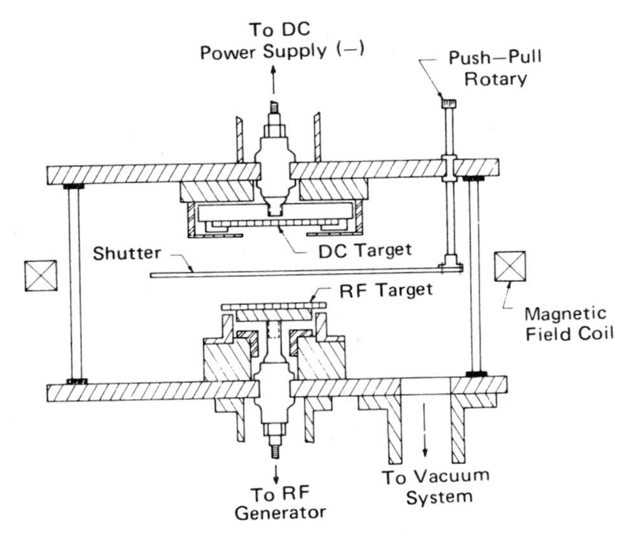

Figure 6-28. DC sputtering system with rf substrate bias (Vossen and O'Neill 1970)

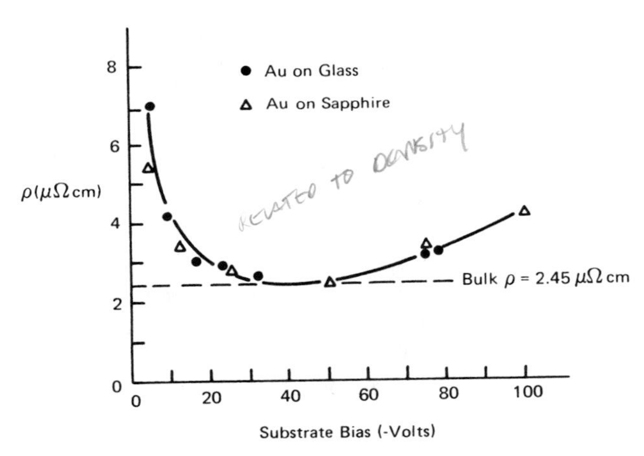

Figure 6-29. The variation of resistivity of 6000Å dc sputtered gold films versus rf substrate bias (Vossen and O'Neill 1968)

Similar behaviour of the resistivity change with bias for tantalum films is shown in Figure 6-30; this figure also illustrates the different effects of dc and rf bias discussed earlier in this section, with only the rf bias managing to achieve bulk resistivity.

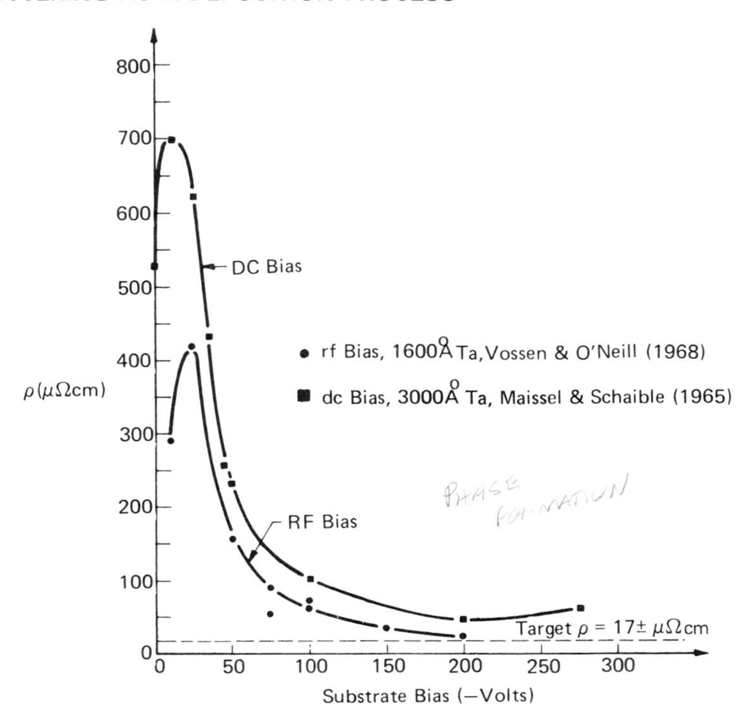

Figure 6-30. Resistivity of Ta films versus substrate bias for both dc- and rf-induced bias
- • rf bias, 1600Å Ta, Vossen and O'Neill (1968)
- ■ dc bias, 3000Å Ta, Maissel and Schaible (1965)

But bias sputtering is not restricted to control of electrical resistivity. It seems as though any and every thin film property can be controlled with this technique. Figures 6-31 to 6-33 show how the electrical properties of sputtered S_iO_2 films, the hardness of sputtered chromium films, and the etch resistance of sputtered silicon nitride films in buffered HF, can all be controlled by bias techniques – and there are countless other examples.

Control of Gas Incorporation

Earlier we saw that the sputtering gas can become incorporated into the growing film. The gas content also is a function of bias, and probably contributes to many other bias effects. Pursuing the work reported earlier, Winters and Kay (1967) measured the argon content of nickel films as a function of bias, and

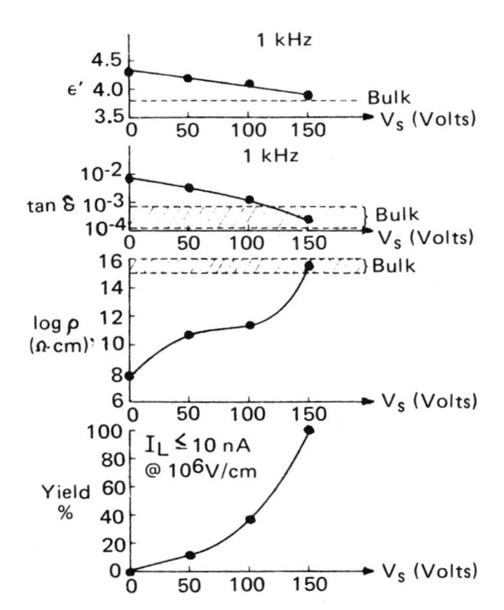

Figure 6-31. The dielectric properties of SiO_2 films rf sputtered with rf-induced substrate bias (Vossen 1971):

(a) relative dielectric constant

(b) dissipation factor

(c) resistivity

(d) capacitor yield based on the criterion shown for capacitor areas of 0.02 cm^2

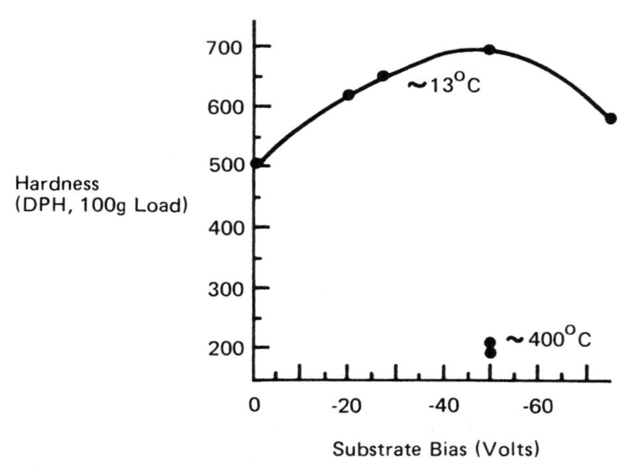

Figure 6-32. Influence of substrate bias voltage and temperature on hardness of sputter-deposited chromium (Patten and McClanahan 1972)

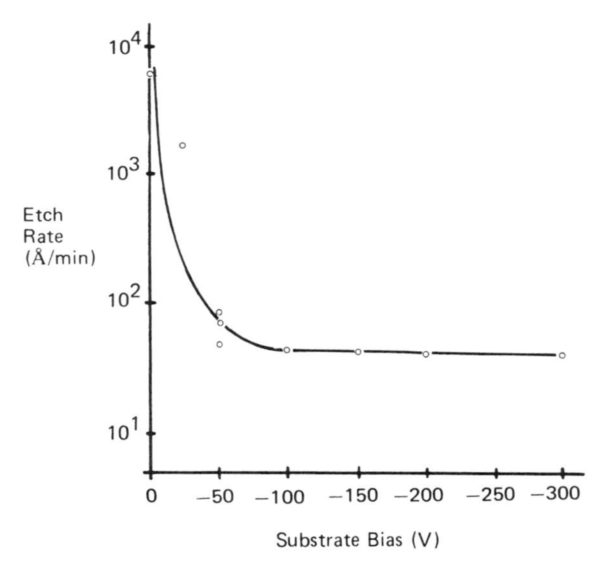

Figure 6-33. The etch rate of reactively sputtered silicon nitride films in buffered HF (Stephens et al. 1976)

their results are shown in Figure 6-34, with the 0-200 eV range clarified in the inset. The decrease in argon content observed is because incoming argon ions at these energies have a high probability of sputtering previously sorbed argon but only a small probability of sticking. However, as the bias voltage is increased, so is the energy of the incoming ions, and they are eventually energetic enough (at ~ 100 eV) to become embedded in the growing film; as soon as this effect dominates the sputtering effect, the argon content starts to rise again.

The growing film is also subject to resputtering by the incoming ions. However, you will recall from the discussion on sputtering target kinetics that the sputtering yield S drops rapidly below 100 eV, so at modest bias voltages of 50V or so, there will generally be little resputtering of the growing film itself; but with more than 100 V bias, resputtering can be significant. We shall return to this point in "Deposition of Multicomponent Materials".

Application of rf Bias

Probably the most common configuration for bias sputtering is with rf on the target and rf bias. This can be accomplished in various ways. A single power supply is normally used, and the rf power is split between the target and sub-

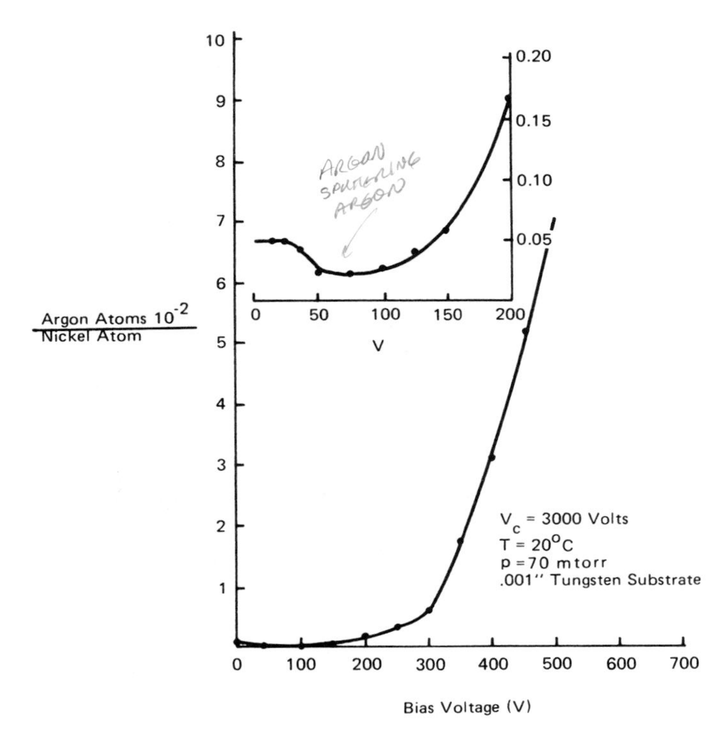

Figure 6-34. Argon concentration of sputtered nickel films, vs bias voltage (Winters and Kay 1967)

strate electrodes either in the matching network (for example with an autotransformer), or less frequently by inductive coupling (the *tuned substrate* system shown in Figure 6-35 and discussed in Chapter 5, "Equivalent Circuits of RF Discharges"), by capacitive coupling (Figure 6-36), or by using a controlled area ratio of electrodes – the *CARE* system discussed in Chapter 5, "Voltage Distribution in RF Systems". By contrast, systems with power more obviously applied to the bias platform are sometimes known as *driven substrate* systems. Keller and Pennebaker (1979) have analyzed the electrical properties of these various systems.

An alternative approach to the driven substrate system with one power supply is to use an rf power amplifier for each electrode with either a single common exciter or individual exciters. In the last case, exciters of nominally the same frequency (usually 13.56 MHz) are used, but they will never in practice be

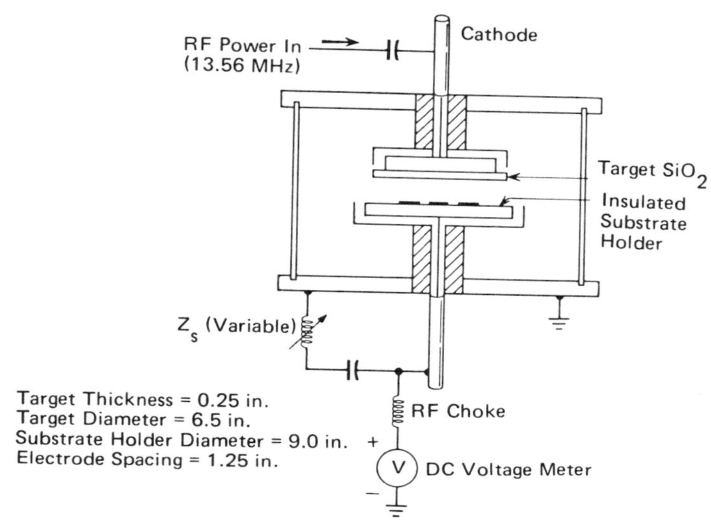

Figure 6-35. Experimental system for rf sputtering with substrate tuning (Logan 1970)

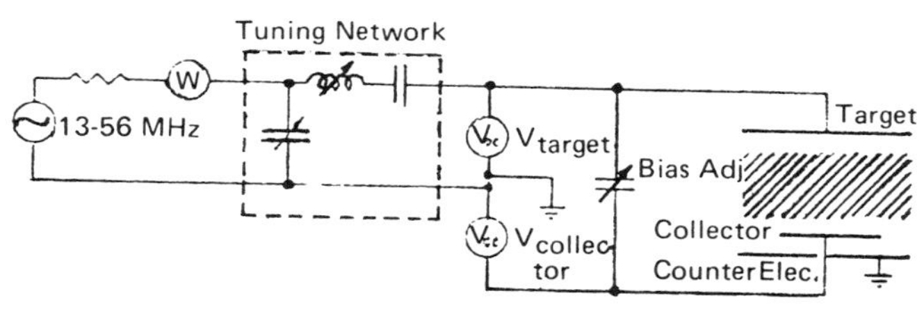

Figure 6-36. Circuit of RF sputtering diode with capacitive collector-to-target coupling (Christensen and Jensen 1972)

exactly the same, so that the phase angle between the two electrodes will continuously and regularly change. In all other cases, there will be a certain phase relationship between the two electrode sheaths that will be a function of operating conditions, cable lengths, and matching network settings. The effects of the phase relationship have been explored by Logan et al. (1977). The significance of substrate bombardment by fast secondary electrons from the target has been noted in the previous section. In rf sputtering with rf bias, these electrons will generally be decelerated by the substrate sheath to an extent determined by the phase of the sheath voltage as they enter. Logan et al. calculated and measured the maximum electron energy (Figure 6-37) and showed also how the threshold

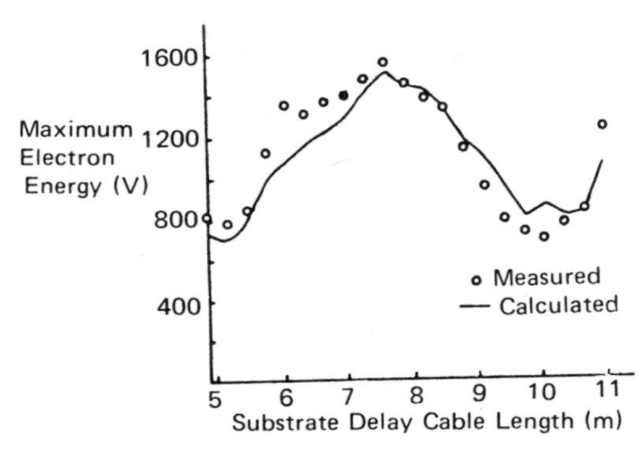

Figure 6-37. Maximum electron energy vs cable length, i.e. phase shift (Logan et al. 1977)

voltage of field effect transistor devices coated with sputtered quartz would be shifted during sputtering (Figure 6-38). The threshold shift increased with the maximum electron energy, and so could be phase controlled (which was achieved by varying cable lengths). As the maximum electron energy increased, higher annealing temperatures were required to anneal out the damage. Other parameters also might change with the phase angle.

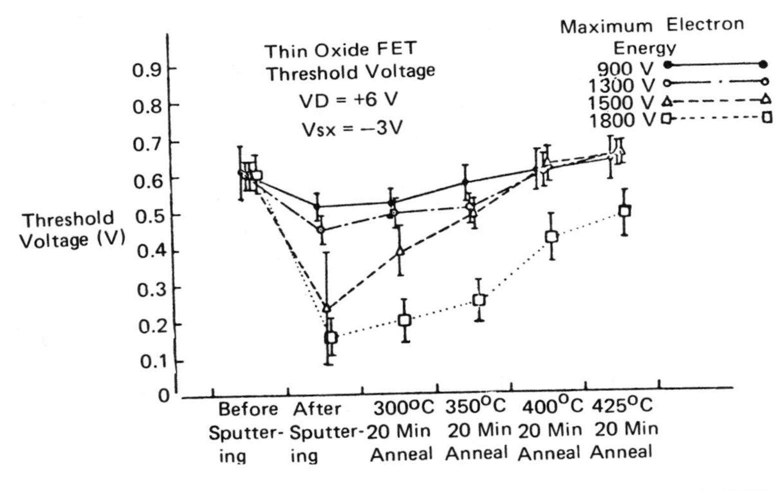

Figure 6-38. FET threshold voltage shift vs electron energy (Logan et al. 1977)

Bias Sputtering Mechanisms

In a previous section, we saw that sorbed argon can be resputtered from the growing film by the bias sputtering technique. It is very common to find explanations for bias sputtering that describe how ". . . . loosely-bonded material can be resputtered from the growing film by low energy ions". This may be an adequate explanation for the resputtering of weakly physisorbed gases, but it does not account for other commonly observed phenomena. As an example, sputtered chromium has a very high affinity for oxygen and tends to oxidize easily in the small amount of oxygen always present in a system; the use of bias sputtering can minimize this oxidation, even though the chromium-oxygen bond is very strong.

Winters and Kay (1972), in an extension of their earlier work on argon incorporation in thin films, studied the composition of films deposited in a gas mixture of argon and nitrogen. As a result of this study, they divided metals into three classes:

a. metals which chemisorb molecular nitrogen and form a nitride, e.g. W.

b. metals which do not chemisorb molecular nitrogen but still form a nitride, e.g. Ni.

c. metals which neither chemisorb nitrogen nor form a nitride, e.g. Au.

Winters and Kay found that bias sputtering caused the nitrogen content of gold to increase, due to the increased sorption of nitrogen when the film was bombarded by energetic N_2^+ ions. However, although this same effect is present with the tungsten and nickel films, it is overshadowed by resputtering of the previously sorbed nitrogen, as it was for the tantalum films in pure argon at low bombardment energies. This rather surprising result is a consequence of the sputtering yield for nitrogen on tungsten being rather large. Winters and Kay maintain that there is adequate evidence to conclude that most chemisorbed gases have large sputtering yields and low thresholds; they suggest that this is a consequence of the poor mass fit in the energy transfer function $4m_1 m_2/(m_1 + m_2)^2$ so that a struck chemisorbed atom cannot transfer its energy to the underlying substrate lattice. However, as we shall be discussing later in this chapter, there are examples in which a very thin metal surface layer, again having a poor mass fit to the substrate, has an anomalously *low* sputtering yield.

So it appears that bias sputtering can lead to either an increase or decrease in the concentration of gaseous species, depending on their surface adsorption characteristics which in turn control their sputtering yields. Winters and Kay also conclude that bias sputtering is unlikely to reduce impurity concentrations by more than a factor of 10, so that the biasing technique is not much of a substitute for good vacuum. However, Figure 6-39 shows how the nitrogen content of metal films (unbiased) changes with the partial pressure of the nitrogen in the argon. Note that the nitrogen content of tungsten, which chemisorbs molecular nitrogen, decreases by less than a factor of 10 over four orders of nitrogen pressure change. So it may well be that, for tungsten and other class (a) metals, that biasing *is* very effective compared with decreasing the partial pressure of the contaminant.

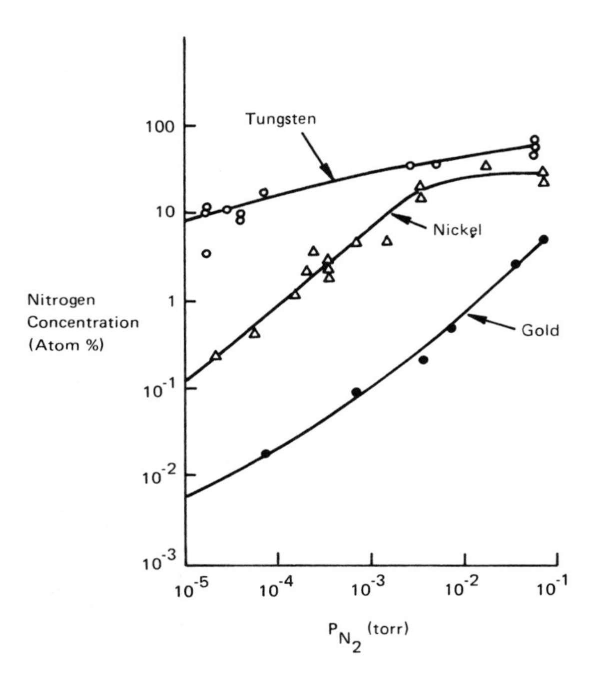

Figure 6-39. Nitrogen concentration in film vs partial pressure of nitrogen; p_{tot} = 70 mtorr, T=300 K, target voltage = 3000 V, bias voltage = 0, and target – substrate distance = 4.5 cm (Winters and Kay 1972)

Winters and Kay make two other points. Firstly, they consider the effect of backscattering in the gas phase which tends to return sputtered material back to its substrate; they conclude that in many instances bias sputtering may be more effective in obtaining high purity films if the sputtering pressure is relatively high. Secondly, they note also that ion bombardment, and hence bias sputtering, also changes the physical appearance of a film. They had previously shown that nickel films became much smoother as the bias voltage was increased from 0 V to 300 V. The topography of the film surface is important in determining its optical appearance, as light is scattered from surface features; we shall discuss more extreme examples of this later in the chapter. In Winters' and Kay's experiments, unbiased films of tungsten, nickel and iron looked dark grey or almost black, whilst biased films appeared metallic.

This last result emphasizes another point. Much of our discussion about bias sputtering so far has been concerned with the effect of bias on gas content. However, gas content is just one of many parameters which contribute towards the nature of a thin film. For example, the gold resistivity change in Figure 6-29 was apparently related to film density and not gas incorporation, and the tantalum resistivity modification in Figure 6-30 was due to a phase change. And other parameters such as the surface roughness or the hardness of a film may be important for a particular application.

Ion bombardment has many effects, some of which we are only now beginning to learn about; for example, in Chapter 7 we shall see how chemical reactions can be accelerated by ion bombardment. Changing the bias voltage will also change the power input to the film. The power due to ion bombardment can be considerable, and the consequent heating may be helpful or, as we shall see shortly, may lead to *grain growth* and other undesirable film features. We have also seen in earlier sections how fast electrons from the target can bombard the substrate. Increasing the bias on the substrate will tend to retard the electrons and lower their bombardment energy. Whilst this is always true in dc systems, the statement has to be restricted in rf systems depending on the phase of the sheath field seen by the fast electrons as they enter the substrate sheath. We discussed phase control in the previous section, and showed how it could be used to influence the threshold shift of FET devices during sputtered quartz deposition.

We would probably be correct to conclude that the control of the gas content of films is a significant role for biasing; but there are likely to be many other effects due to the flux and energy of ion and electron bombardment on the substrate. Some of these might be simple, such as thermal effects, and some might be more subtle influences on the nucleation and growth stages of the film. And in the next few sections we shall see how biasing can be used to control film topography by deliberately encouraging resputtering, but can also cause problems with multicomponent films by resputtering the film and causing a depletion of the component with the higher (or highest) yield.

There is no doubt that biasing can add enormously to the power of sputter deposition. As a broad generalization, it is likely that, no matter what particular film property you are interested in, it can be influenced by the bias sputtering technique. Try it and see!

Analysis of Charged Particle Bombardment at the Substrate

In earlier chapters, the work of Davis and Vanderslice (1963) has been mentioned. They identified the mass and energy of ions bombarding the cathode of a dc discharge, as a means of understanding some of the target sheath phenomena. John Coburn (1970) has used the same approach, but applied to the substrate platform, with and without bias. This application has been extremely fruitful, covering quite a range of topics, and in our energy-conscious world, Coburn has to be congratulated on getting such good mileage out of the technique!

The experimental arrangement is shown in Figure 6-40. Some of the ions bombarding the substrate plane pass through a small aperture into a differentially pumped low pressure region, where they are first energy analyzed and then mass analyzed.

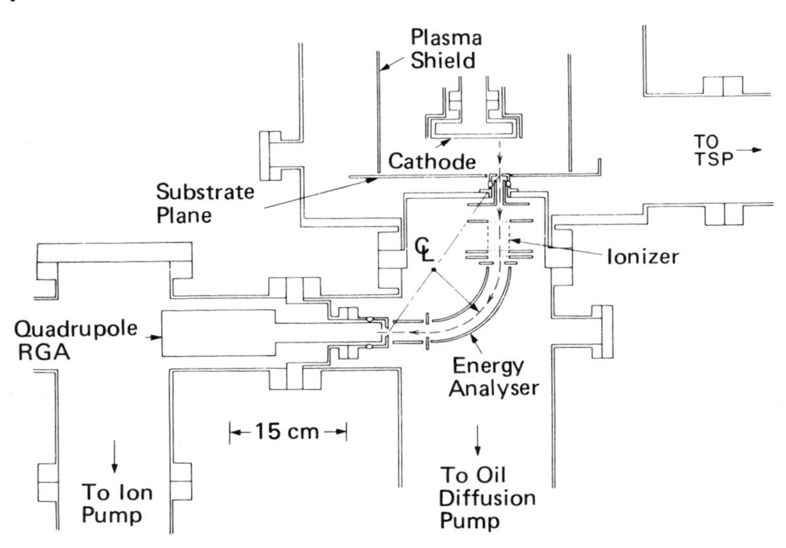

Figure 6-40. Experimental arrangement for the mass and energy analysis of ions bombarding the substrate plane (Coburn 1970)

This arrangement can be used to identify the ion species at the substrate as a function of their energy. Figure 6-41 shows 2 spectra of ions from a system sputtering copper. These spectra differ as a result of the energy analyzer being set to admit different energies. In this case the difference was 1 eV; the bias voltage was -30 V. The figure shows some of the complex and contaminant ions referred to earlier in the chapter.

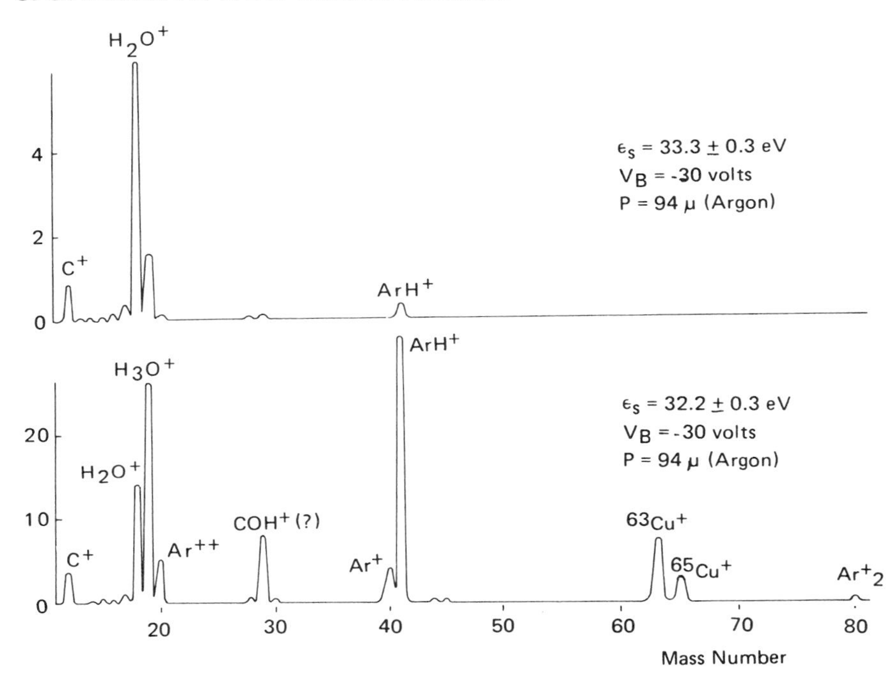

Figure 6-41. Mass spectra of positive ions from a dc discharge prior to sublimation pumping (Coburn 1970). Copper target −1000 V, cathode-bias table spacing 5cm, argon pressure 94 mtorr, bias voltage −30 V.

Upper trace: analyzer set to 33.3 ± 0.3 eV

Lower trace: analyzer set to 32.2 ± 0.3 eV

The mass/energy analyzer can also be used to measure the substrate sheath voltage (plasma potential minus substrate potential) by looking at an ion such as Ar_2^+, which has a low cross-section for collisions in the sheath and a high mass so as to minimize energy modulation by the rf field as it crosses the sheath. The energy distribution will be strongly peaked at the sheath voltage. We have already discussed this application in Chapter 5.

Some of the sputtered target is likely to become ionized, particularly by the Penning mechanism of collision with a noble gas metastable, as discussed in Chapter 2. Copper has an ionization potential of 7.7 eV, well below the 11.6 eV metastable level of argon, and would be expected to be Penning ionized. Figure 6-41 clearly shows copper ions from the target.

The combination of an electron multiplier as an ion detecter and a modern stable low-current amplifier, enables one to look at ion spectra with very high gain. Figure 6-42 is a mass spectrum of ion bombardment on the substrate plane when a target half of copper and half of tantalum was sputtered in argon. In this case the energy analyzer was not used. If the ion fluxes from different mass numbers occur over a very wide range (in this case 6 orders of magnitude), the conventional approach is to take a series of spectra, with each subsequent spectrum having a gain increase of x 10. Spectra for the Cu-Ta target example, taken by this means, are shown in Appendix 6, and are probably easier to interpret in that form. However, the series of spectra can also be taken in a single scan, with a useful savings in time, by using a logarithmic amplifier, as shown in Figure 6-42; actually this figure has a special non-linear logarithmic or *coburithmic scale.*

The largest ion currents in Figure 6-42 come from argon as expected. But the logarithmic amplifier reveals Cu^+, Ta^+, and less abundant argon peaks at lower ion currents. For even smaller currents, less abundant isotopes and complexes appear, followed by contaminants such as Nb^+ and $CH_3{}^+$. And so on. This expanded spectrum (and those in the appendix) show the sensitivity of the technique, and also illustrate one of the mixed blessings common to modern analytical techniques: the closer you look, the more you find.

Mass spectrometer data such as these have to be carefully interpreted, as the ion current magnitude cannot simply be equated with ion abundance; the quadrupole spectrum tends to underestimate higher mass ions, and its sensitivity also depends on ion energy.

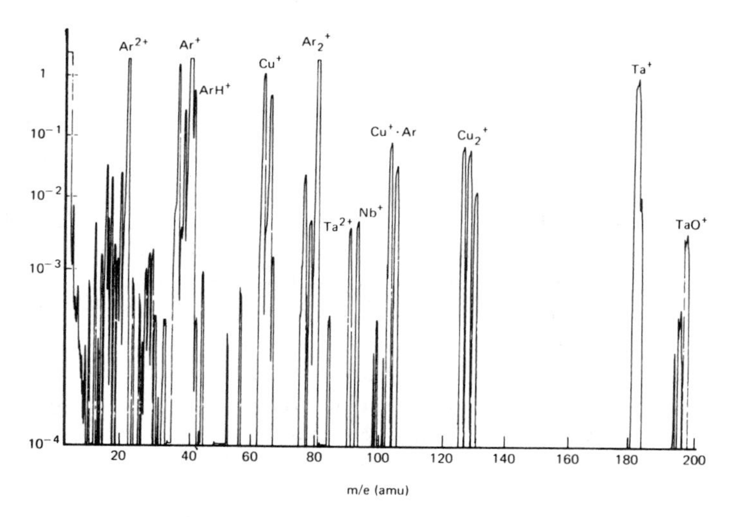

Figure 6-42. Ion mass spectrum at the substrate plane in an argon discharge sputtering copper and tantalum (Coburn 1979)

The mass/energy analyzer has been used by Coburn, by Coburn and Kay and other colleagues, for a variety of applications such as to measure the effects of discharge confinement on plasma potential, as described in Chapter 5, "Experimental Test of the Voltage Distribution Model". The sensitivity of the technique to the composition of a sputtering target makes it useful also as a technique for analyzing the composition in depth of the target as it is sputter etched away. The technique is then called *Glow Discharge Mass Spectrometry (GDMS)*, and is one of a series of techniques for *depth profiling* a sample, as discussed later in "Sputter Etching". In GDMS, one can employ noble gases with higher metastable states so as to encourage Penning ionization of the target components and so enhance the sensitivity of the technique.

I suspect that mass/energy analysis will continue to be useful in glow discharge analysis. Essentially the same technique has been recently used by Komiya et al. (1977) to study vapour deposition using a hollow cathode glow discharge.

Bias Evaporation

So far we have been looking at the use of bias to control the properties of films deposited by sputtering. But the bias technique affects conditions at the substrate mostly, and in principle deposition could be by any technique, for example by evaporation. There are two manifestations of this combination, which are different in application rather than in concept. *Ion plating* is used primarily for high rate metal deposition with very large dc bias applied and is discussed further in the section on ion plating; *bias evaporation* is normally used with rf bias, with similar bias values as in bias sputtering.

Bias evaporation has been used by Vossen and O'Neill (1970) for the evaporation of aluminium. They suggest its use for any materials that recrystallize easily, and which would recrystallize under the energy input from a dc sputtering target. A schematic of the bias evaporation configuration is shown in Figure 6-43. Of course, to obtain any considerable amount of ion bombardment

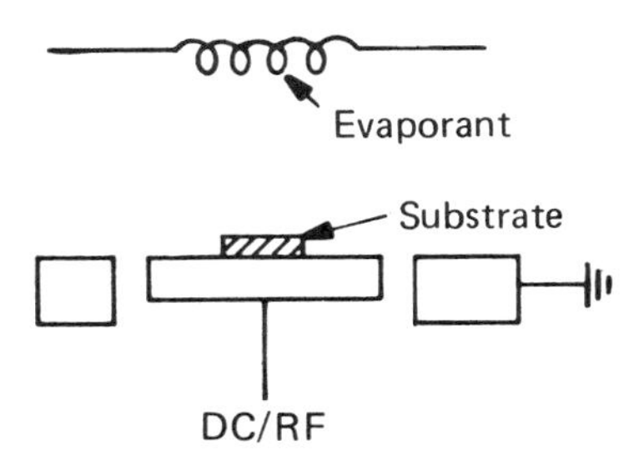

Figure 6-43. Schematic of a bias evaporation system

on the substrate, there must be a plentiful supply of ions, and these are obtained from a local glow discharge with substrate as target or part of the target. To obtain the discharge, the system is filled with argon to a pressure of a few millitorr.

Vossen and O'Neill reported several advantages of this technique over conventional evaporation. These advantages were in terms of being able to achieve almost bulk resistivity, ohmic contacts to silicon without sintering, reduced pinhole density, conformal coverage over severe substrate topography (which is discussed further in the section on that subject), and smoother films. To elucidate the latter, aluminium films generally have a tendency to form 'hillocks' and to show microscopic pitting. With bias, these defects were less evident and reached a minimum at -400 V. For larger biases, the film started to show preferred{111} crystallographic orientation, and the surface roughened as grain growth became evident, probably due to the heating effect of the ion bombardment. However, we must note that some of the improvements were probably a result more of the sputter cleaning of the substrate just prior to deposition rather than of the ion bombardment during it.

Bias Sputtering For Conformal Coverage

Bias sputtering can be used to control *where* a film is deposited, so as to ensure conformal coverage of surface topography.

Coverage of "difficult" surface topography, of which a perpendicular step is the epitome, has long been a problem and is responsible for many device failures, such as an open circuit in aluminium metallization on a silicon chip (Schnable and Keen, 1971).

Conformal coverage is difficult because surface features act effectively as masks and/or present little surface area to the deposition source. Changing the type of deposition source usually alleviates one of these problems only by accentuating the other (Figure 6-44). Vossen (1971) has demonstrated that neither rotating the substrate table nor using planetary rotation offers a proper solution and that the same is true of straightforward sputtering; however, it must be added that some people *are* satisfied with the use of planetary rotation, at least to the extent that it eliminates failures, which is not the same as saying that conformal coverage was achieved.

Seeman (1966) showed how it was possible to uniformly coat the inside walls of deep narrow trenches by dc bias sputtering. Logan et al. (1970) have used resputtering to obtain better edge coverage with sputtered quartz films. Vossen (1971) has shown how biasing can similarly overcome the problem of metallizing steep walls. The basic principle used to cover the side walls is to resputter already deposited material and redeposit this on the side walls. This is facilitated by the sputter ejection pattern being under-cosine (Figure 6-45) for low energy bom-

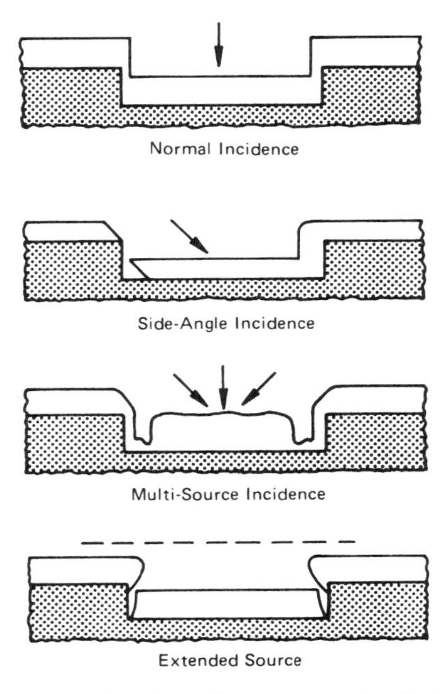

Figure 6-44. Defects arising as a function of source geometry during deposition of a coating to cover a steep step in a substrate (Kern et al. 1973)

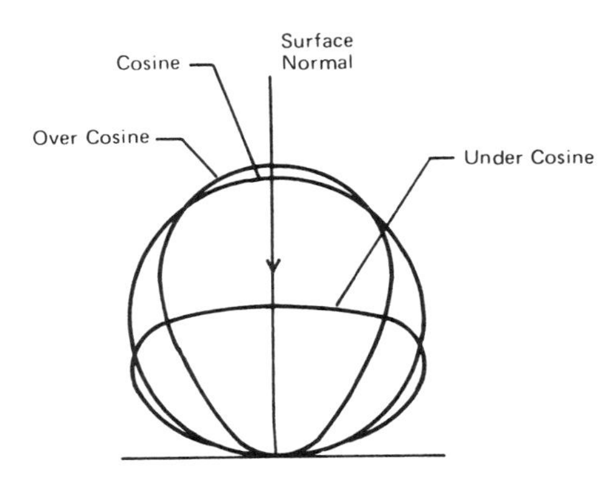

Figure 6-45. Angular distribution of particles from a polycrystalline target as a function of incident ion energy and direction (Kay 1962)

bardment; i.e. there is a tendency for material to be ejected sideways rather than outwards (cosine or over-cosine) as at higher energies. (The cosine law was initially derived from kinetic theory to describe the angular distribution of gas molecules bombarding a surface or leaving it. The result, which follows from an assumption of complete randomness of the motion of individual molecules, states that the number of molecules leaving the surface in a small element of solid angle $d\omega$ at an angle θ to the surface normal, is proportional to $d\omega \cos\theta$. The sputter ejection of atoms from a polycrystalline target is also a quasi-random process and tends to follow the same cosine law – to the extent already noted). As the bias voltage periodically varies, which it does (at mains frequency) for the usual rf or dc bias power supply to a greater or lesser extent (in this case, the greater extent is required – in other cases, we prefer to have a very small ripple), there is thus an effect which Vossen compares to a fireman hosing a wall up and down! There is little tendency for the material already on the side walls to be re-sputtered because the sputtering yield at virtually oblique incidence is practically zero (see later).

Vossen's experiments were performed using silicon substrates into which 6μm high vertical steps had been anisotropically etched. Figure 6-46 shows the varying distributions of platinum dc sputtered onto the silicon test wafer at various rf-induced bias voltages. As the bias voltage is increased, relatively more metal is resputtered onto the side walls. Similarly, for a given bias voltage a greater proportion of material will be resputtered onto the side walls as the deposition rate is reduced. This is clearly a very useful technique for conformal coverage, but the parameters must be carefully chosen. In Figure 6-46, one sees again evidence of grain growth in the metal film. It is probably still true that people try to avoid having sharp corners rather than cope with them. Steep walls from etching processes can be turned into more gradual slopes by using 'taper control layers' (Kern et al. 1973) in wet chemical etching, or by using selective degradation of the photoresist in reactive ion etching, as described in Chapter 7, "Isotropic or 'Anisotropic' Etching?"

Finally in this section, we note the use of the bias technique to achieve not conformal coating, but almost its opposite – a deposited film with a planar surface, regardless of the topography of the substrate. This usage of bias sputtering for *planarization* (Ting et al. 1978) will be clearer after looking at the angular dependence of the sputtering yield, in "Sputter Etching".

Backscattering in Bias Sputtering

In an earlier section ("A Conventional DC Sputtering System – Choosing the Pressure Range"), we saw how molecules from a sputtering target can collide with gas phase molecules and be scattered, in some cases right back onto the target. This is a problem in bias sputtering, because the substrate backing plate

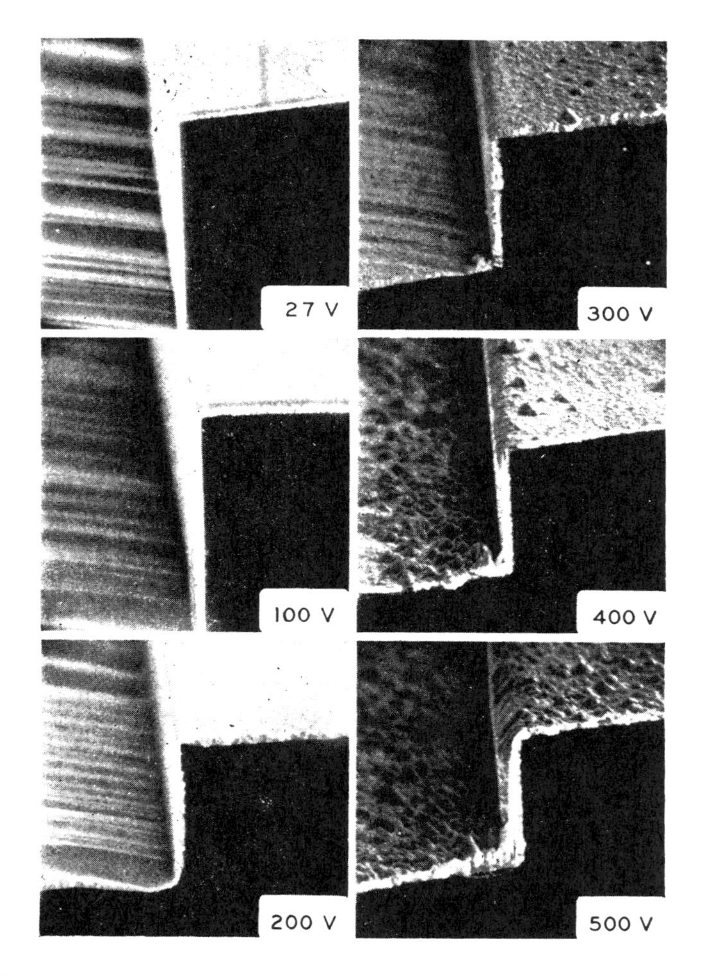

Figure 6-46. The thickness distribution of Pt coatings dc sputtered with various rf-induced bias voltages. The plane surface accumulation rate was held constant at 300 Å/min (Vossen 1971)

(Figure 6-26) will also be subjected to ion bombardment, and the resultant sputtering of the plate and backscattering in the gas phase can lead to backing plate material being deposited onto the substrate and incorporated into the growing film. A similar phenomenon of mixing would occur on composite targets, and would be important in sputter etching if not in sputter deposition. This problem has been investigated by Vossen et al. (1970). They placed 2.4 cm silicon wafers on various 15 cm diameter metal targets, principally Pt, in an rf diode sputtering system, with interelectrode separation of 5 cm. Using a dc level of -700V on the rf target, they found that the platinum concentration on the silicon reached an equilibrium value after about 10 minutes. This equilibrium value ranged from about 1 monolayer at 2.5 millitorr to about 4 monolayers at 20 millitorr, showed little spatial dependence except near the edges of the wafer (Figure 6-47), and increased monotonically with pressure although showing signs of saturation. The equilibrium value is presumably a balance between deposition by backscattering, and resputtering. In this work the target voltages used were more representative of sputter etching than of bias sputtering, but note that although sputtering of the backing plate would be much slower at the lower voltages used in bias sputtering, so also would resputtering from the substrate. Vossen et al. did vary the target voltage over the range 500 V - 1200 V, and found that the platinum coverage on the substrate increased from 2 monolayers to 4.5 monolayers as the voltage was reduced to the lower limit. It would be unwise to extrapolate these results down to the 50 V or so frequently used in bias deposition, because the sputtering yields of materials change so rapidly with energy at such values.

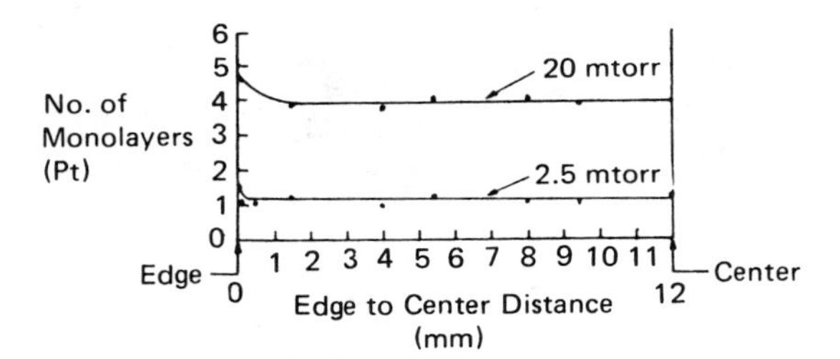

Figure 6-47. Typical wafer profiles; effect of argon pressure (Vossen et al. 1970)

The very clear lesson from these and similar results (Chang et al. 1973) is that one should use a backing plate which is at least compatible with the material being deposited, and may ideally be the same. The same message applies to the comparable situation in both types of diode plasma etching, as discussed in Chapter 7.

We shall return to the subject of backscattering in the section on "Sputter Etching", where we shall also see that there can be some very unfortunate choices of backing plate material.

Deposition of Multicomponent Films

Most of our discussion to date has been concerned with the sputtering of single elements. But often multicomponent films are required, and these may be alloys, compounds, or a mixture of both. They can usually be prepared by sputtering from a single compound target, simultaneously from several different targets, by reactive sputtering, or by a combination of these techniques.

There are four stages that we have to consider, of the transport of an atom from a sputtering target to its incorporation into the thin film. These stages are sputter ejection from the target, transport through the gas, condensation onto the substrate, and stable incorporation into the film. We shall now consider these four stages.

Alloys

Let's consider initially the deposition of an alloy film, and specifically we'll think about deposition from a Ni:Fe target of uniform 80:20 composition. When we first start to sputter the target, we'll be removing the layer of metal oxide that is inevitably on the surface of the target. During this period, the shutter must be interposed between the target and the substrate to prevent condensation onto the latter. After some while, which will probably be in the 5 minute-30 minute time range, the sputtering system will have stabilized and the surface oxide will be removed, revealing the metallic alloy. In dc sputtering, the removal of the oxide will be accompanied by a corresponding change (usually a decrease) in the discharge current, since the secondary electron coefficient of the metal and the oxide differ. Even if rf sputtering is eventually to be used, the use of dc sputtering to determine the time required to remove the surface oxide, is often extremely helpful, and this of course can be applied to single element sputtering as well as multicomponent.

Now the metal alloy surface, initially having an 80:20 composition, is exposed. But nickel has a slightly higher sputtering yield than iron; for 1000 eV argon ions, it's 2.1 for nickel and 1.4 for Fe (see Appendix 6). So if we used ions of this energy, the nickel and iron atoms would be sputter ejected from the target in the ratio 80 *x* 2.1 : 20 *x* 1.4; 86% of the atoms leaving the target would be nickel compared with 80% in the target. But this situation can't last for long, because a result of the preferential ejection of the nickel will be to cause the surface of the target to become enriched with iron. As this iron enrichment occurs, so the sputtering rate of iron atoms increases and of nickel atoms decreases until they are again leaving in the ratio 80:20. Ultimately, in steady state, the departure rate from the target must equal the supply rate, which is the 80:20 alloy from the bulk of the target. In order to achieve this, in the present example, the surface composition would adjust to 72.7% Ni:27.3% Fe. During the period when the surface composition is stablizing, the shutter must continue to be used to prevent condensation onto the substrate.

Two phenomena can prevent the steady state conditions described above, from being reached. The depletion of one element on the target surface (nickel in our example) will set up a concentration gradient of each of the elements at the surface, and this will encourage diffusion. In the case above, nickel will diffuse towards the surface and iron away from it; with enough diffusion, the initial 86:14 sputtering ratio could be maintained. Diffusion is minimized by keeping the target cold enough, and 'cold enough' will vary from alloy to alloy, of course.

Keeping the target cool will also offset another potential problem that can be encountered when sputtering multicomponent materials. If one of the elements or compounds has an appreciable vapour pressure at the target temperature reached, then it will evaporate from the target. Remember from Chapter 1, "Monolayer Formation Time", that a vapour pressure of 10^{-6} torr would correspond to an evaporation rate of about 1 monolayer per second.

When the target has reached steady state, the shutter can be moved back and deposition can begin. Transport of the sputtered material from target to substrate will rarely be straight line travel. Even at the lowest sputtering pressures, collisions with the sputtering gas atoms will take place. As the pressure increases, transport becomes more like a diffusion process and some material is redeposited back onto the target. The redeposition does not affect the alloy composition because the target adjusts itself so that the *net* composition leaving the target is the same as the bulk. However, as the mfp of sputtered material decreases, more of it will eventually land on the walls and proportionately less on the substrate. There is no reason why the target atoms (Ni and Fe in the example) should have the same mfp, and so we should expect that the ratio of the Ni and Fe fluxes at the substrate will not be 80:20. On the other hand, the mfp values shouldn't be very different, so this will be only a small perturbation on the 80:20 value.

The next stage involves the condensation of the material onto the substrate. As we saw in the earlier section on "Thin Film Formation", an atom arriving at the substrate migrates around on the surface for a while, hopping from one adsorption site to another, until it either joins with another migrating atom to form a more stable pair, or it evaporates. The *condensation coefficient* is the proportion of the atoms arriving at the substrate that remains without evaporating, and this will be determined by the arrival rate, the bonding energy between adatoms and substrate, and the substrate temperature. In general, the condensation coefficient of each constituent of a multicomponent target will be different, causing an effective change in the composition of the depositing alloy.

Finally, the growing film will be bombarded by ions and so there will be a certain amount of resputtering of the growing film. As well as sputtering gaseous components from the film, the ion bombardment will sputter the solid components too. Since sputtering increases with increasing ion energy, such resputtering will be particularly effective under bias sputtering conditions. In the deposition of alloys, as at the target, the film will be depleted of the component with the higher sputtering yield. At the substrate, however, there will be no compensating mechanism.

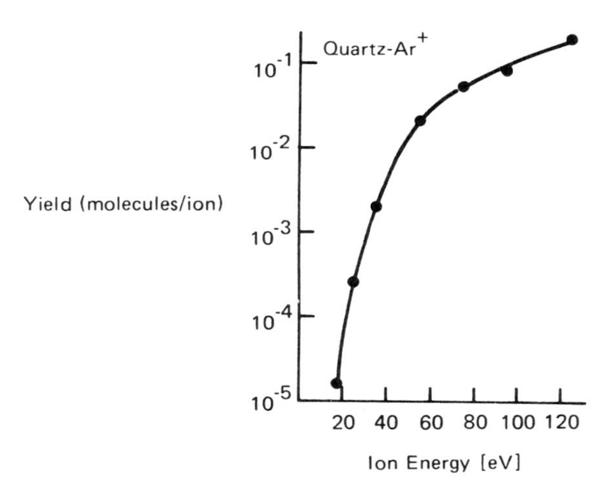

Figure 6-48. Sputtering yield of quartz in argon, in molecules per ion vs bombarding energy (Jorgensen and Wehner 1965)

Earlier in this chapter we have discussed sputtering primarily with rather energetic (~ 500 eV) ions. In bias sputtering, bias voltages of the order of 50 V are more common, and so low energy sputtering yield data becomes of interest. Figure 6-48 shows sputtering yield data for quartz in argon (Jorgensen and Wehner 1965). This data is typical also of the sputtering of metals, for which more yield data, obtained by Stuart and Wehner (1962) using an optical spectroscopic technique, is shown in Appendix 6. What is commonly observed is that the yield, which decreases linearly as the ion energy reduces from 1000 eV to 100 eV, starts to decrease much more rapidly somewhere below 100 eV; n.b. the logarithmic vertical axis in Figure 6-48. An apparent threshold is observed, although this is influenced by the sensitivity of the detection technique. Table 6-2 shows

Table 6-2 Threshold Energies (Stuart & Wehner 1962)

	Ne	Ar	Kr	Xe	Hg
Be.	12	15	15	15	
Al.	13	13	15	18	18
Ti.	22	20	17	18	25
V.	21	23	25	28	25
Cr.	22	22	18	20	23
Fe.	22	20	25	23	25
Co.	20	25	22	22	
Ni.	23	21	25	20	
Cu.	17	17	16	15	20
Ge.	23	25	22	18	25
Zr.	23	22	18	25	30
Nb.	27	25	26	32	
Mo.	24	24	28	27	32
Rh.	25	24	25	25	
Pd.	20	20	20	15	20
Ag.	12	15	15	17	
Ta.	25	26	30	30	30
W	35	33	30	30	30
Re.	35	35	25	30	35
Pt.	27	25	22	22	25
Au.	20	20	20	18	
Th.	20	24	25	25	
U	20	23	25	22	27

the threshold data obtained by Stuart and Wehner, corresponding to a detection limit $\sim 10^{-5}$ atoms/ion. They observed no correlation between thresholds and energy transfer functions, and concluded that a binary sputtering model is inappropriate at these energies, in accord with more recent theories.

In a later section on "Sputter Etching", I shall describe some work by Tarng and Wehner (1972) that shows how the sputtering yield of some thin film materials varies enormously according to the thickness of the film and the nature of the underlying substrate. There is very likely to be a comparable effect in bias sputtering, compounded if one is depositing a multicomponent film. Under such circumstances, sputtering yield data from homogeneous targets is probably of little relevance.

Taking all of these various effects together, we can see that the composition of a sputter deposited alloy film can be quite different from that of the target. However, the composition change should be reproducible for a given set of conditions, and so it becomes mainly a matter of choosing or fabricating a target of appropriate composition.

There are two other points to be noted. Firstly, a sputtering ratio of the two elements in the target holds for a particular ion bombardment energy. If that energy changes, e.g. by changing the target voltage or by changing the pressure, there will be a transition period while the target surface adjusts to a new steady state composition. Secondly, the composition of the deposited film can be changed somewhat by adjusting the bias voltage and other process parameters, but this means may be inconsistent with other process requirements. More flexibility may be attained by using two (or more) quasi-independent targets. One would then need to consider the deposition uniformity problems posed by the geometric constraints of this arrangement, as well as by the beam-like nature of electron and negative ion (if present) bombardment of the substrate. Nevertheless, this can be done successfully, with adequate confirmation from the deposition of multicomponent magnetic films.

Compounds

We now turn to consider the sputtering of a chemical compound target. Is this target sputtered as molecular species or as atoms? The general answer is both. The question has been studied by Coburn et al. (1974), who also cite several other papers where sputtered molecular species were detected.

Coburn et al. studied the rf sputtering of metal oxides. Using the ion sampling and analysis system that was described earlier in the section on bias sputtering, they have looked specifically at the relative numbers of atomic metal ions and dimeric metal oxide molecular ions arriving at the substrate plane. In some cases, ternary and more complex molecular ions were observed but always at very low intensities. Coburn et al. presented their data in terms of $\eta = MO^+/(M^+ + MO^+)$ for a series of metals M. They found that η was strongly influenced by the relevant M–O bond strength, as shown in Figure 6-49. Not surprisingly, more strongly bonded metal oxide ions were less likely to be disassociated. Whilst η is defined and measured for ions, one would expect this ratio to be at least indicative of the corresponding neutral species, and probably numerically not so different. Coburn et al. also found that η decreased significantly with increasing pressure, and since η at zero pressure is likely to be the best representation of the ion ratio leaving the target (there being no collisions in this hypothetical

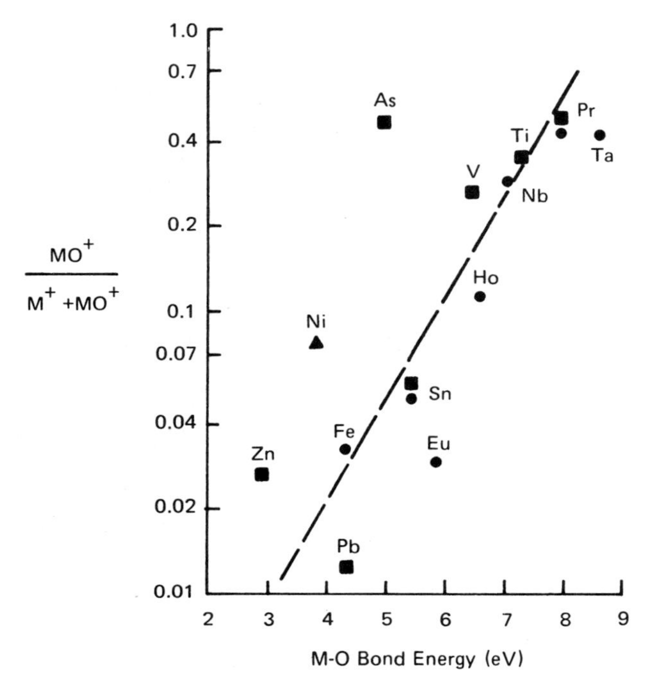

Figure 6-49. The dependence of $MO^+/(M^+ + MO^+)$, i.e. the relative ion flux of the dimer, on the bond energy M-O. Argon pressure 60 mtorr, rf power 100 watts at 13.56 Mhz, target area 20 cm^2 (Coburn et al. 1974)

case), the η values shown would tend to underestimate the relative concentration of the sputtered molecular ion.

Figure 6-49 and other results indicate that the M–O bond strength is not the only factor influencing η. Coburn et al. propose that a molecular sputtering event is more likely to occur if the atom struck by the bombarding ion can transfer enough energy to its molecular partner that they can be ejected as a pair. The energy transfer function (Chapter 1, "Energy Transfer in Binary Collisions") tells us that this is more likely when the atom masses are similar, and in Figure 6-49 it is certainly the lighter metals, with smaller mass ratios compared to oxygen, that lie above the line and exhibit larger molecular ion fluxes than a direct dependence on the M–O bond would suggest. Finally, Coburn et al. demonstrate that the presence of contaminants such as water vapour, has a marked effect on the ratio η.

Restoration of Stoichiometry

Given that a metal oxide target, and by implication any other compound target, is unlikely to be completely sputtered in a molecular form, it is not surprising that the stoichiometry of the resulting thin film will be different from that of the target, usually being deficient in the gaseous or other volatile species. For example, films from a quartz target sputtered in argon tend to be deficient in oxygen. This can be compensated for by sputtering in a mixture of 95% Ar:5% O_2 (Erskine and Cserhati 1978). The oxygen served to fully oxidize the sputtered film and so restore its stoichiometry. It can be particularly effective in doing this because the glow discharge environment provides energetic electrons to dissociate the molecular oxygen into its chemically more active atomic form.

We should note that we do not always want to achieve the stoichiometry of the equivalent bulk material. For example, Mogab et al. (1975) found that silicon nitride films which met their requirements were richer in nitrogen than the stoichiometric Si_3N_4.

Reactive Sputtering – Again

But if we are going to react a small amount of Si or SiO with O atoms, why not a lot? Indeed this can be done, and sputtered quartz films can be fabricated by sputtering a very pure elemental silicon target in a 50% Ar:50% O_2 mixture (Erskine and Cserhati 1978). This is the process of *reactive sputtering*, the chemical combination of the sputtered species and a component in the gas phase, that we usually try to avoid. Reactive sputtering can be used to promote total chemical conversion of the target material or, as in the previous section only to compensate for a deficiency in the film.

Quite small quantities of gas can be very effective and quite critical, so that careful partial pressure control, e.g. by mass spectrometry, is required. Note though, that although partial pressures of reactive gases can be very low, one still needs an adequate supply or flow of reactive gas. This requirement is discussed in more detail in the next chapter on "Plasma Etching". Oxygen and nitrogen are frequently used in reactive sputtering. In the discharge, molecular ions will be formed, and diassociation will lead to atomic ions and atomic neutrals in ground and excited states. Each of these species can play its role in the chemical conversion of the film.

The general situation will be rather complex and it is more fruitful to pursue specific cases. In principle, reactions could take place on the target, in the gas phase, and on the substrate. Target reactions result in one actually sputtering a compound target. These usually have a lower yield than elements, and result in the often observed reduction in deposition rate with added reactive gas.

Gas phase reactions suffer from the same problem of conserving energy and momentun that was discussed in Chapter 2. However, since there is in this case a molecular product, there is a chance that surplus energy will be absorbed into vibrational and rotational transitions, so such reactions should be much more likely than in atomic ion-electron pair recombination. On the other hand, there seems to be a general opinion that more reaction under these circumstances takes place on the substrate.

We have already discussed some of the various bombardments that go on at the substrate. We now add bombardment by reactive positive ions and neutrals to the list. Relevant work by Winters and Kay on the effects of the sorption characteristics of various gaseous and ionic species has already been cited. In the case of electronegative gases such as oxygen, we shall also have bombardment by fast oxygen negative ions formed in the target sheath or on the target itself. Then add substrate bias to confuse things!

There are all sorts of peculiar and interesting effects in reactive sputtering; they usually turn out to be due to some of the effects we have noted. For more practical information, see the review articles by Holland (1956) and by Maissel (1970), and the extensive bibliography by Vossen and Cuomo (1978).

SPUTTER ETCHING

Sputter etching is the name conventionally given to the process of removal of material from a surface by sputter ejection. There are several uses for the process other than as a prerequisite for sputter deposition, as we have already discussed.

Pattern Production

If material is selectively removed from a surface by using a suitable mask, then an etch pattern can be produced. The process of sputtering is fairly universal, so that the variety of wet chemicals used for the same process can be eliminated. Since sputtering results from bombardment by ions that move along electric field lines, and because field lines are always perpendicular to an equipotential surface, then etch profiles are inherently vertical in contrast to the isotropic profiles observed with wet chemical etching (Figure 6-50).

Sputter etching can be carried out in a conventional sputtering system with an in situ glow discharge, or with an externally generated ion beam. Ion beam sources are themselves glow discharge devices and are discussed in a later section. The change with an ion beam system is that the operating pressure can be much lower ($< 10^{-4}$ torr). There is then less chance of sputtered material (from the target or more importantly from the mask or target support) colliding in the gas phase and being backscattered onto the target. There is also a negligible flux onto

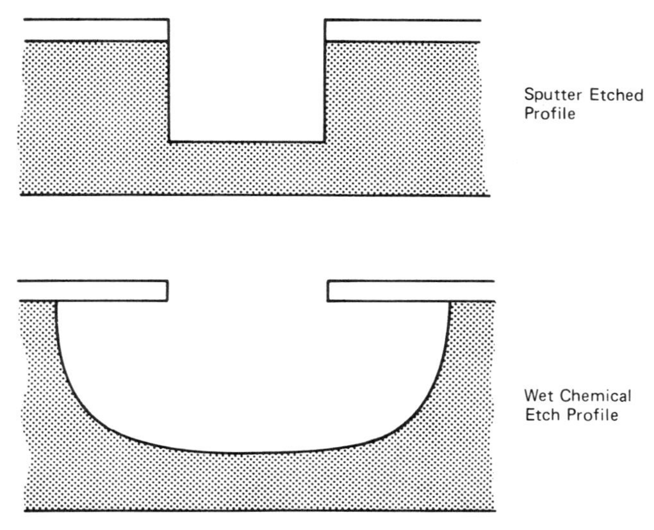

Figure 6-50. Etch profiles produced by sputter etching and by wet chemical etching.

the target of energetic neutrals produced by charge exchange. These fast neutrals are not affected by the electric field and therefore do not follow the field lines onto the target. They could lead to undercutting of the mask. In practice, there is a noticeable difference of directionality between results obtained in sputtering and ion beam systems. The difference is not that great, however; this may be due to pronounced forward scattering of the ions neutralized in the charge exchange process, so that the fast neutrals are more directional than one might at first suspect.

But even for ion beam systems, the sputter etch profile shown in Figure 6-50 is rather idealized. Two prime reasons for this are *mask erosion* and *trenching*. If the mask itself is also etched by the sputtering process, and this is always the case to some extent, then the final dimension of the opening in the mask will be greater than the initial dimension, leading to a tapered profile (Figure 6-51). The propensity of the mask to be sputter etched depends on its thickness and its profile. For etching of small dimensions of the order of microns, lithographically processed patterned photoresist materials are used. To achieve such resolution, the resists have thicknesses $\sim 1\ \mu m$. The polymeric materials used for photoresists tend to sputter rather easily and also degrade under ion bombardment. Sometimes the heating effect of the glow causes *resist flow*, so that its profile changes. These effects combine to make resist masks prone to etch back during sputter etching.

There are much more resistant etch masks. Oxides generally have low sputtering yields, and alumina and magnesium oxide make good masks. But they are not photosensitive, so they must be patterned using more conventional photoresist

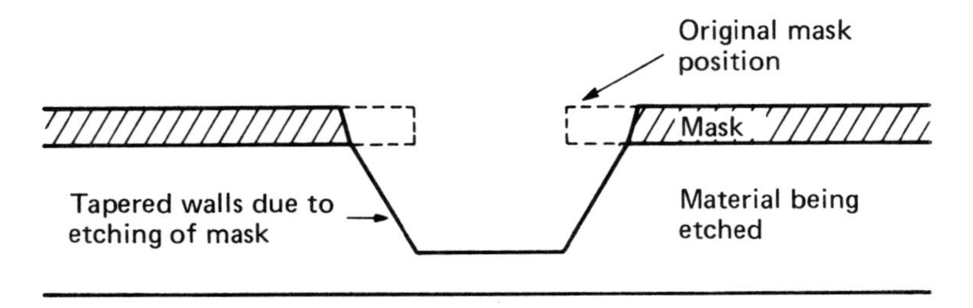

Figure 6-51. Etch profile resulting from mask erosion.

masks. The additional steps of oxide deposition and etching must be weighed against the reduced mask erosion.

Trenching is the enhanced erosion around the foot of an etched wall, leading to the 'molar' shape shown in Figure 6-52. It results from the increased flux of ions at the trenches due to reflection off the side walls of the etch pit, and perhaps also from material sputtered from the side wall onto the base of the pit.

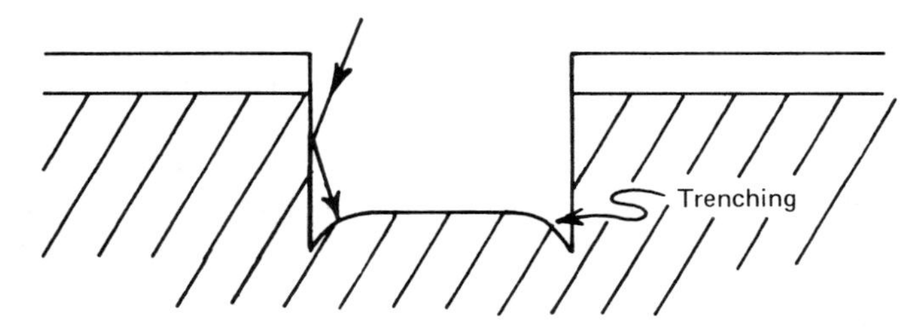

Figure 6-52. 'Trenching' in sputter etching

Several other phenomena control the topography of a sputter etched target. One concerns the redeposition of sputtered material. In the section on bias sputtering, we saw how some material is sputtered sideways, and this can be redeposited onto the side wall of another feature. This effect can be further illustrated by some results of Glöersen (1975). During the sputter etching of aluminium using a photoresist mask (actually in an ion beam system, but the principle is the same), sputtered aluminium is redeposited onto the side wall of the resist. After processing, the resist is chemically removed, leaving a wall of metal around the original resist position.

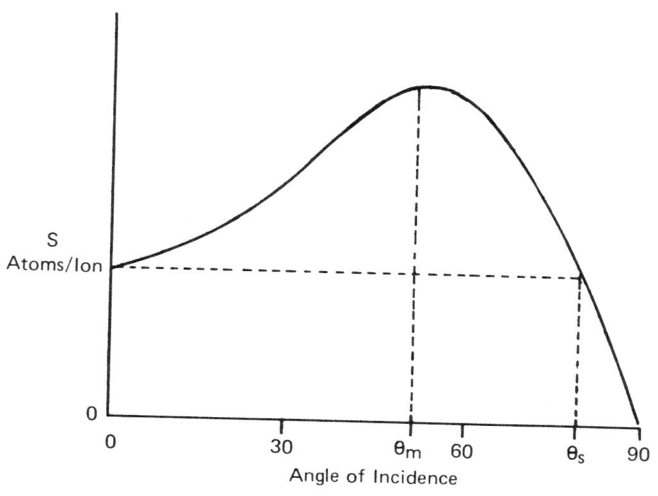

Figure 6-53. Variation of sputtering yield with angle of incidence (with respect to the target normal)

Another phenomenon related to etch topography depends on the angular dependence of sputtering yield. Figure 6-53 is a generalization of the effect; there are many specific examples, e.g. Wehner (1959), Cheney and Pitkin (1965), Glöersen (1975). Normal (in the angular sense) sputter ejection due to normal ion incidence requires 180° reversal of the momentum; this is not very likely and accounts for the low energy efficiency of the sputtering process. As the angle of incidence increases, so does the sputtering yield, again for momentum considerations. This is accompanied by the sputter ejection angular distribution becoming pronounced in the direction of specular reflection. Ultimately the yield goes to zero at 90° incidence, resulting in a maximum yield somewhere between 0° and 90° depending on the target material, and the ion identity and energy. This angular dependence has several consequences. One is the formation of *facets*. Figure 6-54 shows two surfaces being etched, one normal to the ion beam and the other at angle θ. If we consider sections of each surface that present the same cross-sectional area A to the incoming beam, then we can see that the etch rates in the direction of the beam, R_b and R_c respectively, will be proportional to the sputtering yields at the relevant angles of incidence. Although the inclined surface is subjected to a lower flux of ions, this is irrelevant because the volume of material removed depends on the *cross-sectional* area and the etch rate.

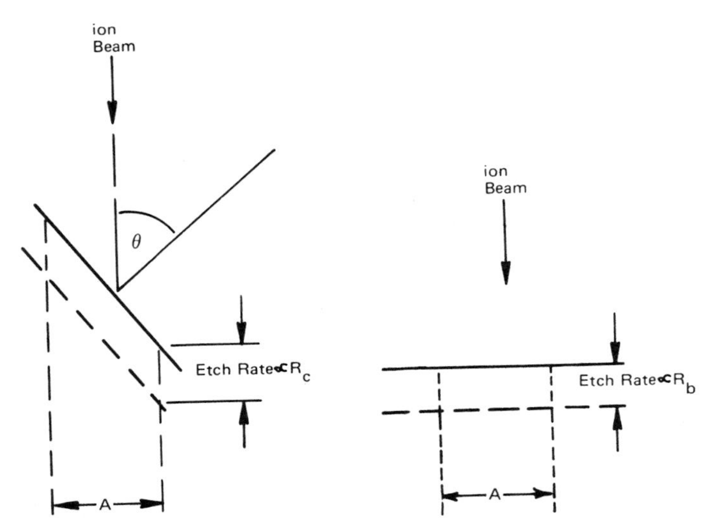

Figure 6-54. The etch rate parallel to the ion beam is proportional to the sputtering yield at the same incidence angle.

The phenomenon of facetting, which is illustrated in Figure 6-55, usually starts on corners which always have some rounding, and therefore present a variety of incidence angles to the incoming ions; in bias sputtering systems, the effect is pronounced due to the field concentrations occurring at sharp corners. A facet will develop for the plane which has the highest sputtering yield and etch rate ($\theta = \theta_m$ in Figure 6-53), as the more slowly etching planes are consumed. The effect is then transferred to the substrate but at a different angle corresponding to θ_m for the substrate material. A second material, photoresist in the figure, is not necessarily required and any artifact on the target surface can promote the facetting.

The mechanism for the *planarization process* (Ting et al. 1978) used to encourage the formation of a planar thin film over a non-planar substrate, is related to the mechanism of facetting, but has the additional complication that etching and deposition are occurring simultaneously

Folks involved in ion beam etching use the angular dependence of yield to take care of trenching and other phenomena, and often use non-normal incidence and planetary rotation to achieve the required results (Lee 1979). Neither of these tricks is possible in glow discharge sputter etching since ion bombardment is controlled by the glow discharge, which follows the target and hence ensures normal ion incidence.

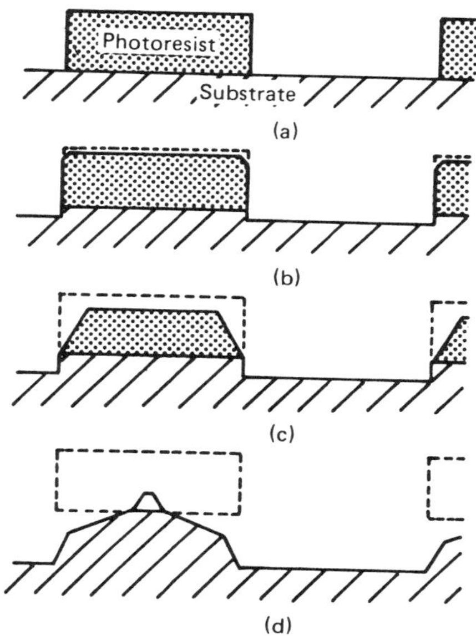

Figure 6-55. A simple model depicting the stages in the ion beam etching of grooves using a photoresist mask. (a) The photoresist cross-sections prior to initiating ion beam etching. (b) The onset of facet formation in the photoresist. (c) The photoresist facet intersects the original substrate surface plane. (d) After etching further, the facet in the photoresist propagates laterally thereby exposing additional substrate to ion beam etching and consequent facetting (Smith 1974)

Etch Topography

A phenomenon related to facetting is that of *cone formation*. The root of this can already be seen in Figure 6-55 where the resist has the shape of a truncated cone. The earliest observations of cones occurred nearly 40 years ago (Guentherschulze and Tollmien 1942), and there are now a large number of reports of cones on a variety of metals and semiconductors (Wehner 1959, Stewart and Thompson 1969, Nobes et al. 1969, Hauffe 1971, Wilson and Kidd 1971, Bayly 1972, Tsong and Barber 1973, Robinson and Kaufmann 1979). Explanations and computer simulations (Catana et al. 1972) of these artifacts are based on the angular dependence of the sputtering yield, and show how structural faults or impurities can develop into cones. Although the phenomenon is frequently referred to as cone growth, this is a misnomer since the cones do not grow but instead result from not etching.

The interest in cone growths is also due to the observation by Wehner and Hajicek (1971) that cones could be developed on pure copper surfaces by depositing a very low flux of molybdenum at the same time the copper is being etched in a conventional sputtering discharge environment. The effect is attributed to the low sputtering yield of the molybdenum and to its high surface diffusion mobility. Also using an in situ glow discharge, we carried out similar experiments using copper and tungsten and found, for example, that cone shape and density are very dependent on the structure of the original copper substrate; Figure 6-56 is of two copper specimens prepared side-by-side under identical etching/deposition conditions, and Figure 6-57 is also of copper but under slightly different conditions. The rounding of the cones which is evident, particularly in Figure 6-57, is probably due to the non-normal incidence of some sputtering particles which results from an in situ glow discharge.

As well as having interesting electrical properties, cones also produce optical effects. For example, a high density of closely spaced cones produces so much light scattering that the surface resembles matt black velvet, a condition that could be easily confused with gross contamination, and has resulted in at least several sputtering targets being dispatched to Valhalla (no, not the IBM location) and diffusion pump oils cursed, in both cases unjustly!

Pertinent to the Cu/Mo cone system, Tarng and Wehner (1972) made an interesting Auger study of the deposition of molybdenum onto various metal surfaces, and its subsequent sputter removal. The equivalent of 12 monolayers of Mo were deposited onto tungsten, copper, gold, and two grades of aluminium under conditions of complete condensation, and were then removed by sputter etching. The surface coverage was monitored throughout by Auger analysis. Results are shown in Figure 6-58 as a function of the amount of molybdenum that would have been removed if the target had been Mo entirely. Mo appears to initially completely cover the W, and is then removed almost as though from bulk Mo. On the other metal substrates, deposition did not produce complete coverage. However, under ion bombardment, the coverage in these cases at first begins to rise as the Mo is spread more uniformly over the surface. There is then a slow decrease in coverage as the Mo is etched away. The sputtering yield of Mo atoms from Cu, Au, and Al is much lower than that of Mo from Mo, especially at low energies; 200 eV argon ions sputter Mo from Al with a yield three orders of magnitude lower than that for bulk Mo or bulk Al.

Some very peculiar surface structures have been reported by Oohashi and Yamanaka (1972) under simultaneous etching/deposition conditions. Rather than use a separate deposition source, their substrates were placed on a tantalum target, so that the tantalum was backscattered onto the substrate whilst the latter was being etched. Figures 6-59 shows some of the 'winding' patterns they obtained. Prof. Yamanaka kindly sent me photomicrographs of Cu, Al, Ag, C, Co, Mn, Au, Fe, Zr, Pt, Ni, Ti and Brass, all sputter etched similarly on a Ta target. Only the Cu showed classic cones, whilst the others displayed all sorts of strange topographies. Goodness knows how these fit into the cone formation theories!

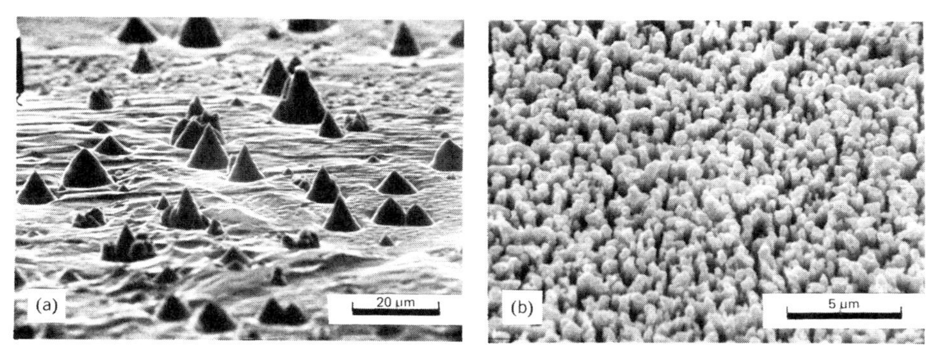

Figure 6-56. Scanning electron micrographs of two copper surfaces, sputter etched with simultaneous low rate deposition of W (Chapman et al. 1973)

(a) Bulk polycrystal Cu.

(b) Evaporated thin film Cu.

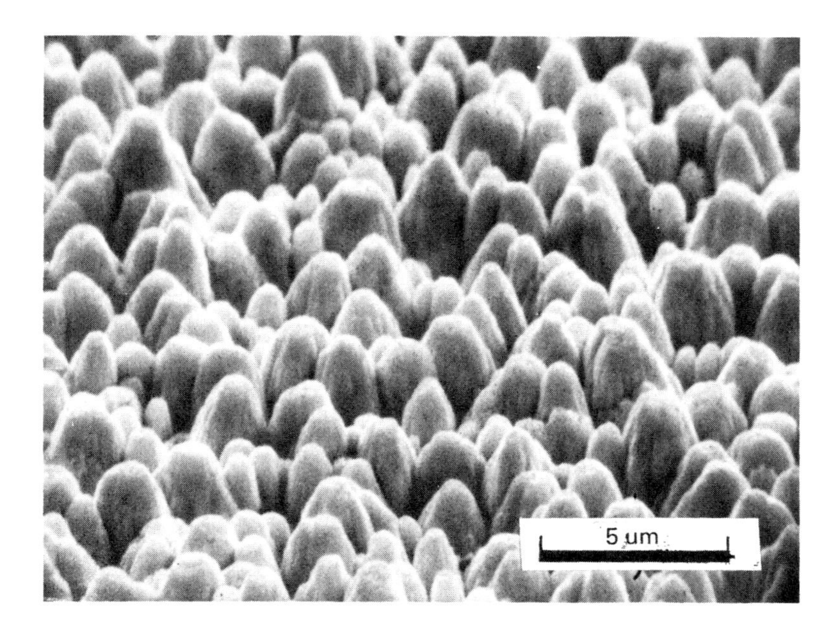

Figure 6-57. As Figure 6-56, but for an evaporated copper film, and under slightly different conditions

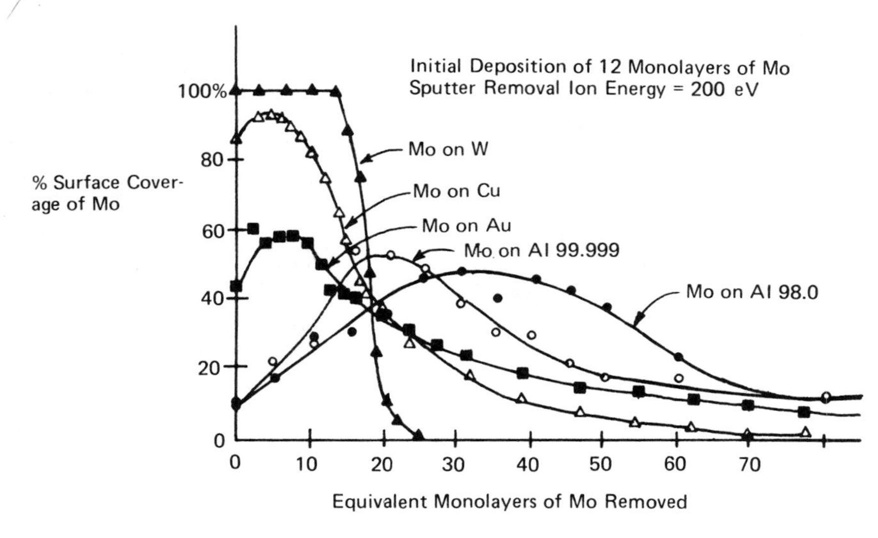

Figure 6-58. Surface coverage of Mo on various metal surfaces, subjected to Ar^+ bombardment. The coverage is shown as a function of time, measured in terms of the number of monolayers of Mo that would have been removed from a pure bulk metal Mo sample in the same time (Tarng and Wehner 1972)

Surface Analysis

There is now quite a range of techniques for analyzing surfaces, each with their specific advantages. They usually analyze a surface layer tens or hundreds of Ångstroms in thickness, according to the penetration and escape depth of the particles involved in the specific analytical process. This is both an advantage in identifying the location of the analyzed section, and a disadvantage because the material beneath the surface layer is not analyzed.

This problem is eliminated by removing the surface layers sequentially, exposing deeper layers. This is usually done by sputtering, and the overall process is known as *depth profiling*. The sputter removal can either be part of the analytical process as in *SIMS* (Secondary Ion Mass Spectroscopy), or quite separate as in *Auger* analysis. In these latter cases, sputtering can either be continuous during analysis, or intermittent with analysis. Profiling techniques have been reviewed by Coburn and Kay (1974).

In all cases, one would like to be able to expose the sample by one atomic layer at a time. It should be clear from the discussions on alloy sputtering in "Deposition of Multicomponent Materials" and the preceding section on "Etch Topography" that layer-by-layer profiling of the substrate cannot be assumed, and this adds enormously to the problems of surface analysis, which have been reviewed by Coburn (1976). Presumably one could, in principle, solve the problem by comparing the material actually leaving the surface with that remaining on it, for example by combining the results of simultaneous SIMS and Auger; but I'm sure that's horrendously difficult and I'm not aware that anyone has succeeded in doing it yet.

Surface Cleaning

Glow discharges can be used in several ways to clean the surfaces of materials, and this is usually done in situ immediately prior to another vacuum process such as deposition by sputtering or evaporation.

Glow discharge cleaning involves placing the substrates to be cleaned in the glow so that they are bombarded by low energy ions and electrons. The precise energy of bombardment will depend on whether the substrates are insulating or conducting, whether the discharge is rf or dc excited (usually it's the latter), and whether the substrates are subject to high energy negative particle bombardment from the target sheath.

Beneficial results seem to result from this technique, particularly if the cleaning is carried out in an oxygen discharge (Hirai et al. 1966). Impurities are desorbed from the surface due to the ion and electron bombardment, or due to the heating associated with these bombardments. With the oxygen discharge, there will also be effective oxidation of organic impurities on the surface by atomic oxygen formed dissociatively in the discharge, and these oxides will generally be volatile. This is the *plasma ashing* process described in Chapter 7. There are also several reports of oxygen glow discharge cleaning improving the adhesion of subsequently deposited films, by modifying the substrate surface. Such effects could be due, though, to residual oxygen in the system forming an intermediate oxide layer at the interface. "Thin Film Adhesion" is discussed further at the end of this chapter.

The same glow discharge process is also used for treating certain polymeric materials prior to subsequent processing, e.g. before deposition of a coating onto a polymer. Apparently the action of the discharge is to change the surface structure of the polymer, creating dangling bonds and the like (Hall et al. 1968, Bersin 1974). Unfortunately I don't know much about this application.

As well as glow discharge cleaning, there is also *sputter cleaning* in which the material being cleaned is made the target of a sputtering discharge. The discharge could be rf or dc, with the former required for insulating materials. In this case, the target is exposed to energetic ions and, in the rf case, also to low energy electrons. In contrast to glow discharge cleaning, sputtering is encouraged and the target is cleaned by removal of its surface, contamination and target too!

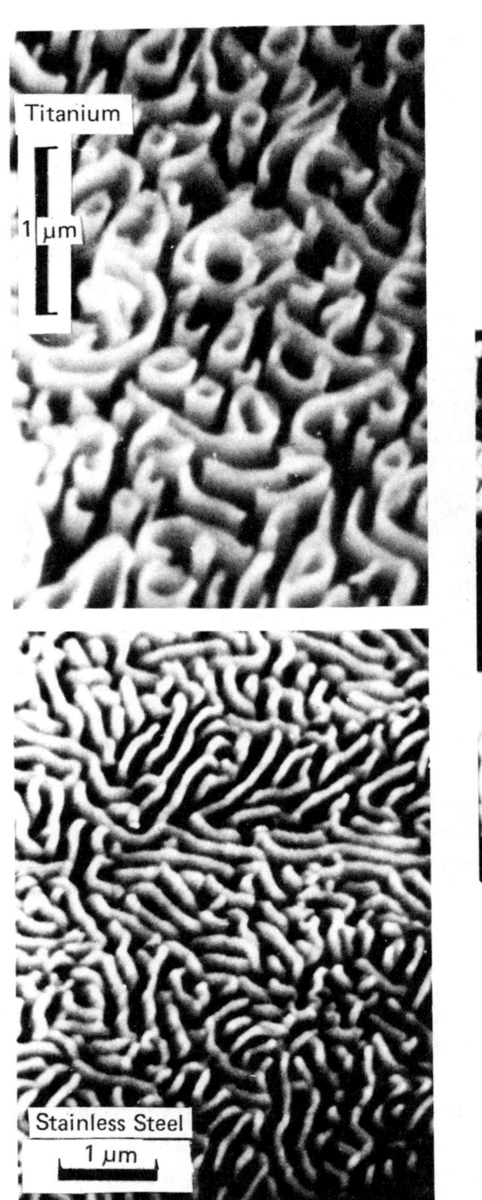

Figure 6-59. Scanning electron micrographs of sputter etched metal surfaces, showing the effect of backscattered tantalum (Oohashi and Yamanaka 1972)

In sputter deposition, sputter etching is commonly used to clean both target and substrate prior to deposition. Target cleaning by this technique is referred to as *presputtering*, which process is also used to heat the system and bring it to steady state, whilst protecting the substrate from deposition with a shutter (described in "Practical Aspects of Sputtering Systems"). The effectiveness of the shutter has been demonstrated by Chang et al. (1973) who showed, using several surface analytical techniques, that rf presputtering of a tungsten target for 10 minutes at about 10 mtorr produced no detectable damage or deposit on single crystal silicon wafers 3 cm away from the W target. A shutter was interposed about 5 mm above the wafers.

Similarly, substrates can be cleaned by sputter cleaning prior to deposition. Although glow discharge cleaning is effective for removing much contamination, it is ineffective in removing the native oxide layers or other compounds from metal or semiconductor surfaces. Removal of these layers is frequently necessary, particularly for electrical applications such as obtaining ohmic contacts. *Sputter cleaning* of substrates is known by just that phrase, or alternatively as *backsputtering*. In the same study already cited, Chang et al. rf backsputtered silicon wafers on a tungsten plate, in argon for 10 minutes at 10 mtorr at a wafer bias voltage of -400 V. In this case, the shutter was about 2.5 cm above the wafers. They found that backsputtering produced an amorphous contaminated layer of Ar, W, C and Si $\geqslant 3$ monolayers, on the wafers. As in the results of Vossen discussed in "Bias Techniques", the W and C were backscattered from the substrate and concentrated near the edge of the wafers.

Clearly, the backsputtering technique could have disastrous consequences if the wrong backing plate is used. In the work of Chang et al., the combination of the use of a tungsten backing plate with the bias sputter deposition of tungsten onto the silicon wafers, was very effective in preventing oxide formation at the wafer-film interface, whilst simultaneously promoting a mixed W-Si interface; tungsten is not an unwelcome contamination in this case.

Finally, the application of a positively biased probe in a discharge will raise the plasma potential, as we already know by now, and so will increase the energy of ion bombardment onto all grounded conducting surfaces in the discharge. This can be a useful cleaning technique, but must be recognized also as a potential source of contamination, since the sputtered material has to end up somewhere, usually on the substrate (Murphy, 1811).

Implications for Bias Sputtering

In the last few sections, we have been considering effects in sputter etching. Although one normally thinks of etching as a subtractive process at a target, we have to remember that the deposition which occurs on a substrate in bias sputtering is in fact the net result of deposition from the target and removal by resputtering due to ion bombardment. The bombardment leads to structural and compositional changes in the growing film; under certain circumstances, it can reduce the net deposition rate to zero or even make it negative – i.e. cause etching of the substrate.

SOME OTHER SPUTTERING CONFIGURATIONS

All of the discussion of sputtering systems so far has been about the cold cathode diode type of system. (Actually bias sputtering systems have three electrodes, but they are not generally referred to as triode systems). These systems need not be planar, and could have wire targets or other cylindrical symmetry as shown in Figure 6-60, which is from the review by Holland (1974). In all of these cases, secondary electrons are generated by ion impact at the target.

Enhancement of Ionization

In our discussion, in "A Conventional DC Sputtering System", about the range of operating pressure of a dc diode system, we found that a limitation was due to lack of electron impact ionization to sustain the discharge. There are two ways of combating this limitation, either by increasing the probability of ionization or by increasing the number of electrons, as follows:

- Using a magnetic field to increase the path length of an electron before it is collected or recombines on an electrode or wall.
- Using the ionization enhancement given by rf excitation.
- Injecting more electrons into the discharge by using a *hot filament* as an electron source.

Various configurations to enhance ionization are shown in Figure 6-61, which is also from Holland (1974).

Of this multitude of sputtering configurations, we have already discussed both dc and rf planar diode systems, with and without bias capabilities. In the following section we shall consider magnetic field effects in diode systems and the various types of magnetron systems. But first, a few words about hot filament triode systems are in order.

Hot Filament Discharges

The essential parts of a hot filament discharge system are the filament itself and the anode. The filament, which is normally tungsten or another refractory metal,

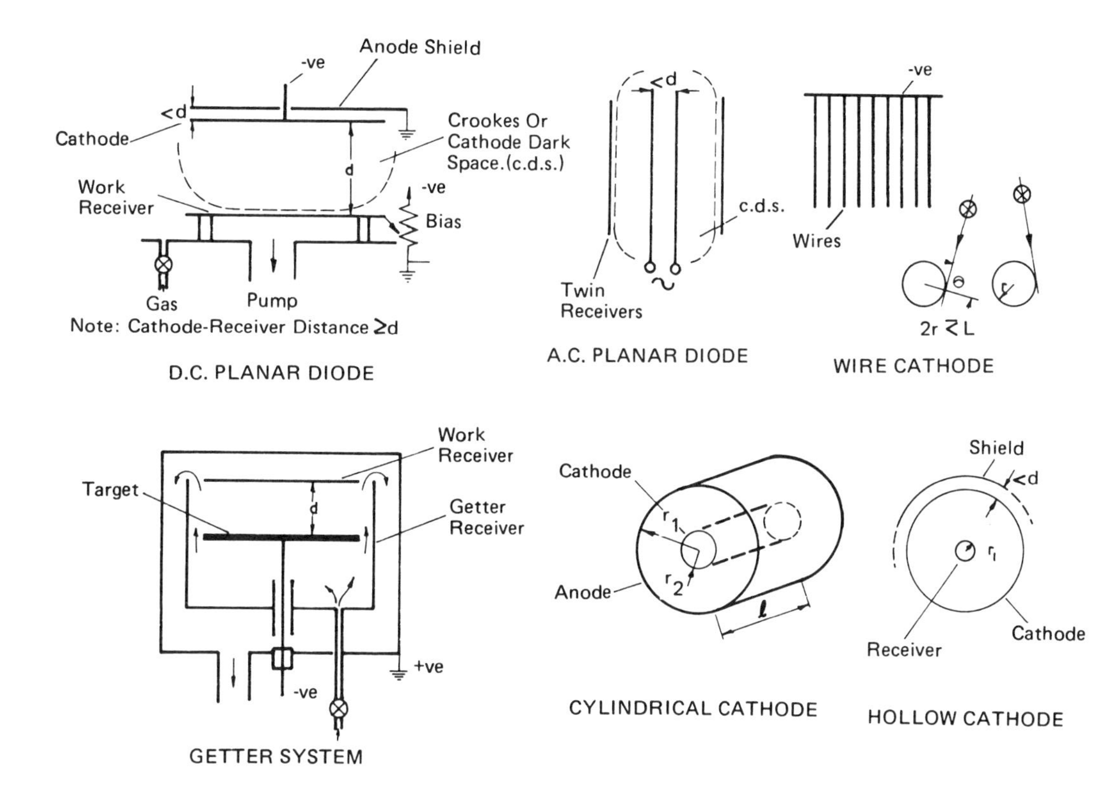

Figure 6-60. Cold cathode glow discharge sputtering systems (Holland 1974)

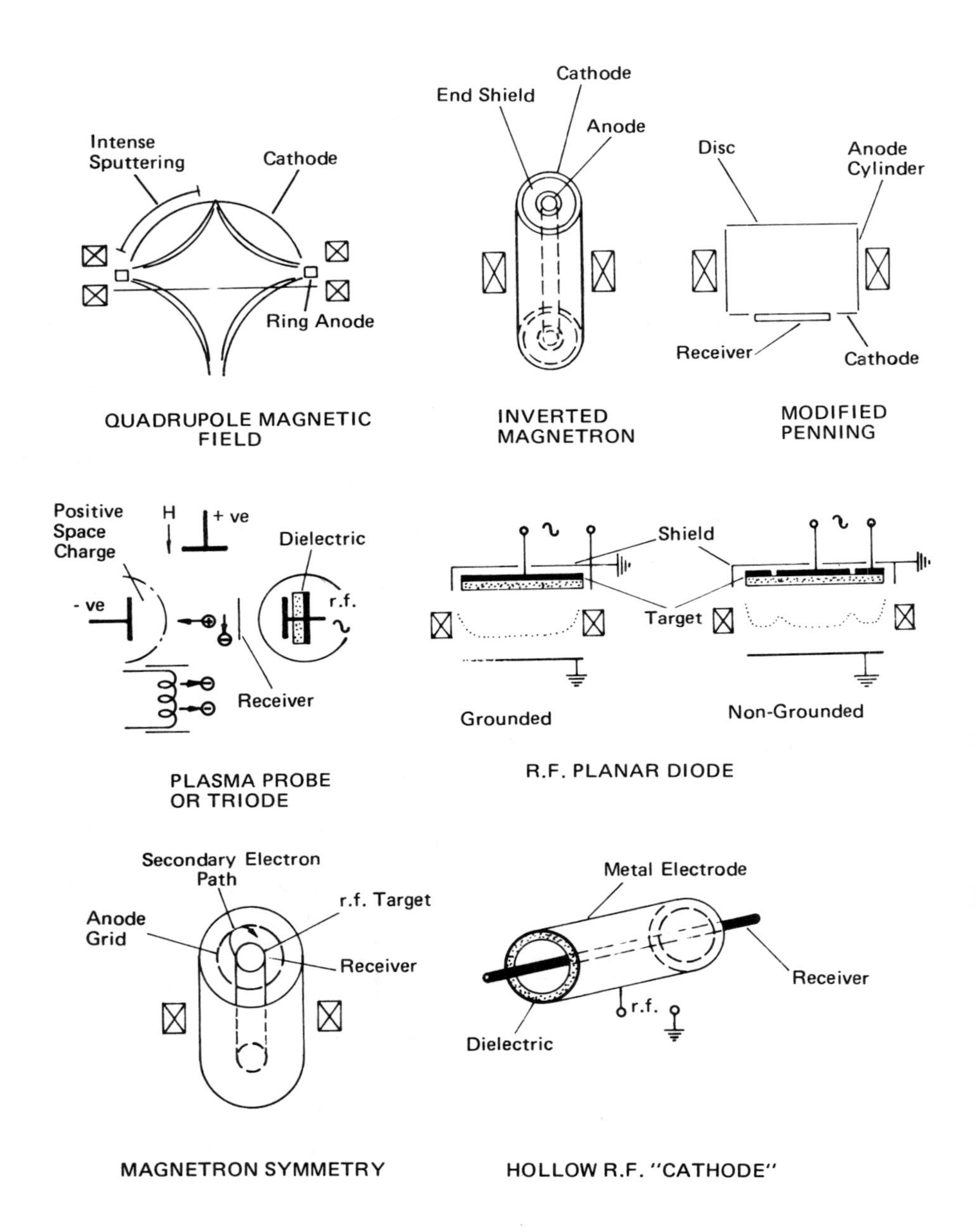

Figure 6-61. High vacuum and plasma probe sputtering systems (Holland 1974)

is heated to incandescence, usually by the resistive heating ($I^2 R$) of a low voltage, high current ac or dc source. At such elevated temperatures, the filament can become a copious source of thermionically emitted electrons. Hence one refers to *hot filament cathodes*, as opposed to *cold cathodes* where the electrons are emitted by secondary emission.

The potential supply of current by thermionic emission is given by the *Richardson-Dushman equation*:

$$j = AT^2 \exp - \frac{e\phi}{kT}$$

where ϕ is the *work function* of the metal and A is a constant equal to 120 amp/ $cm^2 deg^2$.

Usually the emission current from a filament will be severely limited by the electron space charge surrounding it, but with an anode located in the vacuum to withdraw the electrons, and a suitable gas introduced to a pressure of a few millitorr, a glow discharge can easily be generated. This discharge will be qualitatively the same as a cold cathode discharge, with sheaths formed at both cathode (the filament) and anode. What is distinctive about the hot filament discharge is that electron emission from the cathode is primarily by thermonic emission rather than by ion impact. Since the electron supply can be increased almost indefinitely, it is possible to generate low impedance discharges of several amperes with interelectrode voltages of just a few tens of volts. Pioneers in this area were surprised to be able to generate discharges with less than the ionization potential applied across the discharge, but they didn't know about non-uniform electric fields and two-step ionization processes. Hot filament discharges will work down to rather low pressures ($\sim 10^{-5}$ torr), particularly with the aid of a magnetic field, as is discussed in the following section.

The hot filament discharge is converted into a sputtering system by adding a target with an adequate dc or rf potential applied to it. The discharge is then the source of ions, and the potential on the target controls the ion energy and flux onto the target. This arrangement, shown in Figure 6-61d, is commonly known as a *triode sputtering system*. A substrate can be placed in a suitable position when thin film deposition is required.

Triode systems were very popular in earlier days of sputtering, particularly in the 50's and early 60's. The capability of quasi-independent control of discharge and sputtering process made this an attractive configuration, particularly to researchers, as did its ability to operate down to low pressures. These capabilities resemble those of ion beam sputtering. Triodes seemed to lose popularity in the mid 60's only to resurface around 1972 for several applications, partly because of their ability to sputter efficiently at low voltage and so avoid radiation damage, whilst maintaining adequate current density and deposition rate. An extensive analysis of a triode system has been made by Tisone and colleagues (see Tisone 1975). More recently, the hot filament triode system has been used as a high rate sputtering source, and for plasma etching by Chapman and Minkiewicz (1979) and Heiman et al. (1979).

MAGNETICALLY ENHANCED SPUTTERING SYSTEMS

Magnetic field effects are used quite a lot in sputtering systems. They give advantages in sputtering rate and extendability of operating range, and can reduce electron bombardment at the substrate. The problems of uniformity which sometimes arise, can be overcome in many cases.

The primary interaction between a particle of charge q and velocity v, and a magnetic field B, is to produce a force F on the particle of magnitude F = Bqv. The direction of the force is perpendicular to both the magnetic field and the velocity (Figure 6-62) and is better expressed in the vector form:

$$\mathbf{F} = q\ \mathbf{v} \times \mathbf{B}$$

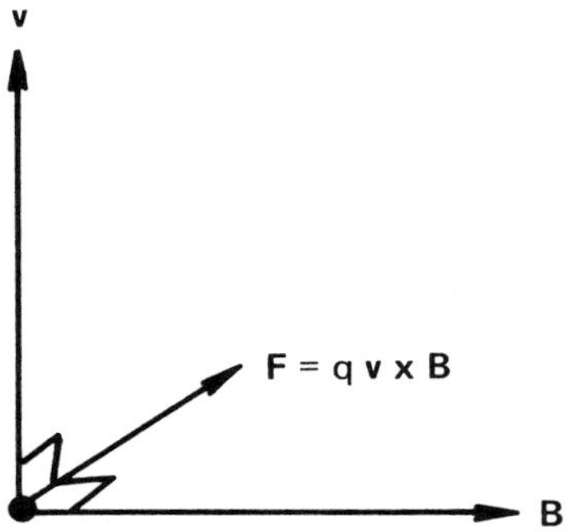

Figure 6-62. The electromagnetic force **F** is perpendicular to both **v** and **B**

This force will produce an acceleration that is inversely proportional to the mass of the charge. For the magnetic fields used in sputtering, which are typically 100 gauss, only the electrons will be affected; the ions are too massive.

The purpose of using a magnetic field in a sputtering system is to make more efficient use of the electrons, and cause them to produce more ionization. In a conventional glow discharge, electrons are soon lost by recombination at the walls. To minimize this loss, there are two ways of using a magnetic field which are rather different in approach:

Axial Magnetic Fields

Axial magnetic fields are used in planar diode glow discharges and hot filament discharges for the purposes of increasing the path length of the electrons before they are collected by the anode, and of keeping electrons away from the vacuum chamber walls and hence reducing recombination. We shall illustrate the effect on the hot filament system introduced in the last section (Figure 6-63a). An electron travelling along the axis of the discharge will be unaffected, since **B** and **v** will be parallel and so their vector product will be zero. But suppose that the electron is travelling at θ to the magnetic field (Figure 6-63b). Then it will be subjected to a force Bev sin θ perpendicular to the field. Provided that it doesn't make any collisions, the electron will then describe circular motion around **B** at a radius given by:

$$\frac{m_e(v\sin\theta)^2}{r} = Bev\sin\theta$$

or

$$\frac{m_e v\sin\theta}{Be} = r$$

Coupled with the velocity $v\cos\theta$ parallel to **B**, the general motion will be a helix (Figure 6-63c).

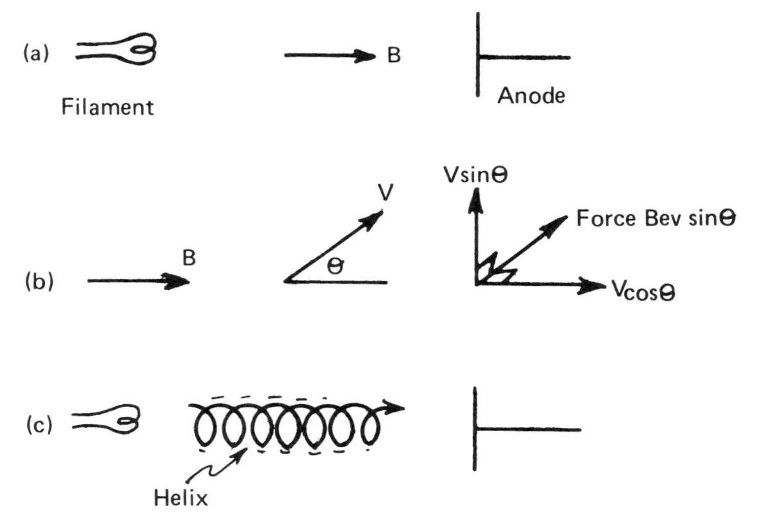

Figure 6-63. Effect of an axial magnetic field

(a) hot filament discharge with axial magnetic field B

(b) force on a particle with velocity v at θ to the axis

(c) net motion of particle is helical.

After each revolution of the helix, the electron will return to the same radial position around the axis of the discharge. An effect of the magnetic field is therefore to reduce the net velocity of the electron towards the wall to zero, hence reducing wall recombination losses. A second effect of the helical path is to increase the total path travelled by an electron emitted from the filament on its way to the anode, thus enabling it to cause more ionization and excitation. These effects are very evident when the magnetic field is applied; the discharge is observed to be confined to a core which narrows with increasing magnetic field and brightens, indicating an increased electron density in the core.

A more complete analysis of this *magnetically confined* or *magnetically immersed* glow discharge would consider the spreading of the glow due to diffusion, balanced against the magnetic confinement. Similar calculations for electron beams are well developed (Hemenway et al. 1967). In the beam case, one has to consider the space charge expansion of the beam, but presumably these models could be developed to deal with the discharge situation.

Using magnetic field confinement in this way, hot filament triode systems, which normally can operate at a few millitorr by way of their effective thermionic electron sources, can extend their operating range down to about 10^{-5} torr. [The magnetic field can also have a considerable effect on the distribution of etching species in a hot filament triode plasma etching system, as shown by Chen et al. (1979).]

Axial magnetic fields are also used in the more conventional cold cathode diode sputtering systems described earlier. Again the effect can be used to lower the operating pressure or, probably more frequently, to increase the ion current and hence sputtering rate, at a constant pressure. Unlike the hot filament system, the substrate is normally on the anode in the cold cathode system. An effect of the magnetic field will be to constrict the discharge and perhaps cause non-uniform deposition on the substrates. One must be aware of these potential deleterious effects. Nevertheless, axial magnetic fields prove to be useful in many cases.

Magnetrons

Magnetron systems take the same philosophy one step further and attempt to trap electrons near the target so as to increase their ionizing effect. This is achieved with electric and magnetic fields that are generally perpendicular.

Consider the electron normally emitted with velocity v from a surface (Figure 6-64a) into a region of magnetic field B (into the paper in Figure 6-64a) and zero electric field. In a similar way to the axial magnetic field case, the electron will describe a semicircle of radius r given by $m_e v/Be$, providing it does not collide en route, and will return to the surface with velocity v. So the effect

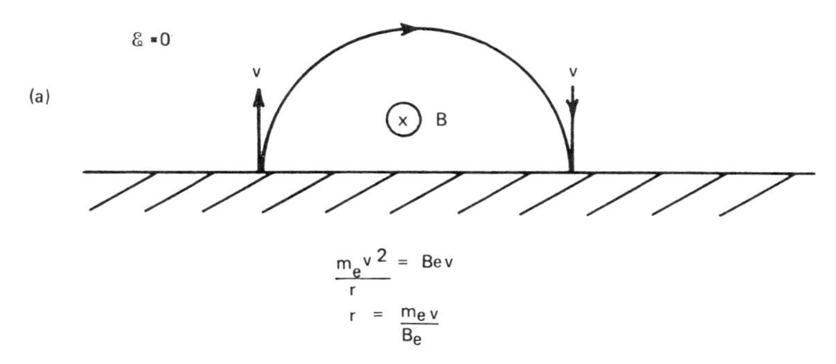

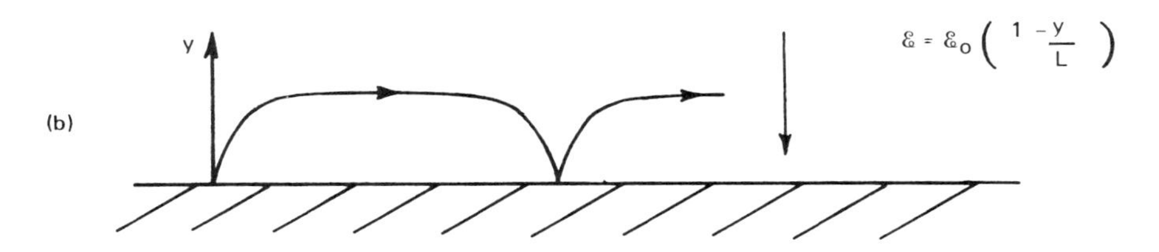

Figure 6-64. Motion of an electron ejected from a surface with velocity v into a region of magnetic field B parallel to the surface:
(a) with no electric field
(b) with a linearly decreasing field.

of the magnetic field is to trap the electron near the surface from which it was emitted.

To examine a situation which is closer to the magnetic sputtering application, Figure 6-64b is of a sputtering target where there is a strong electric field $\mathcal{E}$ in the dark space above its surface, and again a magnetic field B parallel to the surface of the target. Let the electric field $\mathcal{E}$ decrease linearly across the dark space of thickness L; as we have seen earlier, this is a dependence frequently used for dark space fields, and although somewhat questionable, is quite adequate for the purpose of illustration. If y is the dimension away from the target, and the target surface is y = 0, then

$$\mathcal{E} = \mathcal{E}_0(1 - \frac{y}{L})$$

where $\mathcal{E}_0$ is the field at the target. Let x be the distance along the target surface from the point of emission of a secondary electron. This electron will be rapidly accelerated, initially away from the target by the strong normal electric field at the surface. By comparison, we can assume that the electron emission velocity is zero. The equation of motion will then be:

$$\ddot{x} = \frac{Be\dot{y}}{m_e}$$

$$\therefore \dot{x} = \frac{Bey}{m_e}$$

$$\ddot{y} = \frac{e\mathcal{E} - Be\dot{x}}{m_e}$$

After substituting in the y dependences of $\mathcal{E}$ and $\dot{x}$, the resulting differential equation can be solved to yield

$$y = \frac{e\mathcal{E}_o}{m_e\omega^2}(1 - \cos \omega t)$$

where $$\omega^2 = \frac{e\mathcal{E}_o}{m_e L} + \frac{e^2 B^2}{m_e^2}$$

In the absence of the electric field, ω would be equal to eB/m_e. This is known as the *cyclotron frequency* and has the value 2.8 10^6 B Hz, where B is in gauss. In the absence of the emitting surface in this example, and in the previous example of an axial magnetic field, the electron would rotate around the field lines with the cyclotron frequency.

The addition of the electric field on the target changes the orbits from circular to cycloidal, provided the electron stays within the dark space. Since $r = m_e v/Be$, its instantaneous radius of curvature will decrease as it travels further from the target surface. If it enters the negative glow by straying further than L from the target, then it will describe circular motion in the electric field-free region there, before returning to the dark space.

The maximum excursion y_{max} of the electron from the target, in the absence of collisions, can be found by equating the gain in kinetic energy to the loss of potential energy (Green and Chapman 1976):

$$\tfrac{1}{2} m\dot{x}_{max}^2 = e(V - V_T)$$

where V_T is the negative target voltage and V is the potential at y_{max}. But since $\ddot{x} = Be\dot{y}/m_e$ and $\dot{x} = Bey/m_e$, then we can substitute into the energy balance equation to obtain:

$$y_{max} = \frac{1}{B}\left[\frac{2m}{e}(V - V_T)\right]^{1/2}$$

Note that the form of the electric field has not been used in this derivation, so the result holds quite generally, both within and without the dark space.

The net result of all of this is that the electron is trapped near the target, provided it doesn't make any collisions, so the loss process of fast electrons going to the anode and walls is eliminated. If it does make collisions, the trapping won't be so effective, but on the other hand we want the electrons to make lots of ionizing collisions to sustain the glow, and the magnetron action enables them to do this before being lost to the anode.

The application of magnetic fields to enhance ionization in glow discharges seems to date back to Penning (1936), but their effective use in sputtering systems has been more recent. *Magnetrons* are originally valve (vacuum tube) devices for generating or amplifying high frequency signals, operating on the basic principles just discussed, but the name is now also applied to that group of sputtering devices using the same principle of crossed electric and magnetic fields.

Cylindrical Magnetrons

There are several types of magnetrons for practical sputtering applications. The earliest were probably those with cylindrical geometry and an axial magnetic field (Figure 6-65). With the inner cylinder as the target, the arrangement is known as a *cylindrical magnetron* or *post magnetron* (Thornton 1973, Wasa and Hayakawa 1969); this configuration has the ability to coat a large area of small substrates.

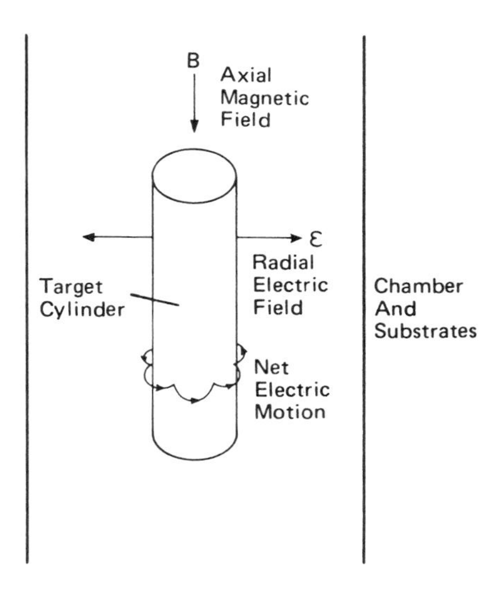

Figure 6-65. Cylindrical magnetron sputtering configuration (Green and Chapman 1976)

Figure 6-66. The extended glow shows that electrons are lost from the ends of the target (Green and Chapman 1975)

When the inside of the chamber becomes the target, the arrangement is known as an *inverted magnetron* (Gill and Kay 1965) and has a capability of depositing uniform thickness films over strangely shaped substrates placed along the axis.

Although electrons are constrained in their radial motion, there is little to restrict their axial motion and they can therefore 'walk' off the end of the target (Figure 6-66). These *end losses* can be prevented by various types of end containment (Thornton and Penfold 1978), to increase the effectiveness of the discharge.

Circular Magnetrons

The magnetron action can be achieved with other geometries, too. Figure 6-67 is a schematic of the *circular magnetron* due to Clarke (1971).

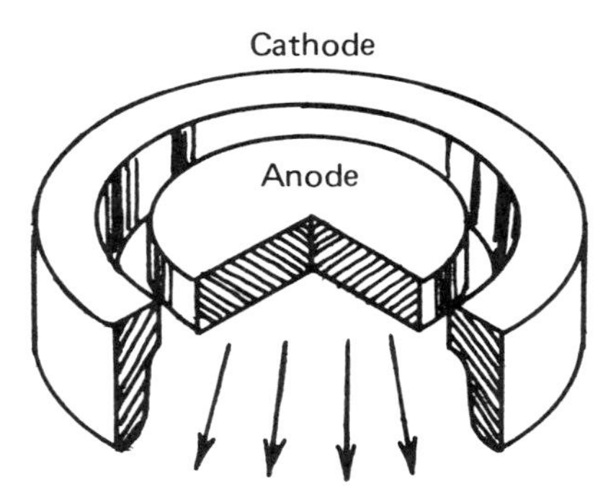

Figure 6-67. Circular magnetron sputtering configuration. Various cathode cross-sections are employed to minimize erosion problems

The operating principle is the same as that of the cylindrical magnetron, but results differ in that, for example, the sputter ejected material has a pronounced forward direction (perpendicular to the anode); substrates are therefore usually placed on a rotating carousel to ensure good uniformity of film thickness. As with the other magnetrons, substrates are protected from the fast electrons which were very evident in the diode configurations, and this can be a big advantage when dealing with bombardment-sensitive substrates, such as some semiconductor devices. A disadvantage of circular magnetrons is that the localized erosion of the target causes deposition rates to change over a period of time, but this can be overcome to some extent by careful choice of the cathode shape. Magnetrons of this type have been reviewed by Fraser (1978).

Planar Magnetrons

The third type of practical sputtering magnetron is the planar magnetron (Figure 6-68), which was first described in the literature by Chapin (1974). As with the circular magnetron, a 'looping' magnetic field is used, and this restricts the sputter erosion of the target to a 'racetrack' area. The geometry of the rectangular version of the planar magnetron makes it suitable as a line source with substrate motion again required for good uniformity. A circular configuration could also be used. The magnetic field causes very effective concentration of the glow (Figure 6-69) and the associated localized erosion of the target is again somewhat of a problem. Planar magnetrons have been reviewed by Waits (1978).

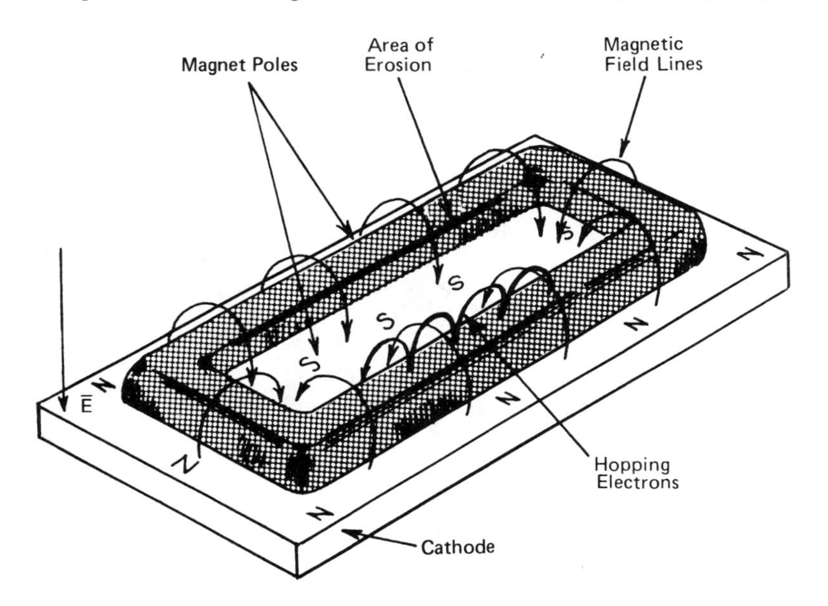

Figure 6-68. Planar magnetron sputtering configuration

General Comments

Magnetrons have proved to be very useful additions to the family of sputtering devices. They offer high deposition rates, with greater than one micron per minute being available if required. They also virtually eliminate bombardment of the substrate by fast charged particles; it is claimed that the heat input is reduced to the heat of condensation of the film (although it is not clear to me

Figure 6-69. The glow in a planar magnetron is magnetically confined (courtesy of Airco Temescal)

how fast neutral argon bombardment – from ions neutralized and reflected at the target – is eliminated).

The disadvantages of magnetrons are of localized erosion of the target (although not in the cylindrical magnetrons), which causes deposition rates to vary with time and requires frequent replacement of the target, and of arcing. Magnetrons seem to be more susceptible to arcing than other types of sputtering system, and this is probably exacerbated by the high current, low voltage power supplies which are required to drive the low impedance discharges. Current densities may be up to 100 mA/cm^2, with 10 or 20 mA/cm^2 at a few hundred volts being quite typical. A 'unipolar arc', as described previously, can be established – for example by a small oxide patch of the cathode surface – and the arc can cause virtually the entire discharge current to be concentrated into the arc spot on the cathode. The huge power dissipation at the cathode causes the target to locally melt. Arcing of this type, although common to all magnetrons, is very clearly shown in the planar magnetron: *race track arcs* appear to be small particles that hop around the race track area of erosion (Chapin 1974) and are possibly evaporated clusters of target material which become charged by thermionic emission and then are driven by the qvB force. Arcing is particularly prevalent in reactive magnetron sputtering, presumably because of insulator formation on the target.

The sputtering of magnetic materials in a magnetron poses interesting problems because they tend to distort or even eliminate the field lines causing magnetron action. There are remedies for this problem, most of them proprietary.

Magnetron systems are usually dc powered, although rf versions are in use. The power dissipated at the target by the large ion currents involved causes considerable problems with target cooling, and these problems are considerably greater with insulators – which usually hinder heat flow as well as electrical flow.

For a general review of magnetrons and magnetic field effects in sputtering, readers are referred to the review by Thornton and Penfold (1978) and, for applications, to the papers presented at the magnetron sessions of the Annual Symposium of the American Vacuum Society – see, for example, Journal of Vacuum Science and Technology, Vol. 14, No. 1 (1977); Vol. 15, No. 2 (1978); and Vol. 16, No. 2 (1979).

ANALYTICAL AND MONITORING TECHNIQUES

There are now a whole range of techniques available for analyzing glow discharges, for process control, and for monitoring deposition rates.

Probe measurements for measuring plasma parameters have already been mentioned, and they are just one of a set of electrical techniques which also includes double probes and propagation of microwave radiation through the glow (Thornton 1978).

Optical techniques are appealing as analytical tools because they do not interfere with the process at all, at least not when used in an emission mode. Spectroscopy of sputtering discharges has been developed by Greene and Sequeda-Osorio (1973) to give spatial resolution within the discharge. Not only the discharge gas but also elements from the targets can be detected in this way (Figure 6-70), which therefore offers a means of process control. Optical techniques have also been used to monitor emission from the growing film, by using a glass rod to transmit the emission to the monochromater (Ratinen 1973). Optical techniques have been reviewed by Greene (1978).

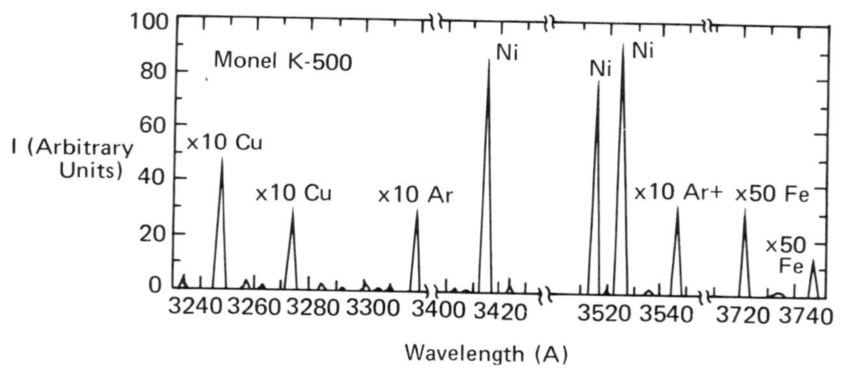

Figure 6-70. A portion of the emission spectrum from the discharge when sputtering Monel K-500 (a nickel-based heat resistant alloy) in argon (Greene et al. 1975)

Mass spectrometry is also a valuable tool. The mass/energy analyzer of Coburn and colleagues has already been mentioned; more generally, mass spectrometry can be used to analyze the sputtering atmosphere and to control, for example, the partial pressure of active components in a reactive sputtering system.

Film deposition rates can be measured in situ by monitoring the change of frequency of a quartz oscillator as the mass of the film changes its characteristic frequency, with a resolution ~ 1Å attainable. Other in situ techniques include monitoring the intensity of a light beam reflected from a growing dielectric film; the reflected beam displays maxima and minima as its optical thickness increases (see Glang 1970). The infra red radiation from the film can be used to monitor its temperature.

ION BEAM SYSTEMS

In the beginning of this chapter, we decided that the ions used for sputtering purposes could be produced by an in situ glow discharge. Most of the discussion in this chapter has been about such systems. We have seen that the environmental conditions necessary to sustain the glow discharge can have a significant effect on the quality of the deposited films.

Ions can also be generated by an external *ion beam source*. This means that the substrate can be located in a virtually field-free high vacuum environment, and this has several implications for the growth of the film. A typical arrangement for ion beam sputtering is shown in Figure 6-71.

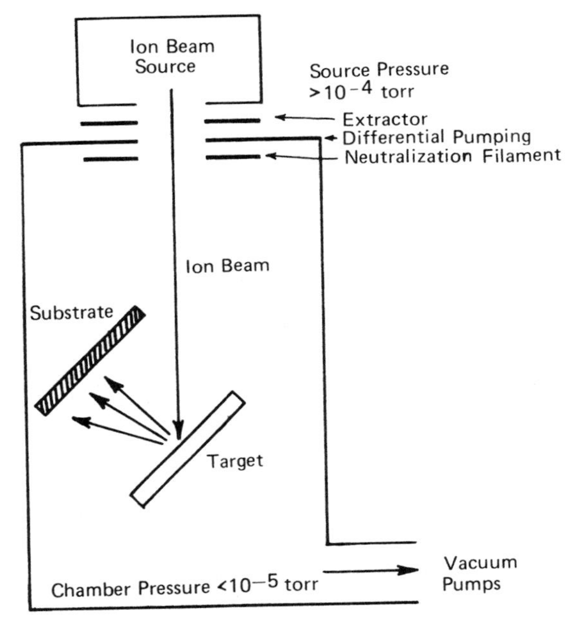

Figure 6-71. Typical configuration for ion beam sputter deposition

Ion Beam Sources

There are various ways of producing ions, but to achieve the current densities required for sputtering, one uses electron impact ionization in hot or cold cathode discharges. There are many source designs (Kaufman 1978), reflecting the wide interest in ion beam technology, of which sputtering is just one part.

The ion beam sources are operated at pressures above 10^{-4} torr. Some of the ions generated are accelerated and extracted by a series of grids. As we have already seen in Chapter 4, "Space Charge Limited Current", the current density j that can be extracted by a voltage V between two planes distance d apart, is given by the Child-Langmuir equation. For collisionless sheaths, this predicts $j \propto V^{3/2}/d^2$. With reasonable values for the separation d between two electrodes, current densities ~ 1 mA/cm^2 at 1000 eV are obtained, and these are comparable to more conventional sputtering systems.

The extraction grids also serve as limiting pumping conductances so that a differential pressure can be set up and the process chamber can be operated at better vacuum, $< 10^{-5}$ torr, where molecular flow conditions obtain.

After extraction, the positively charged ion beam would expand due to Coulomb repulsion, and make the beam non-parallel. This is prevented by using a *neutralization filament* close to the source, consisting of a heated filament that thermionically emits electrons and neutralizes the beam. This also makes it possible to bombard insulators without them charging up.

Ion Beam Sputtering

For the purposes of sputter etching or sputter deposition, a noble gas ion beam is extracted from the ion source and used to bombard a target (Figure 6-71). The target can be used as a sputter deposition source to coat a substrate; this process is variously known as *ion beam sputter deposition, secondary ion beam deposition* (although the depositing material is un-ionized), and ambiguously as *ion beam deposition*. Alternatively the sputter etching of the target can be the main purpose, in which case the process is known as *ion beam etching*, and has been discussed earlier in the chapter.

The various differences between ion beam and glow discharge sputtering have been highlighted by Harper (1978), and can be summarized as follows:

- *For etching*, the control of the process is improved by: the parallel ion beam, the ability to control incidence angle, the ability to control beam current and energy independent of target processes, and the ability to sputter insulators without them charging up. In addition, the low process chamber pressure minimizes redeposition by gas phase scattering, and the monoenergetic ion beam permits study of ion impact processes, sputtering yields, etc.
- *For deposition*, the fact that the substrates are much more isolated from the glow discharge generation process minimizes unwanted heating of the substrate and minimizes fast electron bombardment. Because the substrates are not part of the electrical circuit, it is much easier to incorporate substrate heating, cooling, and process controls. The low operating pressure in the process chamber minimizes gas contamination of the growing film, particu-

larly since the vacuum pumps can operate with maximum conductance to the chamber and so minimize the partial pressure of contaminants. The low operating pressure also reduces energy-attenuating collisions of the sputtered particles en route from target to substrate. Weissmantel et al. (1972) have taken advantage of this environment to make in situ observations of sputter deposited film growth in the electron microscope.

I have not mentioned before (and perhaps should have) that sputtered particles are ejected with higher energies than are evaporated particles. The average energy from an evaporation filament at temperature T is of the order of kT and so would amount to 0.25 eV at 3000 K. For bombardment energies of 1 keV or so, the most likely sputter ejection energy is from 1 – 15 eV (Figure 6-72), with

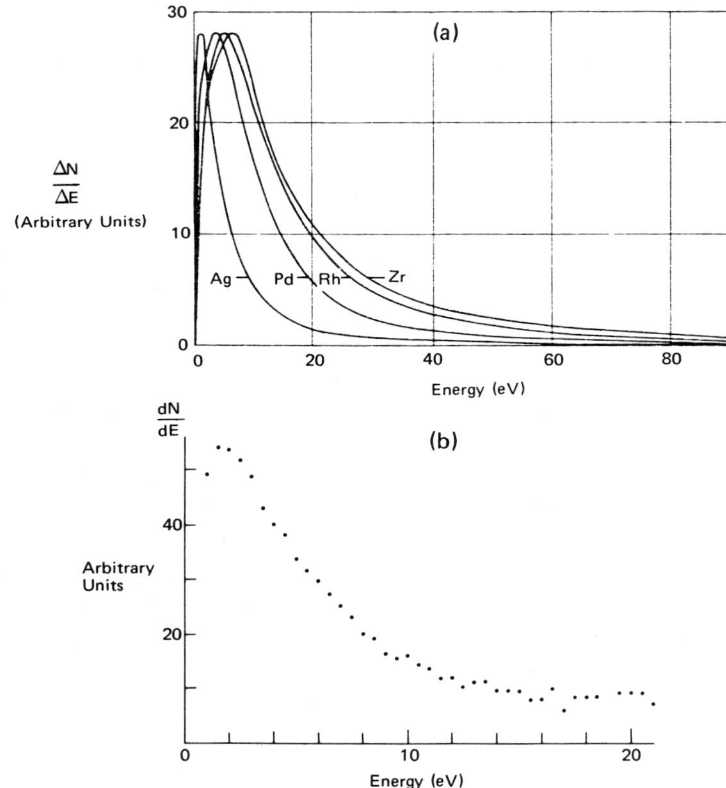

Figure 6-72. (a) Energy distributions for Ag, Pd, Rh and Zr under 1200 eV Kr^+ bombardment (Wehner and Anderson 1970)

(b) Energy spectrum of neutral copper sputtered at 45° incidence and 45° ejection, by 3 keV Ar^+ (Lundquist 1978)

monotonically decreasing numbers of atoms having higher energies. In glow discharge sputtering systems, except at the lowest operating pressures, almost all of the ejection energy of the sputtered atoms will have been dissipated by collisions before the atoms reach the substrate. This will not be so with the pressures extant in an ion beam system. It has been suggested that the higher arrival energy of the atoms will lead to greater adatom surface mobility during the nucleation and growth stages of film formation. Single crystal films deposited in this mode have been observed to have lower epitaxial temperatures (Chapman and Campbell 1969).

Ion beam systems were used in the 60's to gain information about the basic processes in sputtering, and yielded information about sputtering yields, preferential sputter ejection from specific crystal axes of single crystal targets (which helped clarify sputtering mechanisms), sputter ejection energies, etc. Subsequently their use for sputter deposition waned, and work with such systems concentrated on high energy ion bombardment effects, of interest for ion implantation and radiation damage processes. Ion beam etching continued, but it was considered that ion beam sputter deposition was rather impractical in terms of capital cost and limited throughput. More recently, with the availability of large diameter ion beams (Laznovsky 1975) and a need for some specialized sputter deposition applications, there seems to be a renewed enthusiasm for ion beam sputtering processes.

Ion Beam Deposition

The *sputter deposition* process involves the condensation of low energy atoms onto a surface. *Sputter etching* uses high energy atoms or ions; although these may be incorporated in the bombarded surface, the net process is of etching. *Ion beam deposition* (also known as *primary ion beam deposition*) is the formation of a thin film by the direct deposition of ions, with the ion energies chosen to be low enough to ensure net deposition. Whereas the ion beam sputtering processes described above use a noble gas ion beam, ion beam deposition systems use a beam of the material to be deposited, and this is usually a metal; this requires special consideration of the ion source. The practical arrangement would be similar to Figure 6-71, with the target becoming the substrate.

Aisenberg and Chabot (1971) have used this technique to deposit highly adherent diamondlike carbon films, and have subsequently discussed the physics of the technique (Aisenberg and Chabot 1973). Their technique produced a beam of carbon and argon ions and neutrals. It is more representative of the philosophy of the process to use a charged particle beam, which can then be deflected and focused. The beam is also *mass analyzed* to produce a very pure, monoenergetic source of depositing ions. The beam energy chosen is determined by the resulting film quality balanced against self-resputtering as the ion energy increases.

The difficulty of this process is in producing an adequate supply of ions. It is more difficult to produce ions of a solid to start with, and then their flux is limited by space charge limited extraction from the source (low voltage V leads to low current density j, for given d) and when adequate currents can be obtained, there will be problems of beam expansion due to space charge repulsion. It is usually found expedient to extract the ions at a high voltage and then to decelerate the beam. Amano and Lawson (1977, 1978) have used the technique for the controlled deposition of Pb^+ and Mg^+, with deposition rates of 1000 – 3000 Å being deposited in 7 hours from a 10 μA beam at 24 eV.

Although an extremely slow process, ion beam deposition has a potential for very controlled deposition of very pure films, and might provide an alternative to molecular beam systems.

ION PLATING

Ion plating is a vacuum deposition technique that uses a glow discharge to modify the composition of the deposited film. The technique was introduced by Don Mattox in the mid-60's and has been more recently reviewed by him (Mattox 1973a) and by Carpenter (1974). The technique combines evaporation and sputtering. The substrate is made the cathode of a high voltage glow discharge, and this is used to initially sputter clean the surface of the substrate. Material is then evaporated onto the substrate from a resistively heated filament whilst still maintaining the discharge (Figure 6-73). As with bias sputter deposition and ion beam deposition, there is simultaneous deposition and resputtering,

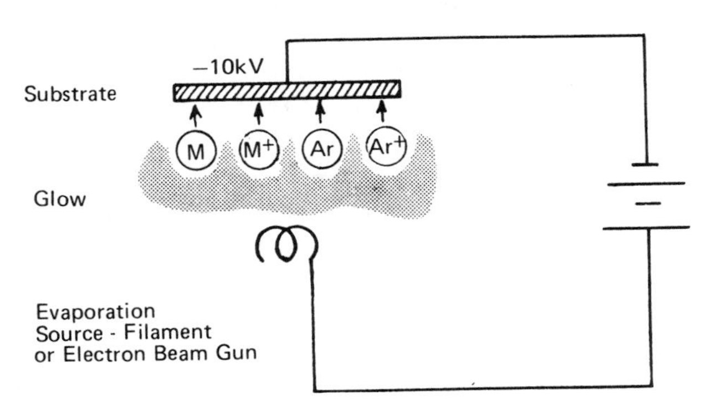

Figure 6-73. Ion plating configuration

with the balance such as to ensure net deposition. Unlike those processes, deposition rates in ion plating are high enough that substantial resputtering can be tolerated. The technique encompasses bias evaporation which was discussed earlier, but is usually differentiated in terms of the accelerating voltages used at the substrate, being very much higher in ion plating than in bias evaporation.

When material from the filament is evaporated into the discharge, some proportion of it will be ionized by the election impact and (possibly) Penning processes that were described in Chapter 2. There are various reports of the degree of ionization attained, usually around a few per cent (Aisenberg and Chabot 1973). But we know that for a noble gas discharge, the degree of ionization is only about 10^{-4}, and apparently ion plating discharges can become self-sustaining without the noble gas, after the process is well underway (Krutenat and Gesick 1970). This can only occur if the evaporation rate is enough to maintain an adequate partial pressure of the evaporant. Under these circumstances, one has the opportunity of observing the emission from a gaseous metal discharge. This phenomenon can also be observed in high rate magnetron sputtering systems, and is very apparent in the colour original of Figure 6-66.

Remembering that sputtering at the substrate can reduce or even eliminate film deposition, then with several thousand volts applied to the substrate at pressures of a few tens of millitorr, film deposition will occur only if the evaporation rate is high. Ion plating is characteristically a high rate deposition process. High voltages are used in order to promote forward sputtering and implantation of the film into the substrate, and the elimination of this interface promotes excellent adhesion, as discussed in the later section on "Thin Film Adhesion". Figure 6-74 shows a depth profile analysis of an ion plated film, and one can see that the interface is spread over a considerable distance.

A secondary effect of the ionization is that the ionized material will follow electric field lines. Combined with the considerable gas phase scattering which occurs at the pressures used, and the high mobility of the deposited material due to the high substrate temperatures produced by the ion bombardment, conformal coverage of three dimensional objects can be achieved.

There is nothing unique about the use of an evaporation source, so that any source of material could be used, as pointed out by Mattox (1973a).

I suspect that ion plating has had a bigger effect on the use of vacuum deposition technology than might at first be clear. With the capability of achieving excellent film durability and high deposition rates up to several mils (1 mil = 25μm) per minute, vacuum deposition is now used extensively in mechanical engineering applications, such as to resurface turbine engine blades and coat rotary engine surfaces (White 1973). High rate sputtering and other evaporation techniques also are being used for similar applications, which are far from the world of semi-conductor fabrication and illustrate the current wide utilization of glow discharge processes.

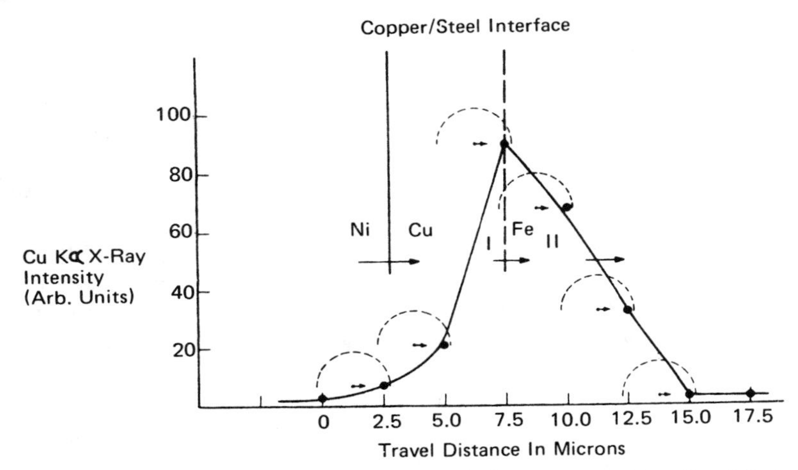

Figure 6-74. Depth profile of ion plated Cu on steel. The decreasing intensity of Cu (measured by electron microprobe) in region II corresponds to penetration into the substrate to a depth of 7-8μm. The nickel layer was used for specimen support (Swaroop and Adler 1973)

ACTIVATED REACTIVE EVAPORATION

Activated reactive evaporation is another glow discharge process used primarily for mechanical engineering applications. The technique has been developed by Rointan Bunshah and colleagues; see Bunshah (1974).

Activated reactive evaporation follows the same principle used in reactive sputtering: a glow discharge is used to dissociate a gas into reactive components (see Chapter 2) which then combine with the growing film. Films of Y_2O_3, TiN, TiC, VC, ZrC, HfC, NbC and TaC (and doubtless many others more recently) have been deposited at high rates (3-12μm/minute) by evaporating the metal in a partial pressure of a few times 10^{-4} torr of O_2, N_2, or C_2H_2. Coupled with high substrate temperatures, extremely durable coatings can be produced.

The configuration for activated reactive evaporation is shown in Figure 6-75. With the application of a field between the source and substrate, an activated reactive ion plating process could also be produced. The structure of thick coatings has been analyzed by Movchan and Demchishin (1969), who classify the structure of coatings into three zones according to the substrate temperature T and the melting point T_m of the coating. The first two zones are divided by $T/T_m = 0.3$, with the transition to the third zone taking place at higher temperature. Many desirable film qualities are achieved at higher temperatures.

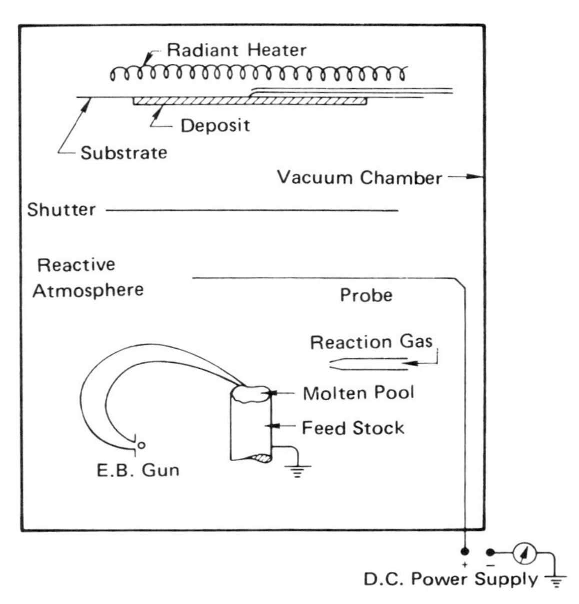

Figure 6-75. Schematic of the experimental arrangement for the activated reactive evaporation process (Bunshah 1974)

THIN FILM ADHESION

The topic of *thin film adhesion* has cropped up several times in the course of this chapter, as it inevitably does in any discussion of thin film deposition. Adhesion is one of the 'unloved' topics in thin film technology. The interaction between the film and the substrate is in itself usually of little interest to those fabricating thin film structures, but unfortunately the quality of this interaction is a vital factor in determining the durability of the device. The thin film is very fragile, and it depends on the substrate for strength. The extent to which it can do this depends largely on the bonding between the two. Inevitably, it turns out that some of the more interesting combinations of materials are incompatible in terms of adhesion. An understanding of thin film adhesion and how it can be controlled is therefore of considerable importance. But even after half a century or so, our knowledge of this topic is extremely limited and largely empirical. A major reason for this is the difficulty of measuring adhesion. Another reason is that confusion often arises over the usages of the words 'adhesion' and related rather undefined terms. In this section, *adhesion* will be used as a generic term describing how a film and substrate remain in contact, and *bonding* will refer to the specific interaction across the interface. As discussed in a review of thin film adhesion (Chapman 1974), from a practical point of view one should really be concerned with the durability of the film-substrate combination, and this is a function of many more things than just the bonding.

There are various types of adhesion:

- The simplest of these is *interfacial adhesion*, in which the two distinct materials, film and substrate, meet at a well-defined interface.

- A second type is *interdiffusion adhesion*, which results from a solid state interdiffusion between the two materials or from solubility of one or both materials in the other. In this case, a discrete interface is replaced by a gradual and continuous change from one material to the other. The 'interdiffusion' can also be promoted artificially.

- A third type is *intermediate layer adhesion*. The film and substrate are bonded together via one or more layers of compounds of the materials with each other and/or with the environmental gases; oxides are particularly common. Again there is no single well-defined interface.

- All these different types of adhesion are further complicated in practice because substrate surfaces are never flat and some degree of *mechanical interlocking* takes place.

Methods of Influencing Adhesion

Although there are some film-substrate combinations which exhibit good adhesion, it usually turns out that a particular film-substrate system to satisfy certain functional requirements is poorly adherent! Sometimes the adhesion can be improved simply – for example, by cleaning the substrate surface so that the film and substrate really do contact: since vacuum deposition is an atomistic or molecular process, this is a real possibility, unlike the contact of two bulk materials. Solvent cleaning usually removes only oils and greases, leaving more tenacious materials such as surface oxides which may prevent interdiffusion. One can remove such deposits by chemical or sputter etching, but the problem then becomes how to deposit the film before the substrate is recontaminated, bearing in mind that a gas monolayer can form in 1 sec at 10^{-6} torr. We shall return to this problem below. Nevertheless, Jacobsson and Kruse (1973) found that for ZnS on glass the critical load in their pull test of adhesion was nearly doubled if the glass was ion bombarded before film deposition. Butler (1970) also found improved adhesion of evaporated copper on glass if the substrate was glow discharge cleaned.

In the case of interdiffusion or compound interfaces, adhesion improvement can often be achieved by substrate heating. One has to be careful, however, that this does not produce adverse effects, such as grain growth. But this can usually be avoided; e.g. reflective adherent films of aluminium on glass can be produced by evaporating a very thin layer of Al at a substrate temperature of 250° C, which is then reduced to 150° C for depositing the reflecting layer. Sometimes very litle heating is required; gold can be evaporated onto silicon (provided the oxide surface is removed) at 50° C to produce excellent adhesion via a diffuse interface extending many atomic layers (Narusawa et al. 1973).

Intermediate layer adhesion occurs naturally for some materials, such as aluminium on glass (Bateson 1952). In other cases, intermediate materials may be introduced to advantage; Colbert et al. (1948a) have reported that metal films, strongly adherent to glass and other silica-containing substrates, can be produced using very thin intermediate layers of metallic oxides, sulphides, sulphates, selenates and phosphates, silver chloride, and magnesium fluoride. Many of these materials were produced in situ by oxidation of an evaporated material in an oxygen-containing glow discharge. Presumably, reactive deposition could also be used. A further patent (Colbert et al. 1948b) deals with the use of intermediate layers of lead compounds. In another specific example, the use of an intermediate layer of SiO_2 seemed even more effective in improving the adhesion of ZnS on glass than predeposition ion bombardment of the substrate (Jacobsson and Kruse 1973).

Intermediate oxide layers can also be achieved by depositing an oxygen-active metal (A) onto an oxide surface (BO), promoting the reaction A + BO → AO + B at the interface. Materials with large heats of oxide formation, such as titanium, molybdenum, tantalum, and chromium, are effective. A further metal layer, adherent to the intermediate material, can then be deposited. This is the basis of several multilayer systems, such as Ti - Au, Ti - Pd - Au, and Ti - Pt - Au (English et al. 1972).

Another version of the intermediate layer philosophy would seem to be particularly useful where substrate heating is restricted, e.g. with diffused semiconductor devices. The approach can be illustrated by its application to producing adherent tungsten on silica. A metallic cation having a different valence from the substrate silicon cation, such as aluminium or phosphorus, is introduced substitutionally into the surface layers of the substrate by diffusion or ion bombardment. This produces unsatisfied bonding of the oxygen anions in the substrate, and the tungsten can now be deposited by evaporation or sputtering and will chemically bond to the substrate (Cuomo et al. 1972).

The formation stages of a thin film, its initial nucleation and subsequent growth, should have an influence on the adhesion and cohesion failure mechanisms of a thin film – for example, those due to the presence of microcracks and voids. This aspect of adhesion and several others have been discussed by Mattox (1963-1973). One may expect that glow discharge and ion bombardment cleaning would have a further influence on nucleation and growth via the creation of deep adsorption sites and other surface defects. Similarly, bias levels and several other deposition parameters will have a marked effect on film microstructure.

It is often stated that sputtered films are more adherent than evaporated films because the sputtered atoms are ejected from their target source with much more energy than those from an evaporation source as discussed in "Ion Beam Sputtering". However, the small mean free path in a glow discharge sputtering system probably means that this ejection energy is largely dissipated by collisions before material reaches the substrate. What are probably more important are the fast electrons from the target, and the field existing at the substrate due to the plasma potential which, as we have seen, is typically +50V with respect to the anode in rf sputtering and less in dc. Substrate bombardment by secondary target electrons may be quite significant since these electrons are responsible for a good deal of substrate heating, which will promote interfacial reaction and interdiffusion as well as influence nucleation; the associated charge will also have a marked effect on nucleation. The sheath field at the substrate will cause positive ion bombardment there, mostly by sputtering gas ions but also by sputtered target atoms which may become ionized by the Penning mechanism. This bombardment will tend to resputter material from the growing film and may also lead to forward sputtering and implantation (as discussed later), particularly if the energy of bombardment is increased by the application of a negative potential to the substrate; i.e. by using the biasing technique. However, one must ensure that the bias does not prevent adhesion-promoting oxide formation which might otherwise naturally occur (Maissel and Schaible 1965).

Mattox has also been responsible for the introduction of ion plating and its use in promoting good adhesion. As we saw in a previous section, ion plating basically consists of bombarding the substrate with high energy gas ions to sputter clean the surface and then of evaporating material through the plasma onto the substrate whilst still maintaining the ion flux, at least for an initial period. As a result, the substrate remains clean during deposition (cf. predeposition cleaning discussed earlier), its surface temperature is high (promoting interfacial reaction and interdiffusion), the growth of the film is modified because of surface bombardment, and some of the evaporated material becomes ionized, is accelerated onto the substate, and may become implanted. Hence, similar to ion implantation, interdiffusion adhesion may exist as a nonequilibrium state even when there is no mutual solubility. Mattox (1969) has suggested the term *pseudodiffusion* for this situation. Swaroop and Adler (1973) have reported a graded diffuse interface of 7 to 8μm depth for ion plated copper on steel.

Similar thinking has led to the use of high energy ion bombardment to enhance adhesion. Perkins and Stroud (1970) have produced extremely adherent contacts by evaporating 300 Å of Ni onto glass and then bombarding with 100 keV argon ions. This is probably at least partly due to nickel atoms being forward sputtered into the glass, which produces interdiffusion; one might also expect a similar mechanism to exist in ion plating and maybe even in bias sputtering to a lesser extent as a means of implanting un-ionized film atoms. Applications of ion bombardment to enhance adhesion have been reviewed by Stroud (1972).

Finally, a technique for increasing the adhesion of thin metal films to glass after deposition is due to Stuart (1969) and is reported by Butler et al. (1971). The back of the substrate is coated with a conductor and an electric field is applied; for copper, this is 1000 V/mm for 10 min at 200° C. The enhanced adhesion is thought to be due to an assisted diffusion of copper ions into the substrate. In the case of aluminium, an alumina layer grows at the interface as though the aluminium were being anodized with the glass as a solid electrolyte. Similar mechanisms may exist at the surface of an insulating substrate in bias sputter deposition.

CONCLUSION

In this chapter we have looked at some of the variations of the sputtering process. It is very clear, by looking at the current literature, that the range of applications of sputtering and the variations of the basic process, is extremely wide. However, these variations can usually be traced to the basic processes described in this chapter.

There have been no dramatic recent developments in sputtering, but there has been a continual improvement in the control of the process. There is also a considerable broadening in the fields of application: e.g. metals can be treated by an ion nitriding process in which the metal becomes the target of a nitrogen-hydrogen discharge so that the surface is hardened (Hudis 1973); the Davis and Vanderslice (1963) mass/energy analyzer was used to analyze the discharge. Continuity.

Over the next few years, we can expect to see a continuation of these trends.

REFERENCES

S. Aisenberg and R. W. Chabot, J. Appl. Phys. **42**, 2953 (1971)

S. Aisenberg and R. W. Chabot, J. Vac. Sci. Tech. **10**, 104 (1973)

O. Almen and G. Bruce, Nucl. Instrum. and Methods **11**, 257 (1961)

J. Amano and R. P. W. Lawson, J. Vac. Sci. Tech. **14**, 831 (1977)

J. Amano and R. P. W. Lawson, J. Vac. Sci. Tech. **15**, 118 (1978)

D. J. Ball, J. Appl. Phys. **43**, 3047 (1972)

S. Bateson, Vacuum **2**, 365 (1952)

E. Bauer and H. Poppa, Thin Solid Films **12**, 167 (1972)

A. R. Bayly, J. Matls. Sci. **7**, 404 (1972)

R. W. Berry, P. M. Hall, and M. T. Harris, *Thin Film Technology*, Van Nostrand (1968)

R. L. Bersin, 32nd Tech. Conf. Soc. of Plastics Engineers, p. 72, San Francisco (1974)

R. F. Bunshah, in *Science and Technology of Surface Coating*, ed. B. N. Chapman and J. C. Anderson, Academic Press, London and New York (1974)

D. W. Butler, J. Phys. E **3**, 979 (1970)

D. W. Butler, C. T. H. Stoddart, and P. R. Stuart, in *Aspects of Adhesion* **6**, ed. D. J. Alner, Univ. of London Press, London (1971)

R. Carpenter, in *Science and Technology of Surface Coating*, ed. B. N. Chapman and J. C. Anderson, Academic Press, London and New York (1974)

G. Carter and J. S. Colligon, *Ion Bombardment of Solids*, Elsevier (1968)

C. Catana, J. S. Colligon, and G. Carter, J. Matls. Sci. **7**, 467 (1972)

C. C. Chang, P. Petroff, G. Quintana, and J. Sosniak, Surface Science **38**, 341 (1973)

J. S. Chapin, Research/Development, p. 37, January (1974)

B. N. Chapman and D. S. Campbell, J. Phys. C. (Solid State Phys.) **2**, 200 (1969)

B. N. Chapman and M. R. Jordan, J. Phys. C. (Solid State Phys.) **2**, 1550 (1969)

B. N. Chapman, J. J. O'Neill Jr., and J. L. Vossen, unpublished results (1973)

B. N. Chapman, J. Vac. Sci. Tech. **11**, 106 (1974)

B. N. Chapman and J. C. Anderson, eds., *Science and Technology of Surface Coating*, Academic Press, London and New York (1974)

B. N. Chapman, D. Downer, and L. J. M. Guimarães, J. Appl. Phys. **45**, 2115 (1974)

B. N. Chapman and V. J. Minkiewicz, submitted to J. Vac. Sci. Tech. (1979)

M. Chen, V. J. Minkiewicz, J. W. Coburn, B. N. Chapman, and K. Lee, submitted to Appl. Phys. Letters (1979)

K. B. Cheney and E. T. Pitkin, J. Appl. Phys. **36**, 3542 (1965)

O. Christensen and P. Jensen, J. Phys. E **5**, 86 (1972)

O. Christensen and M. Brunot, Le Vide, Les Couches Minces **165**, 37 (1973)

O. Christensen, Thin Solid Films **27**, 63 (1975)

P. Clarke, U.S. Patent 3 616 450 (1971)

J. W. Coburn, Rev. Sci. Instrum. **41**, 1219 (1970)

J. W. Coburn and E. Kay, J. Appl. Phys. **43**, 4965 (1972)

J. W. Coburn and E. Kay, CRC Critical Reviews in Solid State Sciences **4**, 561 (1974)

J. W. Coburn, E. Taglauer, and E. Kay, Japan J. Appl. Phys., Suppl. 2, pt. 1, 501 (1974)

J. W. Coburn, J. Vac. Sci. Tech. **13**, 1037 (1976)

J. W. Coburn, unpublished results (1979)

W. H. Colbert, A. R. Weinrich, and W. L. Morgan, British Pats. 605 871 – 605 874 (1948a)

W. H. Colbert, A. R. Weinrich, and W. L. Morgan, British Pat. 605 889 (1948b)

J. Comas and E. A. Wolicki, J. Electrochem. Soc. **117**, 1197 (1970)

J. J. Cuomo, R. F. Mayadas, and R. Rosenberg, US Patent No. 3 704 166 (1972)

J. J. Cuomo, R. J. Gambino, J. M. E. Harper, J. D. Kuptsis, and J. C. Webber, J. Vac. Sci. Tech. **15**, 281 (1978)

W. D. Davis and T. A. Vanderslice, Phys. Rev. **131**, 219 (1963)

F. d'Heurle, L. Berenbaum, and R. Rosenberg, Trans. Met. Soc. AIME **242**, 502 (1968)

D. DiMaria, L. M. Ephrath, and D. R. Young, J. Appl. Phys., to be published (1979)

G. Dupp and A. Scharman, Z. Physik **192**, 284 (1966)

A. T. English, K. L. Tai, and P. A. Turner, Appl. Phys. Letters **21**, 397 (1972)

J. C. Erskine and A. Cserhati, J. Vac. Sci. Tech. **15**, 1823 (1978)

D. B. Fraser, in *Thin Film Processes*, ed. J. L. Vossen and W. Kern, Academic Press, New York and London (1978)

R. A. Gdula, Intl. Electron Devices Mtg. IEEE, Washington, DC (1977)

W. D. Gill and E. Kay, Rev. Sci. Instrum. **36**, 277 (1965)

R. Glang, in *Handbook of Thin Film Technology*, ed. L. I. Maissel and R. Glang, McGraw Hill, New York and London (1970)

P. G. Glöersen, J. Vac. Sci. Tech. **12**, 28 (1975)

F. A. Green and B. N. Chapman, unpublished results (1975)

F. A. Green and B. N. Chapman, J. Vac. Sci. Tech. **13**, 165 (1976)

J. E. Greene and F. Sequeda-Osorio, J. Vac. Sci. Tech. **10**, 1144 (1973)

J. E. Greene, F. Sequeda-Osorio, and B. R. Natarajan, J. Appl. Phys. **46**, 2701 (1975)

J. E. Greene, J. Vac. Sci. Tech. **15**, 1718 (1978)

A. Guentherschulze and W. Tollmien, Z. Physik **119**, 685 (1942)

M. I. Guseva, Sov. Phys. Solid State **1**, 1410 (1960)

J. J. Hanak and J. P. Pellicane, J. Vac. Sci. Tech. **13**, 406 (1976)

J. R. Hall, C. A. L. Westerdahl, A. T. Devine, and M. J. Bodnar, Tech. Report 3788, Feltman Res. Lab., Dover, N. J. (1968)

J. M. E. Harper, in *Thin Film Processes*, ed. J. L. Vossen and W. Kern, Academic Press, New York and London (1978)

W. Hauffe, Phys. Stat. Sol.(a) **4**, 111 (1971)

N. Heiman, V. J. Minkiewicz and B. N. Chapman, submitted to J. Vac. Sci. Tech. (1979)

C. L. Hemenway, R. W. Henry, and M. Caulton, *Physical Electronics*, Wiley, New York and London (1967)

T. W. Hickmott, Appl. Phys. Letters **15**, 232 (1969)

H. Hirai, K. Ando and Y. Maekawa, Mem. Fac. Eng., Osaka City Univ. **8**, 103 (1966)

L. Holland, *Vacuum Deposition of Thin Films,* Chapman and Hall, London (1956)

L. Holland, in *Science & Technology of Surface Coating*, ed. B. N. Chapman and J. C. Anderson, Academic Press, London and New York (1974)

M. Hudis, J. Appl. Phys. **44**, 1489 (1973)

R. Jacobsson and B. Kruse, Thin Solid Films **15**, 71 (1973)

G. V. Jorgensen and G. K. Wehner, J. Appl. Phys. **36**, 2672 (1965)

M. Kaminsky, *Atomic and Ionic Impact Phenomena on Metal Surfaces*, Academic Press, New York (1965)

H. R. Kaufman, J. Vac. Sci. Tech. **15**, 272 (1978)

H. R. Kaufman and R. S. Robinson, J. Vac. Sci. Tech. **16**, 175 (1979)

E. Kay, Adv. Electronics and Electron Phys. **17**, 245 (1962)

E. Kay, Trans. Conf. and School on Sputtering, Pebble Beach, Calif. (1969)

J. H. Keller and W. B. Pennebaker, IBM J. Res. Develop. **23**, 3 (1979)

R. Kelly, Proc. Int. Conf on Ion Beam Modification of Materials, Budapest (1978)

W. Kern, J. L. Vossen, and G. L. Schnable, 11th Annual Proceedings, Reliability Physics Conf., Las Vegas, 214 (1973)

H. R. Koenig and L. I. Maissel, IBM J. Res. Develop. **14**, 168 (1970)

S. Komiya, K. Yoshikawa, and S. Ono, J. Vac. Sci. Tech. **14**, 1161 (1977)

R. C. Krutenat and R. Gesick, J. Vac. Sci. Tech. **7**, S40 (1970)

N. Laegreid and G. K. Wehner, J. Appl. Phys. **32**, 365 (1961)

W. Laznovsky, Research/Development, p. 47, August (1975)

K. D. Leaver and B. N. Chapman, *Thin Films*, Wykeham Publications, London (1971)

R. E. Lee, J. Vac. Sci. Tech. **16**, 164 (1979)

B. Lewis and D. S. Campbell, J. Vac. Sci. Tech. **4**, 209 (1967)

W. Little, H. W. Fowler, and J. Coulson, *The Shorter Oxford English Dictionary on Historical Principles*, Clarendon Press, Oxford (1959)

J. S. Logan, IBM J. Res. Develop. **14**, 172 (1970)

J. S. Logan, F. S. Maddocks, and P. D. Davidse, IBM J. Res. Develop. **14**, 81 (1970)

J. S. Logan, J. H. Keller, and R. G. Simmons, J. Vac. Sci. Tech. **14**, 92 (1977)

T. R. Lundquist, J. Vac. Sci. Tech. **15**, 684 (1978)

W. H-L. Ma and T. P. Ma, Intl. Electron Devices Mtg. IEEE, Washington, DC (1977)

L. I. Maissel and P. M. Schaible, J. Appl. Phys. **36**, 237 (1965)

L. I. Maissel, in *Handbook of Thin Film Technology*, ed. L. I. Maissel and R. Glang, McGraw Hill (1970)

J. T. Maskrey and R. A. Dugdale, Brit. J. Appl. Phys. **17**, 1025 (1966)

D. M. Mattox and J. E. McDonald, J. Appl. Phys. **34**, 2493 (1963)

D. M. Mattox, Sandia Laboratories Report No. SC-R-65-852 (1965)

D. M. Mattox, J. Appl. Phys. **37**, 3613 (1966)

D. M. Mattox, Trans. SAE **78**, 2175 (1969)

D. M. Mattox, J. Vac. Sci. Tech. **10**, 47 (1973a)

D. M. Mattox, Proc. 27th Annual Freq. Control Symp., Fort Monmouth, N.J. (1973b)

C. J. Mogab, P. M. Petroff, and T. T. Sheng, J. Electrochem. Soc. **122**, 815 (1975)

B. A. Movchan and A. V. Demchishin, Phys. Metals and Metallography **4**, 83 (1969)

S. Murphy (1811), original source unknown (John Vossen has suggested the J. Irreproducible Results)

T. Narusawa, S. Komiya, and A. Hiraki, Appl. Phys. Letters **22**, 389 (1973)

C. A. Neugebauer, *Handbook of Thin Film Technology*, ed. L. I. Maissel and R. Glang, McGraw Hill (1970)

M. J. Nobes, J. S. Colligon, and G. Carter, J. Matls. Sci. **4**, 730 (1969)

T. Oohashi and S. Yamanaka, Japan. J. Appl. Phys. **11**, 1581 (1972)

T. Oohashi and S. Yamanaka, private communication (1973)

P. W. Palmberg, J. Vac. Sci. Tech. **9**, 160 (1972)

J. W. Patten and E. D. McClanahan, J. Appl. Phys. **43**, 4811 (1972)

F. M. Penning, Physica **3**, 873 (1936)

J. G. Perkins and P. T. Stroud, AWRE Aldermaston Report No. 09/70 (1970)

H. Ratinen, J. Appl. Phys. **44**, 3817 (1973)

R. S. Robinson, J. Vac. Sci. Tech. **16**, 185 (1979)

G. Schnable and R. S. Keen, Adv. in Electronics and Electron Physics **30**, 79 (1971)

J. M. Seeman, 1st Symposium on the Deposition of Thin Films by Sputtering, Univ. Rochester, p. 30 (1966)

P. Sigmund, Phys. Rev. **184**, 383 (1969)

H. I. Smith, Proc. IEEE **62**, 1361 (1974)

A. W. Stephens, J. L. Vossen, and W. Kern, J. Electrochem. Soc. **123**, 303 (1976)

A. D. G. Stewart and M. W. Thompson, J. Matls. Sci. **4**, 56 (1969)

D. J. Stirland, Appl. Phys. Letters **8**, 326 (1966)

P. T. Stroud, Thin Solid Films **11**, 1 (1972)

P. R. Stuart, private communication (1969)

R. V. Stuart and G. K. Wehner, J. Appl Phys **33**, 2345 (1962)

B. Swaroop and I. Adler, J. Vac. Sci. Tech. **11**, 503 (1973)

M. L. Tarng and G. K. Wehner, J. Appl. Phys. **43**, 2268 (1972)

J. A. Thornton, SAE Report #730544, SAE, New York (1973)

J. A. Thornton and A. S. Penfold, in *Thin Film Processes*, ed. J. L. Vossen and W. Kern, Academic Press, New York and London (1978)

J. A. Thornton, J. Vac. Sci. Tech. **15**, 188 (1978)

C. Y. Ting, V. J. Vivalda and M. G. Schaefer, J. Vac. Sci. Tech. **15**, 1105 (1978)

T. Tisone, Solid State Tech., p. 34, December (1975)

P. Townsend and J. C. Kelly, *Sputtering and Ion Implantation*, Academic Press, London and New York (1976)

I. S. T. Tsong and D. J. Barber, J. Matls. Sci. **8**, 123 (1973)

J. L. Vossen and J. J. O'Neill Jr., RCA Review **29**, 566 (1968)

J. L. Vossen and J. J. O'Neill Jr., RCA Review **31**, 276 (1970)

J. L. Vossen, J. J. O'Neill Jr., K. M. Finlayson, and L. J. Royer, RCA Review **31**, 293 (1970)

J. L. Vossen, J. Vac. Sci. Tech. **8**, S12 (1971)

J. L. Vossen and J. J. Cuomo, in *Thin Films Processes*, ed. J. L. Vossen and W. Kern, Academic Press, New York and London (1978)

R. K. Waits, in *Thin Film Processes*, ed. J. L. Vossen and W. Kern, Academic Press, New York and London (1978)

K. Wasa and S. Hayakawa, Rev. Sci. Instrum. **40**, 693 (1969)

G. K. Wehner, J. Appl. Phys. **30**, 1762 (1959)

G. K. Wehner, General Mills Report No. 2309 (1962)

G. K. Wehner, US Patent 3 021 271 (1962)

G. K. Wehner and G. S. Anderson, in *Handbook of Thin Film Technology*, ed. L. I. Maissel and R. Glang, McGraw Hill (1970)

G. K. Wehner and D. J. Hajicek, J. Appl. Phys. **42**, 1145 (1971)

Chr. Weissmantel, O. Fiedler, G. Mecht, and G. Reisse, Thin Solid Films **13**, 359 (1972)

G. W. White, Research/Development, p. 43, July (1973)

R Widmer, J. Vac. Sci. Tech. **15**, 1197 (1978)

I. M. Wilson and M. W. Kidd, J. Matls. Sci. **6**, 1362 (1971)

H. F. Winters, Adv. in Chemistry Series No. 158, *Radiation Effects on Solid Surfaces*, ed. M. Kaminsky, American Chemical Society (1976)

H. F. Winters and E. Kay, J. Appl. Phys. **38**, 3928 (1967)

H. F. Winters and E. Kay, J. Appl. Phys. **43**, 794 (1972)

H. F. Winters, J. W. Coburn, and E. Kay, J. Appl. Phys. **48**, 4973 (1977)

Chapter 7: Plasma Etching

In contrast to physical sputtering, plasma etching relies on the *chemical* combination of the solid surface to be etched, with an active gaseous species produced in a discharge.

PLASMA ASHING

The idea seems to have first been applied to semiconductor processing during the 1960's as *plasma ashing*, also known as *plasma stripping* (Figure 7-1). This is a technique for the removal of photoresist materials which, being organic, consist essentially of carbon and hydrogen. Solid carbon is converted to gaseous carbon monoxide and carbon dioxide by oxidation in an oxygen discharge. These gases, together with other volatile products of the reaction, such as water vapour

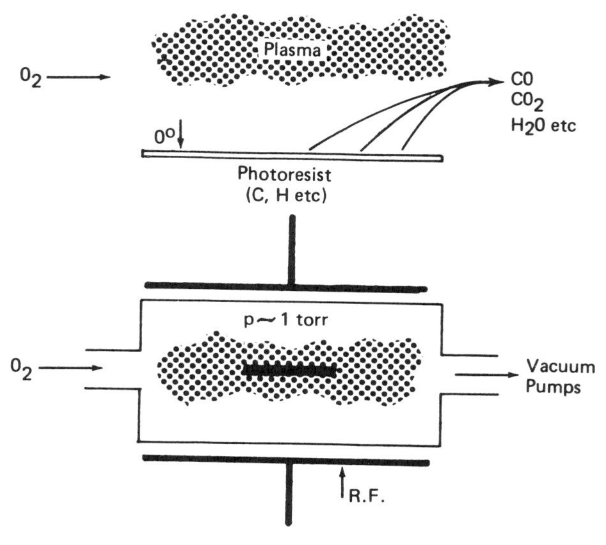

Figure 7-1. The principle and implementation of plasma ashing

(Reichelderfer et al. 1977) are then pumped away by the vacuum system. The purpose of the glow discharge is to convert molecular oxygen, which does not react with photoresists at room temperature, into active oxygen atoms which do react readily. The creation of atomic oxygen in the discharge is achieved by electron impact dissociation (Chapter 2, "Some Other Collision Processes"):

$$e + O_2 \rightarrow O + O + e$$

Some of the dissociative events will be accompanied by ionization, so that the discharge can be sustained. Plasma ashing is now a widely practiced technique, though its implementation has not been without problems: for example, inorganic impurities in the resist that do not form volatile oxides will remain on the wafer surface after etching (Hughes et al. 1973), whereas with wet chemical etching, they would have been washed away with the solvents. This problem is dealt with by meticulous attention to the purity of the photoresist, or by the incorporation of a small quantity of organohalide with the oxygen, in order to plasma etch the inorganics (Jacob 1974, 1976).

PLASMA ETCHING

Given that carbon can be etched by the formation of volatile oxides, it is a short logical step to look for etching processes for other materials by the formation of volatile reaction products in a glow discharge. This is the process of *plasma etching* (Irving et al. 1971), which exists today in a variety of forms and under a variety of names – *plasma etching, plasma assisted etching, reactive sputter etching, reactive ion etching.* These all nevertheless rely on the same basic principle. Development has evolved primarily in and around the semiconductor industry and hence on related materials. Probably the most frequent applications of the process are the etching of silicon and silicon oxides in a discharge of CF_4 to form volatile SiF_4 (Figure 7-2).

There is another path in the genesis of the plasma etching process. Chemical gas phase reactions sometimes occur spontaneously, and can usually be accelerated by heating. There is a range of *gas etching* techniques that do not involve a

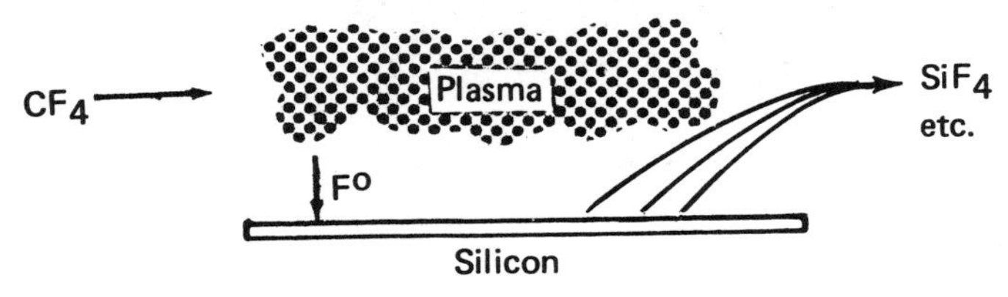

Figure 7-2. The principle of plasma etching

discharge, but instead rely on heating to activate the process. An example would be the etching of sapphire in sulphur tetrafluoride or sulphur hexafluoride (Manasevit and Morritz 1967). However, these processes require a substrate temperature above 1150°C. In semiconductor processing one usually likes to keep wafers cool so as not to disturb diffusion profiles or melt metallization. Plasma etching clearly has an advantage in this respect. It is inherently more controllable, since temperature-controlled chemical reactions tend to have a rate dependence varying exponentially with the temperature (for similar reasons that the Maxwell-Boltzmann distribution discussed in Chapter 1, had an exponential temperature dependence).

The plasma etching technique is quite new and not well understood at present, but this does not seem to be hindering its implementation. Only the etching of silicon materials in fluorine-based gases has been the subject of any reasonably well-defined experiments, and even in that materials system there are many unresolved questions, apparent anomalies, and disagreements on interpretation. Of the various reactor systems (see below), the low pressure diode system is probably best understood because of its ubiquity in sputtering, but that experience cannot always be extended to application with the more complex gases used in plasma etching.

ISOTROPIC OR 'ANISOTROPIC' ETCHING?

Before we look at some of the details of the plasma etching process, let's look at the prime reasons for the impetus it has achieved as a semiconductor manufacturing technique.

One reads that there are cost savings associated with the process (Zafiropoulo 1976) compared with the more conventional wet chemical processes, and there are certainly safety considerations, too: silicon nitride can be etched, using wet chemicals, by hot phosphoric acid at 160°C – not too pleasant for the operator; the equivalent plasma etching process involves CF_4, virtually inert, inside a process chamber.

But the principle motivation behind the adoption of plasma etching is the ability of some reactors to produce fine resolution in small (~ 1 μm) devices, achieved by etching in a directional manner, as follows:

As an example, suppose that one wishes to open a 'window' in a layer of quartz, to expose a feature on a silicon wafer (Figure 7-3c). This could be done by applying photoresist and patterning it by the usual processes so that the opening in the resist is the same size A as the desired window (Figure 7-3a). The oxide layer can now be etched by wet chemical processing to obtain an etched hole with the characteristic isotropic profile (Figure 7-3b). Provided that the process is

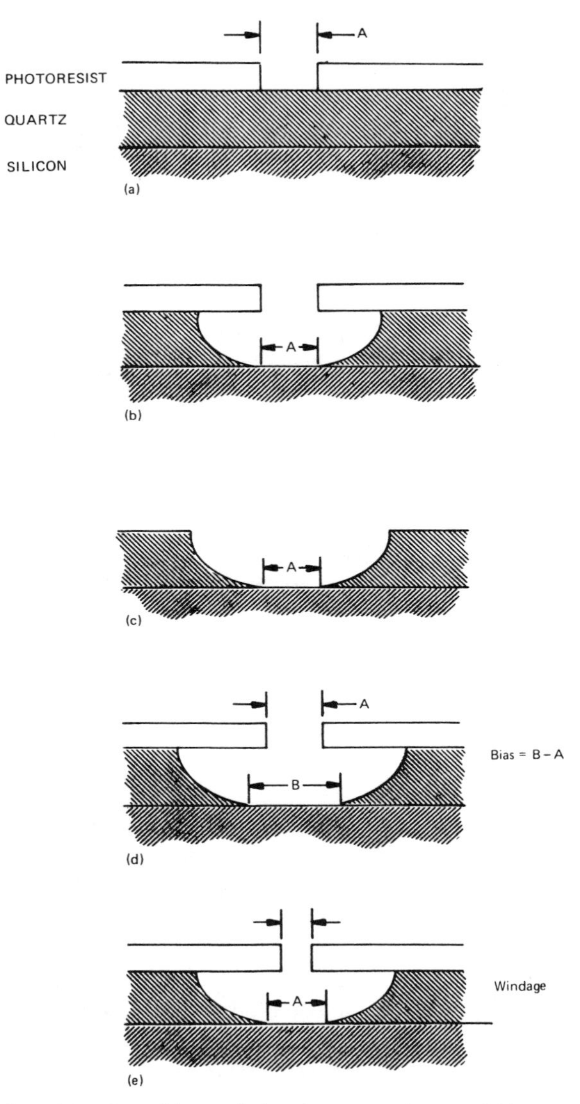

Figure 7-3. Steps in etching a window in quartz; the use of bias and windage

stopped at the correct time, a window of the required size will be achieved; the photoresist mask is then removed (Figure 7-3c). But now we have a practical constraint: any practical process has to cope with a large number of features being etched simultaneously on each chip, with a hundred or more chips on a wafer, and maybe many wafers in each batch. There is then a problem of uniformity. The problem is two-fold, resulting from non-uniformity in the etching process and non-uniformity in the wafer structure being etched, for example in the thickness of the oxide layer. As a result, there will be a distribution of window sizes (Figure 7-3d) and to ensure that all exceed the minimum requirement with a safety margin, there will be an 'overetch' time on the nominal. By using the statistics of the window sizes, we can calculate a *bias* (Figure 7-3d) – the extent to which the etched window size exceeds the photoresist opening (and sometimes defined as half this value) – and the *tolerance*, the statistical variation of the bias. We would like to have both of these parameters as small as possible.

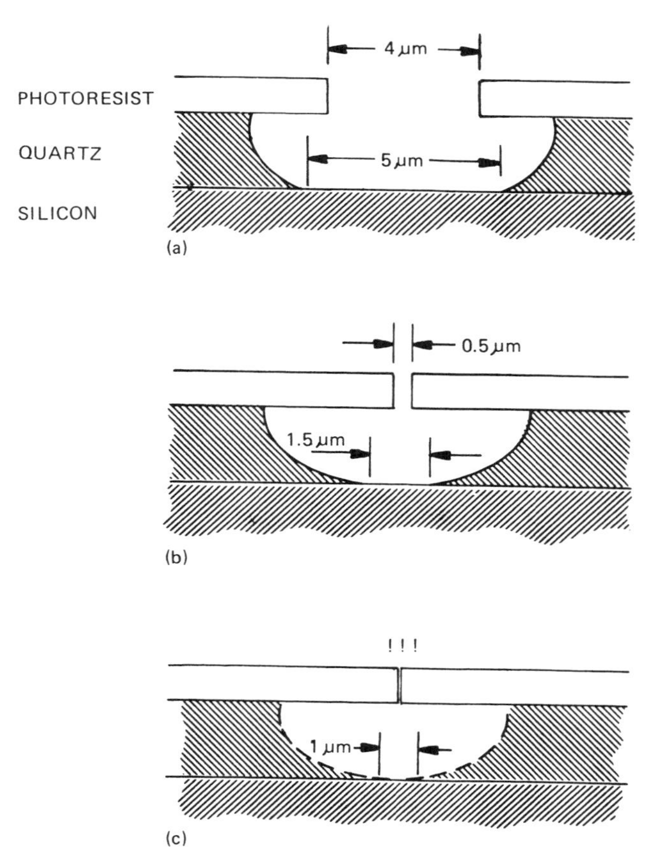

Figure 7-4. The limits of use of windage in small geometries

One might at first think that a small (or better, zero) bias is not necessary since all we need do is make the photoresist opening accordingly smaller; this is known as putting *windage* in the mask (Figure 7-3e). This used to be true, but no longer. Suppose that the process bias is 1 μm, so that 5 μm windows are achieved by 4 μm resist openings (Figure 7-4a). That's alright, but with the same bias, 1.5 μm windows would require 0.5 μm resist openings (Figure 7-4b). Such a dimension in the resist would not only require very close control of the lithography process, but would also represent a poor 'return on investment' – all the problems of sub-micron lithography to achieve super-micron windows. And, carrying the argument further, a 1 μm window would require a resist opening of . . . . (Figure 7-4c)!

Both bias and tolerance can be considerably reduced if *'vertical' etch walls* (or *'directional'* or *'parallel'* or whatever) can be created (Figure 7-5), and this is attainable in some plasma etching processes (Bondur 1976). With such a capability, bias and tolerance could both become zero, since the oxide (in this case) would not be etched beneath the mask, regardless of overetch time (Figure 7-5). But even now there are further constraints:

- The resist openings would all have to be the same size – it's difficult for the plasma system to cope with that non-uniformity, since it is controlled by earlier processing steps.
- The process must be selective, so that etching stops when the relevant interface is reached.
- The resist itself must not etch (referred to as a *nonerodible mask*).
- A vertical wall must be acceptable for subsequent processing.

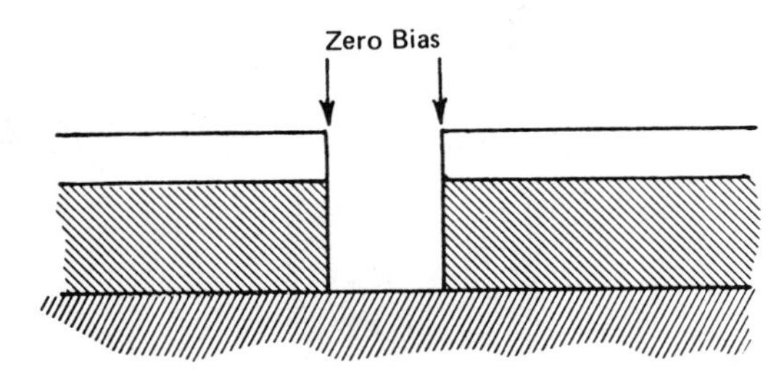

Figure 7-5. 'Directional' plasma etching

Selectivity can be achieved in plasma etching, at least to the extent of obtaining a high ratio of etch rates between some materials. Usually the resist does etch somewhat; this can be discouraged when required, but sometimes it is taken advantage of. Figure 7-6a shows a patterned resist on an oxide layer. Suppose the etching process is directional, as described previously for the non-erodible mask. Now suppose that the resist material and the oxide etch at exactly the same rate, i.e. an *etch rate ratio* (ERR) of 1:1. Then the materials would be indistinguishable to the etching discharge and the shape of the resist would be reproduced in the oxide (Figure 7-6b). So a contoured wall is achieved in the oxide even though the etch process itself is directional. It is therefore necessary to distinguish between the *derived wall profile* and the *inherent directionality* of the etch process. By the same token, an ERR of 1 (photoresist): 2(oxide) would produce oxide walls twice as steep as those in the resist (Figure 7-6c) and with an ERR of 2:1, half as steep. But if the mask is allowed to etch, the capability of

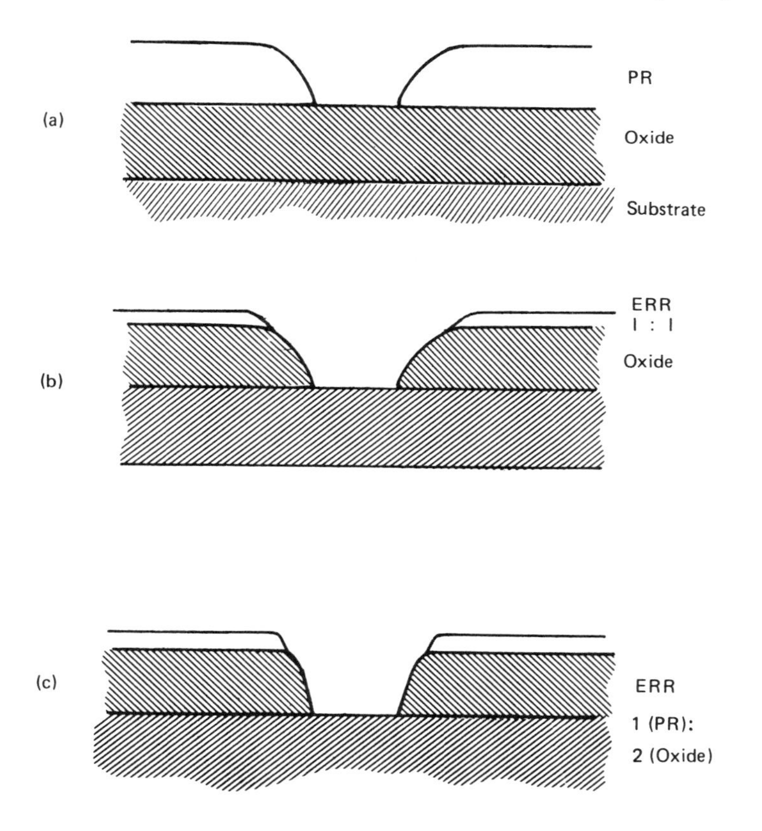

Figure 7-6. The attainment of contoured etch walls with directional etching

(a) Pre-etching: patterned photoresist in place
(b) Resulting profile for 1:1 ERR
(c) Resulting profile for 1(PR):2(Ox) ERR

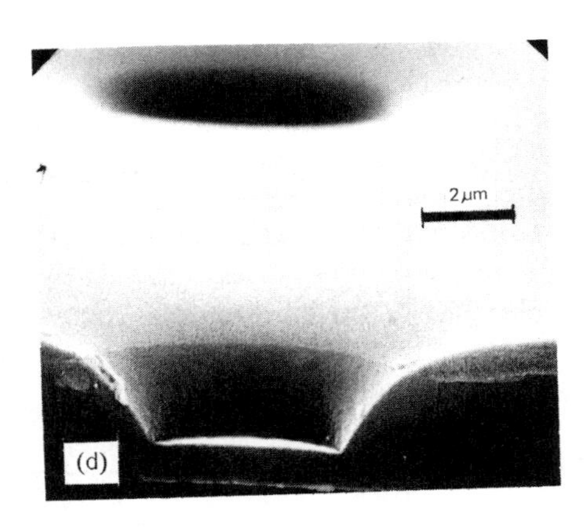

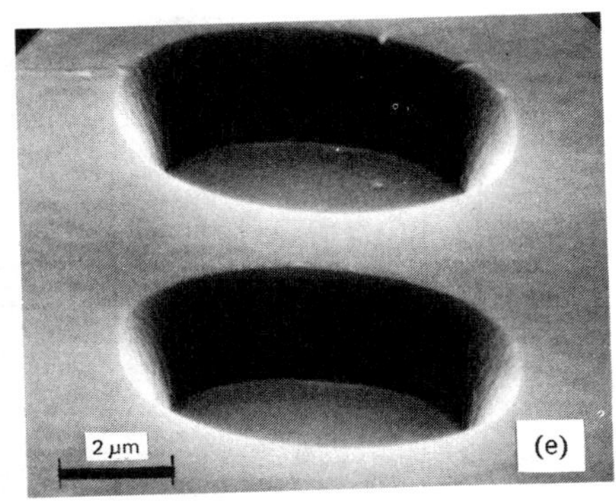

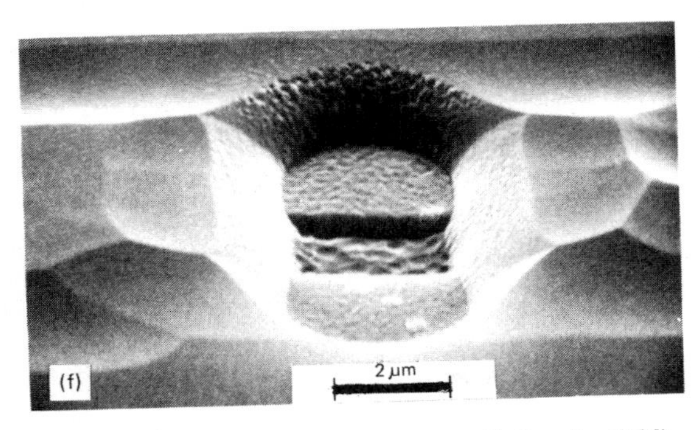

(d) Etched via; photoresist still in place (courtesy G. Brooks, IBM)
(e) Via string; photoresist removed (courtesy G. Gati, IBM)
(f) Etched via with resist removed. Shows wall profiles, texturing due to reactive ion etching, aluminium lines in base of via, and the topography of 'planarized quartz' (see Chapter 6) (courtesy G..Gati, IBM)

producing zero bias and tolerance is lost, although a more acceptable derived wall profile results. As usual, it is a matter of compromise.

Figure 7-6 shows examples where the selective degradation of the resist has been used to obtain a suitable wall profile. Silicon wafers are *metallized*, usually with aluminium or an aluminium alloy, in order to make electrical contact between various parts of the circuit. An electrical insulator is first deposited over all the wafer – a *blanket* coating. The insulator is then etched selectively, using a lithographically patterned photoresist mask, to open up *windows* in the insulator down to those parts of the wafer to be interconnected. The metallization is then put down by evaporation or sputtering to form an electrical *via*. The problem is to ensure good coverage of the side walls (see Chapter 6, "Bias Techniques"). One way of solving this problem is to taper the sidewall of the etch pit, so that the side wall subtends a larger angle at the deposition source. Figure 7-6 shows how a suitable resist: insulator etch rate ratio can achieve this.

A word about nomenclature. *Isotropic etching* means what it says: etching proceeds at the same rate in all directions, resulting in the characteristic linear-circular profile (Figure 7-7a). By definition, anything that is not isotropic is anisotropic. However, in the plasma etching literature, 'anisotropic' has become a misnomer and is generally used to refer to the very directional process that can result in vertical wall profiles (Figure 7-7b). So, at least in the current literature, 'anisotropic', 'directional', and 'vertical' etching are all used synonymously – and none without ambiguity.

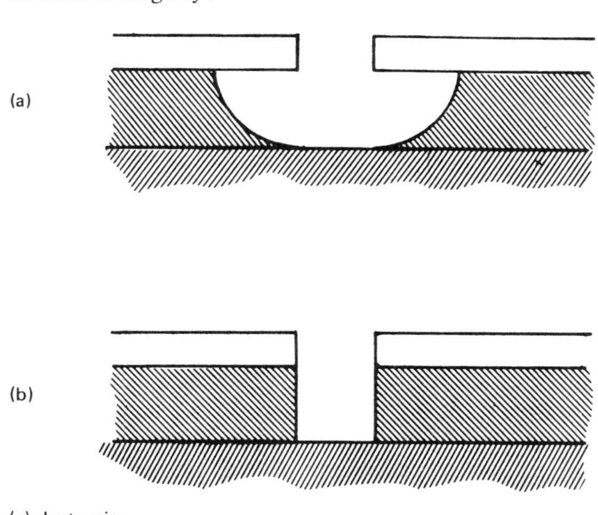

Figure 7-7. Plasma etching profiles

REACTOR SYSTEMS

There are many types of reactors currently available (Figure 7-8). They are all glow discharge systems, but vary considerably in terms of excitation frequency (5 kHz – 5 GHz), operating pressure (1 mtorr – 1 torr), electrode arrangements (internal, external, capacitive, inductive), etch rates (10 Å – 10,000 Å/minute) and etch profiles (isotropic to 'vertical').

There are three principal generic types of reactor. The *barrel* reactors (so named because of the shape of their chambers, although the other two types also have barrel shaped chambers, albeit short and fat) are characterized by high

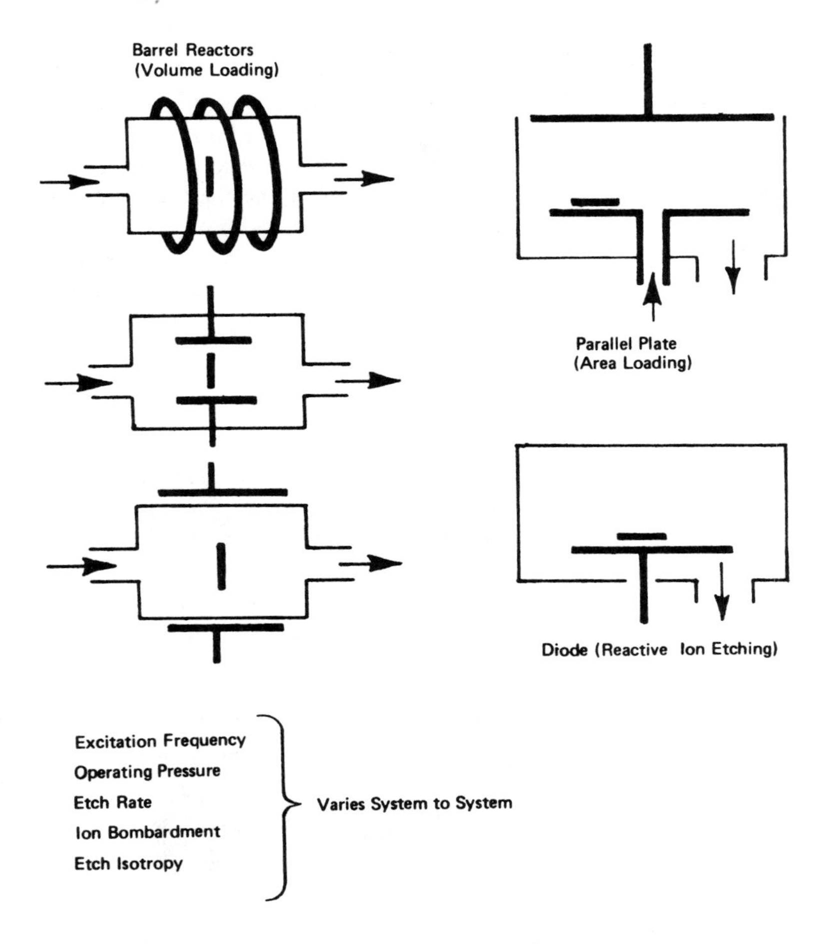

Figure 7-8. Plasma etching equipment

operating pressure (~ 1 torr) and the wafers sitting in the glow, usually on an insulating (often quartz) holder. They are also known as *volume loading* reactors (Poulsen 1977), presumably because of their ability to accommodate a considerable number of wafers in their volume. RF power, usually at 13.56 MHz frequency, is applied to the systems via internal or external capacitive or inductive coupling. This type of reactor is used for the plasma ashing process, and was at one stage the most widespread type of plasma etching reactor. However, it does have the disadvantage that it produces etch profiles which, whilst not always being completely isotropic, always undercut a non-erodible mask.

There is then a set of reactors known variously as *planar diodes, parallel plate* reactors, *area loading* reactors, or *Reinberg* reactors. [Alan Reinberg (1973, 1974) introduced this design of reactor, which was used initially for plasma deposition, and seems capable of performing the dual role (Reinberg 1979).] These reactors are characterized by fairly high operating pressures (100 mtorr – 1 torr), with the wafers placed on the grounded lower electrode; often the gas flow is through the centre of this electrode. Both directional and isotropic etching can be achieved.

Finally, there is the type of system known as a *reactive ion etching* (RIE) reactor, or *low pressure diode*. This is characterized by low operating pressure (10 mtorr – 150 mtorr) with the wafer placed on the non-grounded electrode (Bondur 1976). This type of system is essentially the same as a sputtering system and its use for plasma etching seems to have evolved from its use in sputtering. Etching in this type of reactor is almost always inherently directional.

These are the main types of reactors, although there are several other configurations being explored or in use; two of these are discussed later ("Two More Reactors").

ETCHING MECHANISMS

Plasma etching is a comparatively new technique and more development is required to expand the versatility of the process and exploit its potential.

Development and understanding of processes go hand in hand. The etching of silicon compounds in fluorine-based gases has been the subject of most study. Any understanding of the process had to explain four basic phenomena (Figure 7-9):

- Directional etching can often be achieved (Bondur 1976), and:
- with the addition of H_2 to CF_4,
- the addition of O_2 to CF_4,
- or the increase of substrate area (the 'loading effect'),

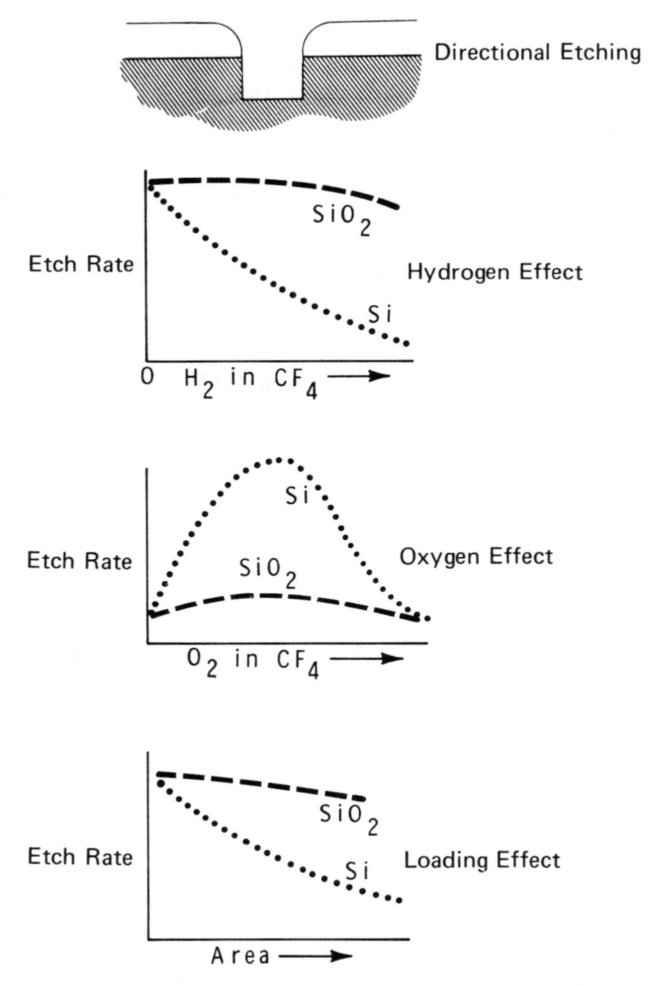

Figure 7-9. Basic phenomena in plasma etching

the etch rate of silicon varies considerably, while that of SiO_2 changes comparatively little (Ephrath 1979, Hosokawa et al. 1974, Mogab et al. 1978, Mogab 1977).

A few key experiments have provided the basis for understanding the process, and this basis is currently being strengthened and extended. Mass spectrometry (Figure 7-10) has identified the overall reaction products (Coburn and Winters

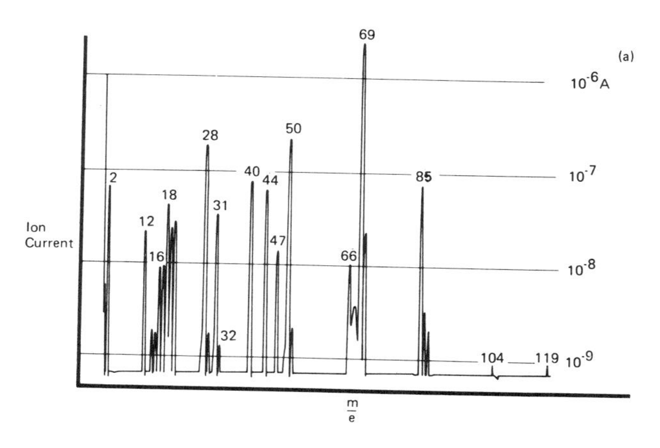

Figure 7-10. Mass spectroscopy analysis of the effluent gas from the etching of SiO_2 in CF_4 (Chapman et al. 1979)

1978, Chapman et al. 1978). For the etching of silicon in CF_4, the effluent contains silicon tetrafluoride, SiF_4, and C_2F_6 as well as unreacted CF_4. For the etching of SiO_2 in CF_4, or of Si in a CF_4/O_2 mixture, the dominant silicon product is again silicon tetrafluoride, this time accompanied by carbon monoxide and dioxide, COF and COF_2, as well as CF_4 and O_2 (Table 7-1). Note that the only significant overall silicon product is SiF_4, silicon tetrafluoride.

Table 7-1 Reaction Products – The Effluent Gas

a) Si in CF_4:

- CF_4
- SiF_4
- C_2F_6 etc.

b) SiO_2 in CF_4 or Si in CF_4/O_2:

- CF_4
- O_2
- CO
- CO_2
- SiF_4
- COF_2 etc.

Mauer et al. (1978) have pointed out that elemental silicon and the non-volatile partially fluorinated products SiF, SiF_2, and SiF_3, are likely to be sputtered from the reaction surface. Coburn and Kay (1977) have used ion mass spectroscopy (i.e. without ionizer) to observe ions in the discharge, and they certainly observed SiF^+ and $SiF_3{}^+$. However, most of these are likely to have been formed by the electron impact dissociation of SiF_4 in the glow, rather than of SiF and SiF_3, particularly in the low flow rate-high residence time conditions used by Coburn and Kay.

Non-volatile partially fluorinated silicon compounds are most likely to be redeposited onto the walls and other internal surfaces. At these new sites, they are likely to become further fluorinated to form volatile silicon tetrafluoride, so that the intermediate step has no overall effect; there is a chance, though, that they may be buried in the growing film on the walls, or converted to a compound (such as silicon oxide if there is some oxygen in the system) that is etch-resistant at the low bombardment energy site. In this case, some silicon atoms will have been removed from the surface by less than the four fluorine atoms necessary to form silicon tetrafluoride, and this could complicate a quantitative analysis of the etching process (for example, see "Gas Flow Rate Effects" later in the chapter), although quantitatively this is apparently a minor effect.

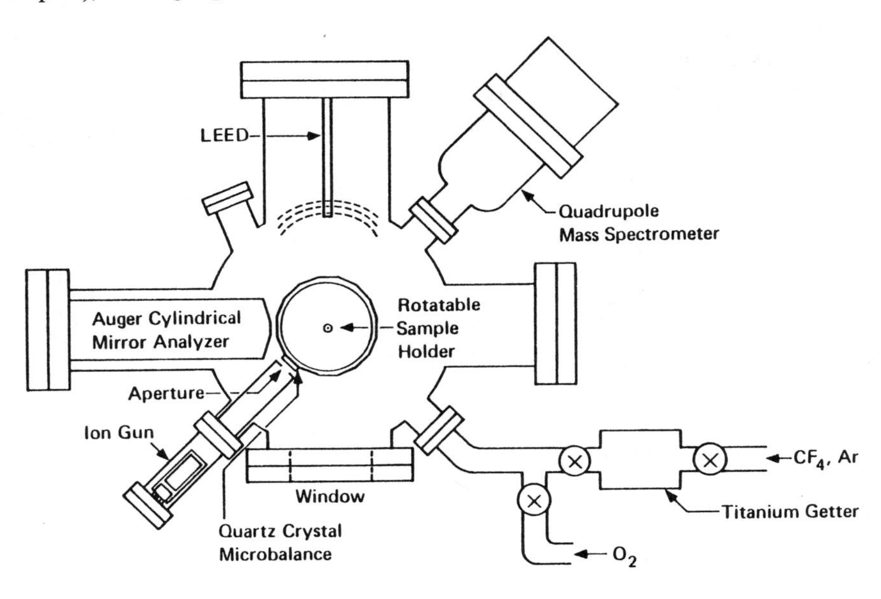

Figure 7-11. Schematic diagram of the experimental system of Coburn et al. (1977), showing the ion gun, Auger surface analyzer, and microbalance etch rate monitor

Although significant quantities of highly unsaturated carbon fluorides were observed, there were no signs of $Si_xC_yF_z$ compounds – and none were expected because Si has a greater affinity for fluorine than does carbon. It appears that we are therefore justified in concluding that the only really significant silicon product of the overall process, is silicon tetrafluoride. This appears to be essentially confirmed by the more recent results of Smolinsky and Flamm (1979).

Coburn et al. (1977) bombarded a silicon surface with ions, measuring the etch rate and subsequently analyzing the surface of the silicon in situ (Figure 7-11). Using argon ions and measuring etch rate as a yield of silicon atoms per incident ion, they observed the typical energy dependence of sputtering yield shown in Figure 7-12. They then used CF_4 gas for their source of ions. Again using mass

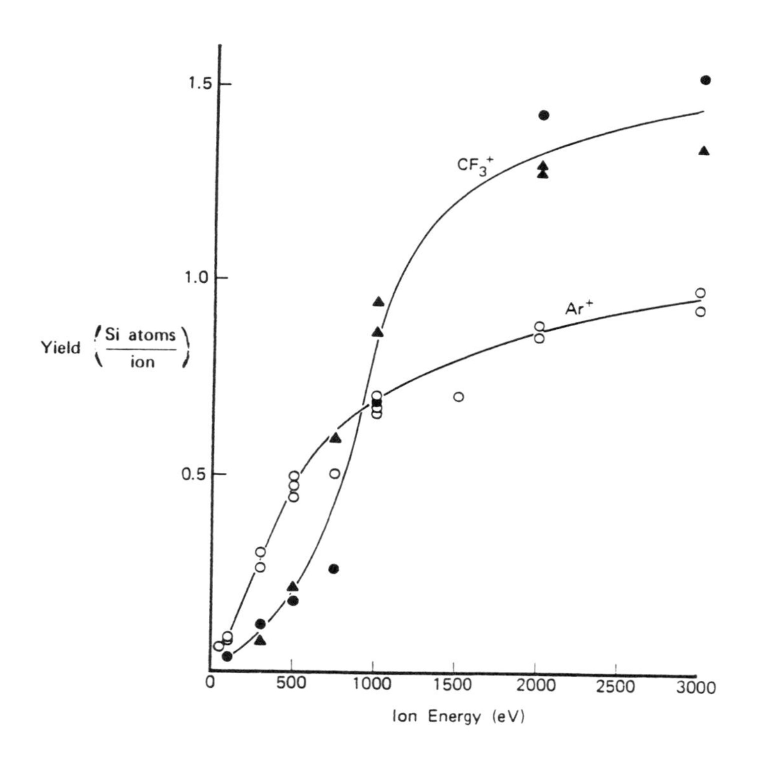

Figure 7-12. Silicon yield curve (Si atoms/ion versus ion energy) for Ar^+ ions and $CF_3{}^+$ ions. In the case of $CF_3{}^+$ ions, these yields are quasi-steady-state (high dosage) yields. The experimental results have been normalized to agree with published yields (Coburn et al. 1977)

spectrometry, one finds that CF_4 gas ionizes to form CF_3^+ ions in about 90% of cases:

$$CF_4 + e \rightarrow CF_3^+ + F + e$$

The other cases lead to CF_2^+ and CF^+; there is no CF_4^+ parent ion. Using this predominantly CF_3^+ ion source to bombard a silicon surface, Coburn et al. found that the behaviour of the fluorocarbon ions at bombardment energies in excess of 1500 eV could be due to physical sputtering only – being somewhat more effective than argon because of the larger ion masses involved. Since CF_3^+ could help to form, at most, only three-quarters of a silicon tetrafluoride molecule, this result is not surprising. What is surprising is that the CF_3^+ yield curve dropped below that of argon at lower energies (Figure 7-12) and that carbon was then found on the silicon surface (Figure 7-13). The steady state concentration of carbon increased with decreasing ion energy, reaching complete coverage for ions of < 500 eV (Figure 7-14), which is an ion energy greater than would normally be encountered in a practical plasma etching system. It appears that the

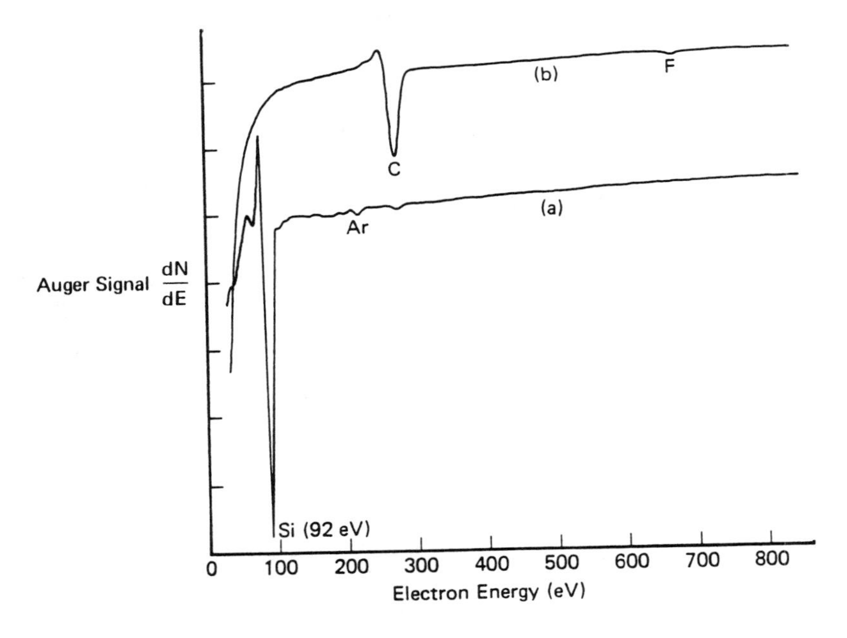

Figure 7-13. Auger electron spectrum from

(a) Si surface cleaned by 500 eV Ar^+ ion bombardment
(b) Same surface as (a) after bombardment with 5 10^{16} CF_3^+ ions/cm^2 of 500 eV energy (Coburn et al. 1977)

energy of impact causes the ion to dissociate, depositing carbon on the silicon surface. The etching problem is thus compounded, since now both carbon and silicon have to be removed from the wafer surface.

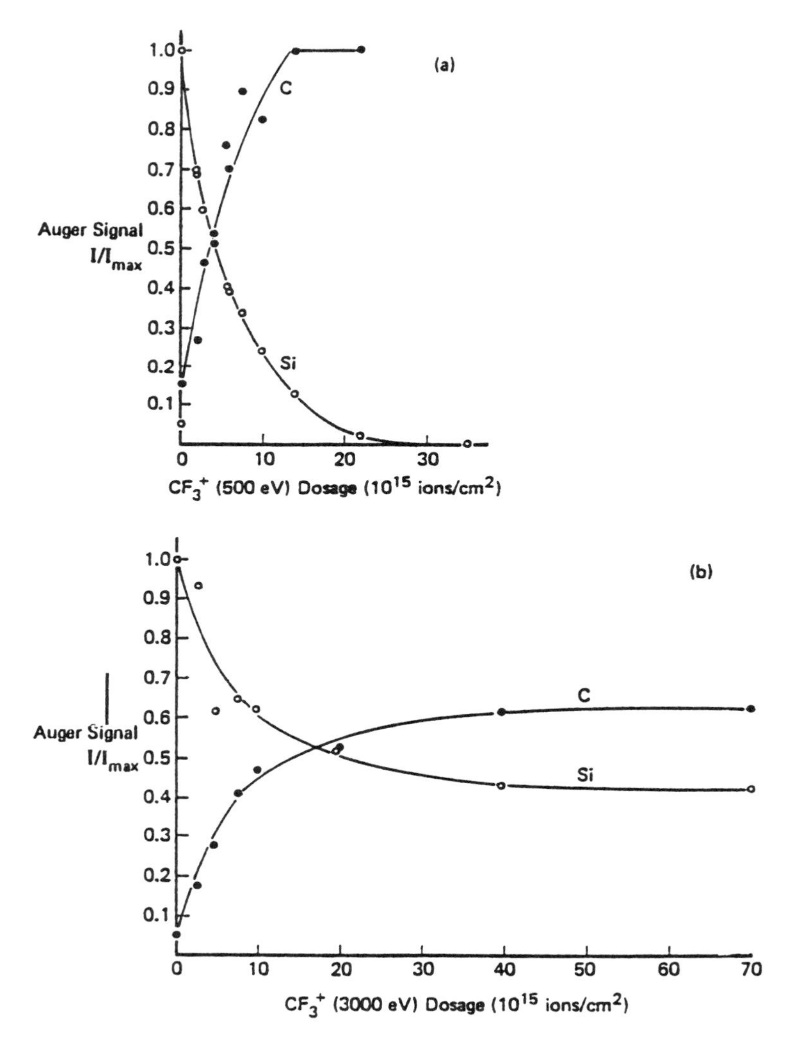

Figure 7-14. Auger signal from a silicon surface (normalized to the maximum value) for Si and C as a function of CF_3^+ ion dosage for

(a) 500 eV CF_3^+ ions and
(b) 3000 eV CF_3^+ ions (Coburn et al. 1977)

Additional parts of the puzzle have been furnished by the optical emission spectroscopy measurements of Harshbarger et al. (1977), which showed how the fluorine and carbon monoxide concentrations in a plasma etching system increased with the addition of oxygen, and how these correlated with the etch rate of silicon (Figure 7-15). Similar measurements showed how the addition of hydrogen caused the fluorine concentration to drop (Figure 7-16), with a corresponding decrease in the silicon etch rate and an increased production of HF.

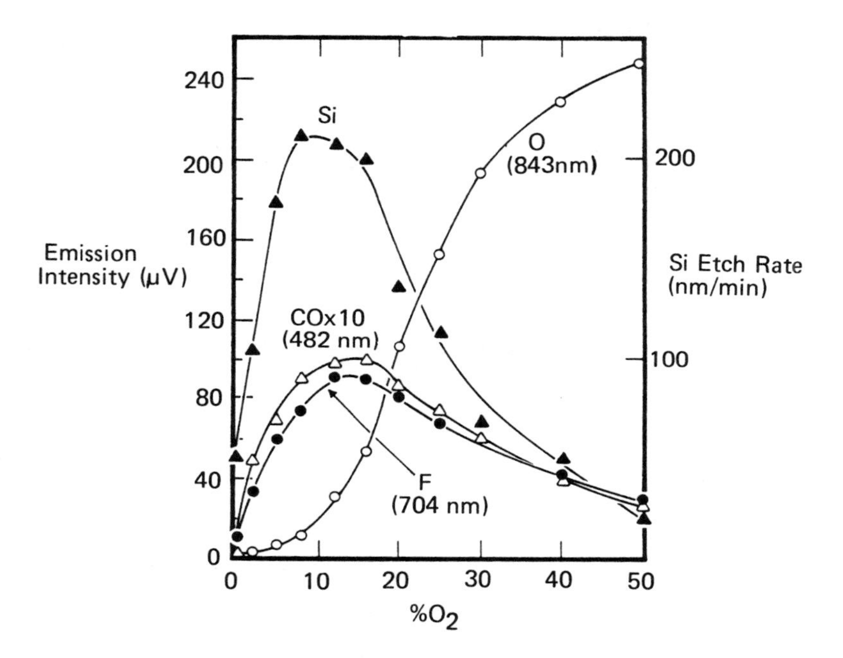

Figure 7-15. Emission intensity (left) and Si etch rate (right) versus % of oxygen in the CF_4 etching gas. The emission intensity for O, F, and CO were obtained by measuring the optical emission lines at 843 nm, 704 nm, and 482 nm, respectively (Harshbarger et al. 1977)

The general picture that emerges (Coburn and Winters 1978) is that one has to remove not only the silicon or silicon oxide surface, but also the carbon that accumulates on all surfaces within the system. The effect of oxygen is to facilitate the carbon removal by the formation of carbon oxides. This carbon would otherwise be removed as C_2F_6, requiring the consumption of fluorine. So oxygen both exposes the silicon or oxide surface to the etching species, and also increases

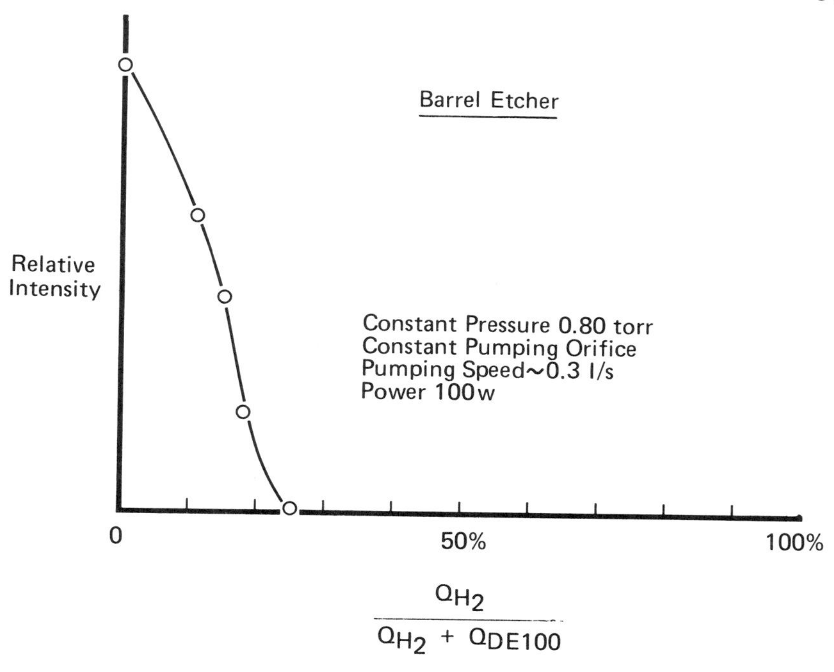

Figure 7-16. Variation of the intensity of the fluorine 7037Å emission line in a high pressure barrel etching system versus the proportion of hydrogen molecules in the input gas (Chapman et al. 1977)

the fluorine concentration in the discharge and flux at the wafer (Table 7-2). Hydrogen has just the opposite effect: it has a high affinity for fluorine and readily forms HF. The concentration and supply of fluorine atoms is thus depressed, so the silicon and carbon on the wafer surface are etched more slowly by a fluorine-deficient gas (Table 7-2).

Three of the observed phenomena (Figure 7-9) can then be understood (Table 7-3). The change in silicon etch rate by addition of hydrogen or oxygen is fairly straightforward. Silicon dioxide etches differently from silicon because of its 'built-in' oxygen supply which is liberated as etching proceeds. So the etching of oxide is always accompanied by an 'oxygen effect', and additional oxygen is somewhat superfluous. For similar reasons, the effect of the addition of hydrogen when oxide is being etched in CF_4 is countered by the liberation of oxygen, and so a great deal of hydrogen is needed to suppress oxide etching.

The loading effect in silicon results from having a limited supply of etching species to deal with whatever area of silicon is presented to be etched, with a

Table 7-2 The Effects of Added Gases

Hydrogen — Reduces fluorine concentration by combination to form HF

Oxygen — Increases fluorine concentration by:

(a) Combining with carbon from reaction gas, which would otherwise require fluorine.

(b) Reacting with CF_3 to liberate F

consequent etch rate-silicon area inverse relationship (Mogab 1977). When etching increasing areas of oxides, the effect of the accompanying liberation of oxygen tends to offset the consumption of etching species.

We still have to account for the directional etching phenomenon. There was a suggestion in the optical spectroscopy results that atomic fluorine was responsible for the plasma etching of silicon. Chapman and Minkiewicz (1976) found

Table 7-3 Interpretation of Observed Phenomena

Hydrogen — Removes fluorine from the reaction chamber

Oxygen — Increases fluorine concentration

	Si	SiO_x
H_2	Scavenges F, etch rate decreases	H_2 effect offset by liberated O_2 effect
O_2	F concentration increases, so does etch rate	Already has a private supply of O_2
Loading	F atoms divided over increasing Si area, etch rate decreases	Liberation of more oxygen with increased SiO_2 area helps overcome F depletion

that silicon could be etched by molecular fluorine gas without a plasma, but only at a very low etch rate ~ 100 Å/minute in 1 torr of gas, corresponding to a reaction probability (for each molecular 'bounce' on the silicon surface) $\sim 10^{-6}$, compared with an estimated value $> 10^{-2}$ for the active species in an RIE system, calculated from flow rate behaviour (q.v. below). Quartz was not etched significantly by molecular fluorine until exposed to short wavelength light, which also increased the silicon etch rate. This gives an example of how effective a plasma can be in creating chemically active species, with the implication in this case that molecular fluorine, already very reactive by most standards, can have its reactivity enhanced by more than four orders of magnitude by the plasma. Although the considerable activity of fluorine atoms was already known (Jones and Skolnik 1976), the synergism with the plasma was less clear.

Comparable experiments using atomic fluorine became feasible when Winters and Coburn (1979) found that XeF_2 gas could be used as a source of atomic fluorine. XeF_2 is dissociatively chemisorbed onto the surfaces of silicon and silicon compounds; the xenon then evaporates off, leaving two independent chemisorbed fluorine atoms. Winters and Coburn found that silicon could be etched at room temperature without radiation or any other external energy input, with a reaction probability of $\sim 10^{-2}$, which could then be enhanced by simultaneous electron or ion bombardment (Coburn and Winters 1979a, 1979b). The effect of simultaneous XeF_2 and ion bombardment was greater than the sum of each (chemical or sputtering) individually (Figure 7-17).

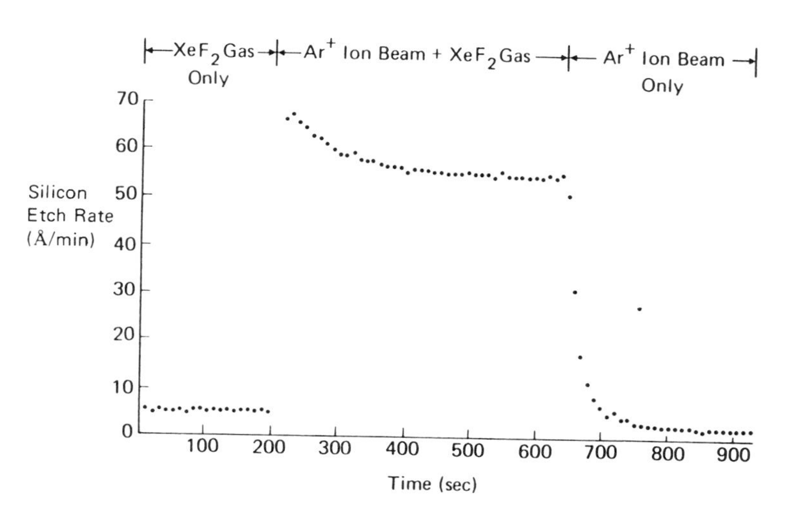

Figure 7-17. Ion-assisted gas-surface chemistry using Ar^+ and XeF_2 on silicon (Coburn and Winters 1979a)

As with molecular fluorine, Winters and Coburn found that silicon oxide was not etched by XeF_2, although surface analysis revealed that there was a limited fluoridation of the surface, indicating that the dissociative chemisorption had again taken place. Etching did however proceed if the oxide sample was simultaneously exposed to a beam of electrons or ions (Figure 7-18). This provides a means for etching a pattern into silicon compounds by 'direct write' electron beam lithography, without the use of a photoresist mask (Chapman and Winters 1978).

The experiments using XeF_2 seem to confirm, for etching of silicon and silicon oxide in fluorocarbon gases, mechanisms implied in some earlier results: if

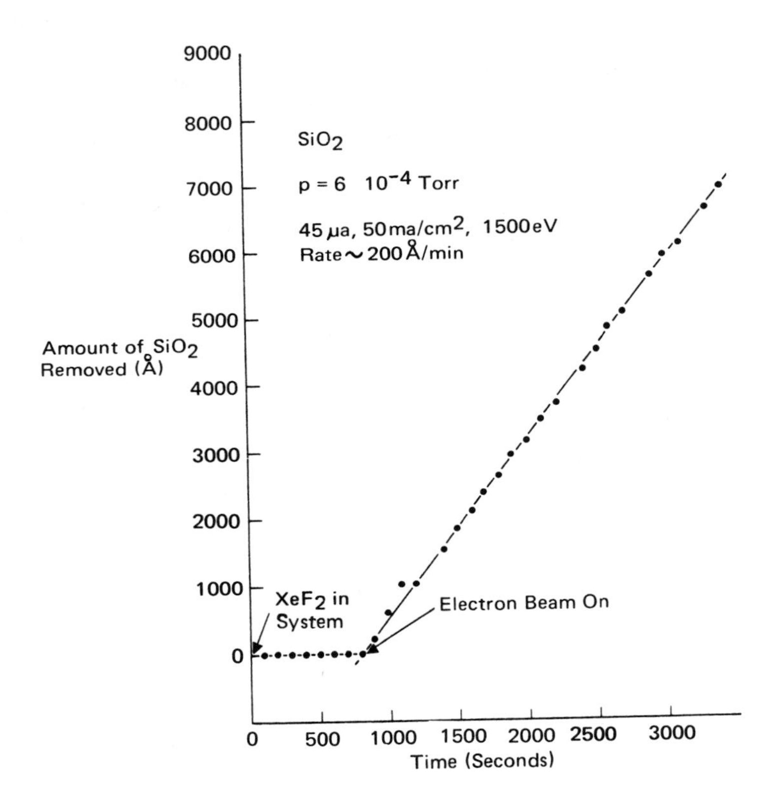

Figure 7-18. Electron-assisted gas-surface chemistry using 1500 eV electrons and XeF_2 on SiO_2; p (total) = 6 10^{-4} torr, with most of the ambient gas being xenon. Neither exposure to XeF_2 nor an electron beam produces etching by itself. Simultaneous exposure produces an etch rate of ~ 200 Å/min (Coburn and Winters 1979a, 1979b)

one protected silicon oxide from ion bombardment and electron bombardment, for example by moving it from target to counterelectrode in a reactive ion etching system at ~ 100 mtorr (Viswanathan 1977), the etch rate dropped precipitously, whilst the etch rate of silicon was much less affected. In addition, oxide was always etched directionally in an RIE system, whereas silicon could sometimes be undercut. In a barrel reactor, silicon was always etched quasi-isotropically. The inference made is that oxide etches only with the assistance of (directional) ion or electron bombardment, while silicon has both a purely chemical and a radiation-enhanced component. Lehmann and Widmer (1978) have demonstrated how the bending of the electric field lines at the edge of the cathode in an RIE system leads to corresponding distortion in the directional etching of silicon oxide in that position.

Results obtained in a triode plasma etching system (Chapman and Minkiewicz 1979), where the energy of ion bombardment can be continuously varied (described further in "Two Other Reactors"), tend to confirm the preceding model for silicon. However, one needs to explain how silicon oxide can be etched rapidly in a barrel reactor, even inside an etch tunnel where ion bombardment is minimal (Figure 7-19). In these barrel etchers, the oxide etch rate does increase with time as the reactor temperature increases, unlike RIE systems where ion bombardment appears to dominate temperature effects (Figure 7-20). It appears that the silicon oxide – fluorine reaction can also be activated by heating as well as by bombardment, and this is not surprising. This has been confirmed by the experiments of Flamm et al. (1979): thermal oxide was etched downstream, and

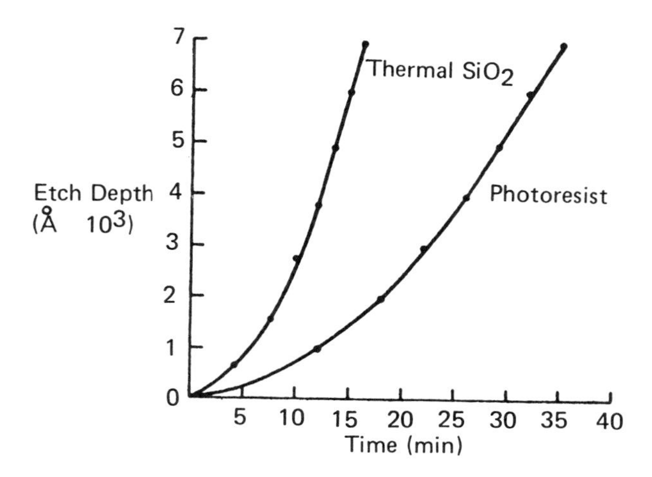

Figure 7-19. Etch depth vs time for thermal SiO_2 and photoresist within an etch tunnel, in CF_4/O_2 at 1.2 torr and 400 W (Clark 1976)

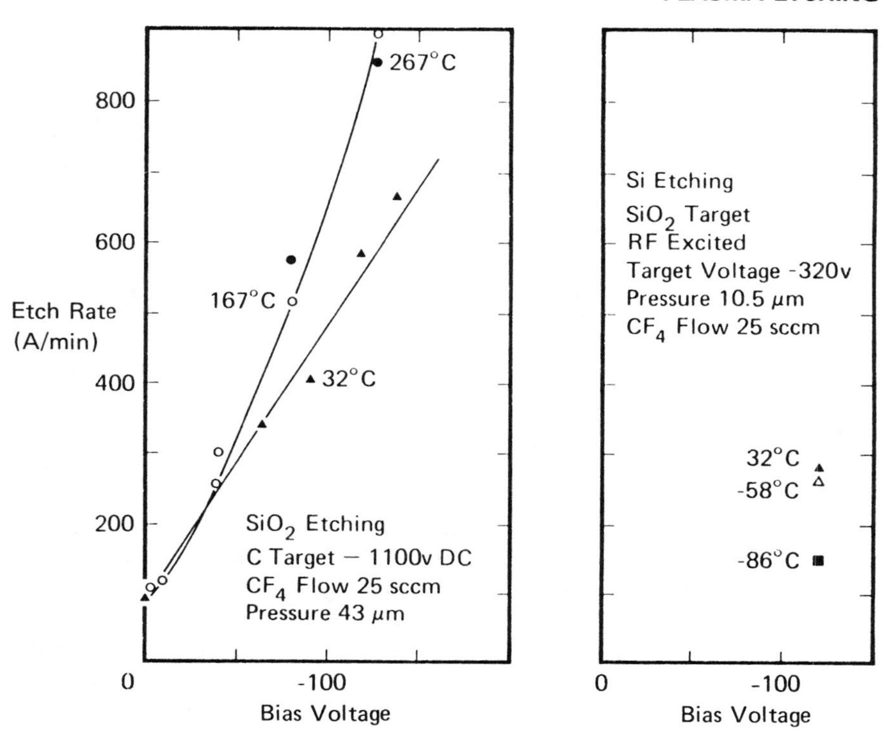

Figure 7-20. Temperature effects in the triode plasma etching of Si and SiO_2 in CF_4

shielded, from a glow discharge of molecular fluorine used as a source of atomic fluorine. The reaction probability of the atomic fluorine ranged from 9.9 10^{-6} at 250K up to 8.5 10^{-5} at 365K. In a similar study, Flamm (1978) found that CF_3 radicals also etched SiO_2 with a reaction probability ~ 10^{-5}, and these radicals would therefore also be effective in the etching of SiO_2 by CF_4 in barrel etchers.

Mogab (1979) has discussed the use of gas mixtures, unfortunately not specified, to overcome the loading effect. The rationale for their operation includes mechanisms for etching which are somewhat similar and somewhat different from the preceding. The gas mixture requires an etchant species A and a recombinant species B. The role of B is to make the active lifetime of A so short that unless A etches very close to where it is generated, it does not etch at all. This eliminates the 'sharing' that would otherwise lead to a loading effect.

When the ratio of the gas species is changed, so does the etch profile. A preponderance of the etchant species A leads to an isotropic profile. Increasing the concentration of the recombinant species B reduces the etch rate, eventually to zero on the side walls of the etch pit; however, at the base of the pit where radiation increases the effectiveness of the etchant species, the reaction proceeds so that directional etching is achieved. This is clearly consistent with the model of radiation-enhanced etching discussed earlier.

It seems that a great deal of progress has been made towards understanding the etching mechanisms of silicon and silicon oxide in fluorine-based gases, but further refinements are required, particularly in the areas of heating effects, assessment of the relative magnitudes of etching by different species, and the activation of these processes.

SELECTIVE ETCHING AND PLASMA POLYMERIZATION

In the previous section, we saw that the addition of hydrogen to a CF_4 discharge caused the etch rate of silicon to decrease whilst the silicon oxide etch rate remained fairly constant. This is evidently a way of *selectively etching* the oxide, by obtaining a favourable *etch rate ratio* (ERR) between the two materials. The attainment of selectivity is of considerable practical importance. We might, for example, wish to etch through an insulating silicon dioxide layer but stop etching when an underlying silicon emitter region is reached.

As we saw in the previous section, the action of the H_2 is to combine with F radicals to form HF, and so 'tie up' fluorine which would otherwise etch the silicon. The carbon depositing on the silicon from the impact-dissociated fluorocarbon ions is then able to hinder etching of the silicon. Evidently one needs to control the F:C ratio in the system, and adding hydrogen is one way of doing this. Other gases should also work, and solid surfaces can scavenge the fluorine too, particularly if they are subjected to ion bombardment. Cathode materials such as carbon or silicon are quite effective at reducing the F:C ratio and achieving selective etching. One could also produce essentially the same result by using gases containing less fluorine per carbon, and this is the basis of several selective etching processes using gases such as C_2F_6, C_3F_8, or C_5F_{12} etc.

Given that *some* hydrogen depresses the silicon etch rate, why not add so much that the etch rate drops to zero? The reason is that the silicon etch rate would actually become negative, i.e. we would start to deposit a coating on the silicon. The coating would be a polymer, and the deposition process is referred to as *plasma polymerization*. The polymer is of the form $(CF_2)_n$, or a related form; it is caused by the association of fluorocarbon radicals on surfaces or, if the pressure is high enough, in the gas phase. Polymerization is not always unwelcome:

plasma polymerized coatings are being explored as protective coatings (Dittmer 1978), and they can also have interesting electrical properties (Hetzler and Kay 1978). Plasma polymerized ethylene coatings have been produced by Morosoff et al. (1978). There are many examples and applications in the literature, and apparently plasma polymerization is another of the useful applications of glow discharges.

But polymerization has an uneasy role in selective plasma etching. On the one hand, we try to encourage some carbon to form on the silicon surface by limiting the F:C ratio. On the other hand, if we go too far, polymerization occurs and etching stops everywhere. The achievement of high etch rate ratios is therefore an exercise in trying to get as close to the onset of polymerization, or the *polymer point* as it is known, without actually depositing polymer on to the wafer.

The balance between etching and polymerization seems to be a very sensitive one which can be 'pushed' either way. Ion bombardment appears to have a major effect, with increased bombardment pushing the balance towards etching. Indeed, one has the situation in RIE systems where etching is occurring on the cathode, and polymerization on the walls. Mauer and Carruthers (1979) have made a study of how the various plasma process parameters control the onset of polymerization.

Selective etching of silicon dioxide: silicon has been reported by many groups of people, with reported etch rate ratios varying from 5:1 up to 50:1. All use the same basic process of controlling the F:C ratio.

Achieving the selective etching of silicon over silicon dioxide is an easier matter, and experiment is in accord with our model. The silicon etch rate is enhanced by adding oxygen to the CF_4 to remove any deposited carbon from the silicon surface. The silicon dioxide etch rate is decreased by shielding it from ion bombardment and keeping it cool. Etch rates in excess of 50:1 can be achieved.

GLOW DISCHARGE ASPECTS OF THE REACTORS

In Chapter 5, we have discussed the background for assessing the differences between the various types of reactors. It appears from the discussion in the previous section that ion bombardment is very important in plasma etching, and heating effects play a role, too. The main division, then, is between the barrel reactors on one hand, and the two types of diode reactors on the other.

In the barrel reactors, the wafers are immersed in the glow, virtually decoupled from ground (Figure 7-21). The surface potential of these wafers is thus not very different from that of the surrounding glow, and so they are subjected only to modest ion bombardment, probably only a few volts or at most tens of volts

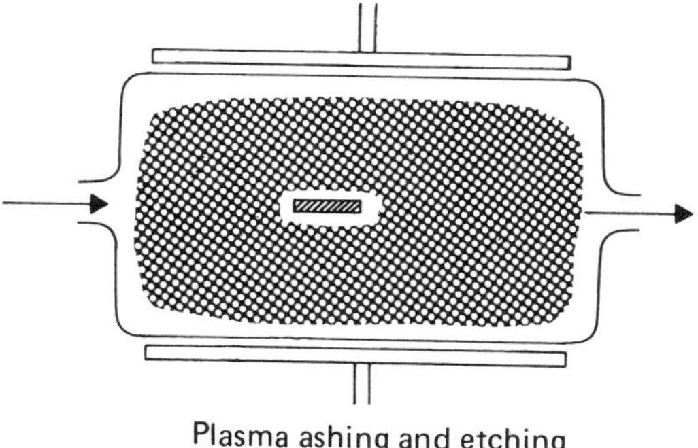

Figure 7-21. Plasma ashing and etching

(Vossen 1979). Inside an etch tunnel (Figure 7-22), the ion *flux* is drastically reduced too. Because of their thermal isolation, wafers would be expected to heat up readily, and have been found to do so.

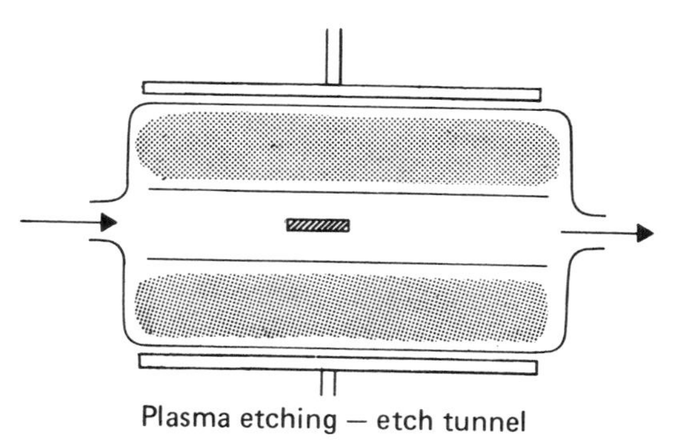

Figure 7-22. Plasma etching with etch tunnel

Planar diodes (Figure 7-23) have the wafer placed on the *grounded* electrode (often temperature controlled) and this has misled some people to expect only low energy ion bombardment there. However, these systems have large electrodes of similar area, and the analysis of Chapter 5 leads us to expect nearly equal sheath voltages at each electrode, having magnitudes of approximately one

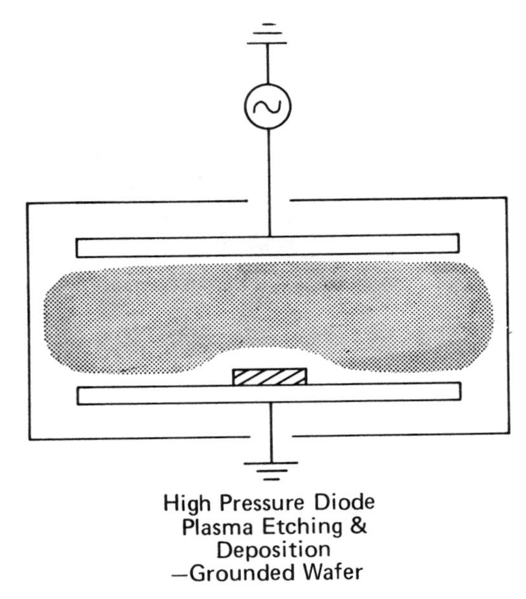

Figure 7-23. High pressure diode plasma etching and deposition-grounded wafer

quarter of the applied rf peak-to-peak voltage (Chapter 5) and so typically 75 V – 250 V for practical plasma etching processes. This is confirmed by electrical measurements and by observation of the dark spaces formed at the sheaths. The energy of ion bombardment on an electrode is determined by the sheath voltage and by collisions in the sheath, as discussed in Chapters 4 and 5.

Reactive ion etching systems (Figure 7-24) are essentially converted sputtering systems. In sputtering, intense ion bombardment at the target is required, and nowhere else. This is achieved by making the target area much less than the area of the grounded chamber and baseplate. As a result, the sheath at the walls is very small, but the sheath at the target, and hence at the wafer, is about half the rf peak-to-peak, amounting typically to -300 V.

The different operating pressures of the two diode systems should not influence ion bombardment energies very greatly, because the larger number of collisions per unit length at the higher pressure is offset by the consequently much thinner dark space sheath. The two effects would be expected to balance in a dc system, but the detailed pressure-dark space thickness relationship is not known for rf systems (at least, not by me). Therefore, it appears that in a planar diode reactor, the sheath voltages and ion bombardment energies are smaller, but not so much smaller, than in an reactive ion etching system. This means that in both systems one must be aware of the potential problems posed by sputtering and backscattering, which we have discussed in Chapter 6.

However, there is another distinction between these two types of planar diode reactor, and this follows from their different operating pressures. Recombination of active species with other gas phase species, both on the walls and in the gas phase, becomes more common with increasing pressure, and may be a major effect in high pressure planar diodes. This, and the fact that any other gas phase process will similarly become more likely with increasing pressure, may explain why the enhancement of silicon etching by the addition of oxygen to CF_4, is quantitatively more effective in higher pressure systems.

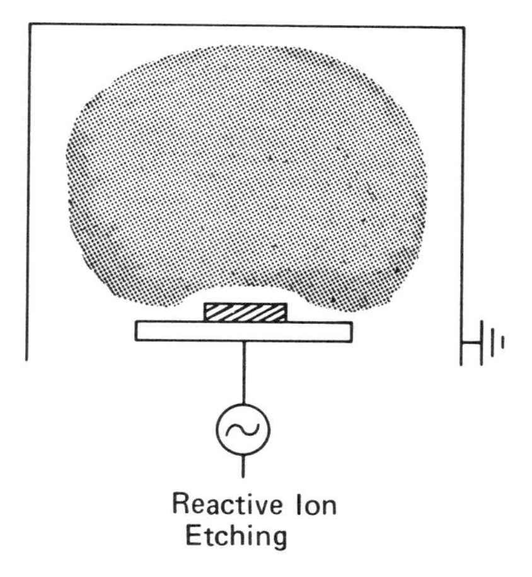

Figure 7-24. Reactive ion etching

TWO MORE REACTORS

Although the etching process is enhanced by the various radiation phenomena that take place in a discharge, there are some deleterious effects, too. Photoresist materials are easily degraded under ion bombardment and heating, setting a limit to the amount of rf power that can be applied to a system, and hence a limit to the etch rate. And certain devices, such as MOS structures, are sensitive to the etching environment (De Maria et al. 1979). The observation of these various interactions has led to the development of two other types of etching reactors in which the design philosophies are somewhat different:

Chemical Dry Etching

The *chemical dry etching system* (Horiike and Shibagaki 1975), shown in Figure 7-25, physically separates the reaction region from the discharge, which is excited by microwaves. The wafers are thus shielded from virtually all radiation and bombardment associated with the glow. Some results for polysilicon and silicon nitride etching are shown in Figure 7-26. Predictably,

- SiO_2 is hardly etched at all, resulting in etch rate ratios of $< 1{:}50$ with respect to polysilicon. Note that this is consistent with a model requiring ion-enhanced chemistry for SiO_2 etching.
- Etching of silicon is isotropic, because of the absence of (directional) bombardment by ions or electrons.

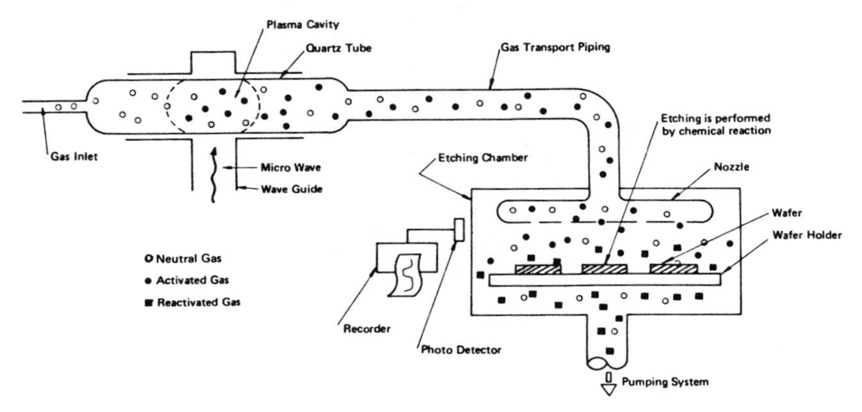

Figure 7-25. Chemical dry etching

A further interesting feature of this reactor is that monitoring of the etching process and endpoint determination can be carried out by observing (in the absence of the bright discharge glow) the weak optical emission due to the reaction at the wafer surface, the phenomenon known as *chemiluminescence* (Chuang 1979).

Triode Plasma Etching

A rather different rationale has been used by Chapman and Minkiewicz (1979) in the development of the *triode plasma etching system*. Since bombardment seems to be necessary for directional etching, and since directional etching is necessary for fine resolution, then the control of ion bombardment is fundamental in this design. Wafers etched in a diode system (high or low pressure) are placed on an electrode and the energies of ion bombardment to which they are subjected, are determined by the needs of sustaining the discharge and of generating reactive species, rather than by a need for a specific ion bombardment energy. This limitation can be avoided by maintaining the discharge with two electrodes, and then introducing the wafer on a third electrode that can be biased so as to control, quasi-independently, the energy and flux of ions on the wafer (Figure 7-27). One practical version of this design is shown in Figure 7-28 and is essentially the same as a bias sputtering system (Chapter 6). Another version (Chapman 1979) uses a hot filament discharge (see Chapter 6) and enables controlled etching to be achieved without any high voltages (> 100 V) anywhere in the processing chamber. We also learn from the hot filament system that ion-enhanced etching of Si and SiO_2 at first proceeds with increasing rate as the ion energy increases, but becomes more constant above about 200 eV.

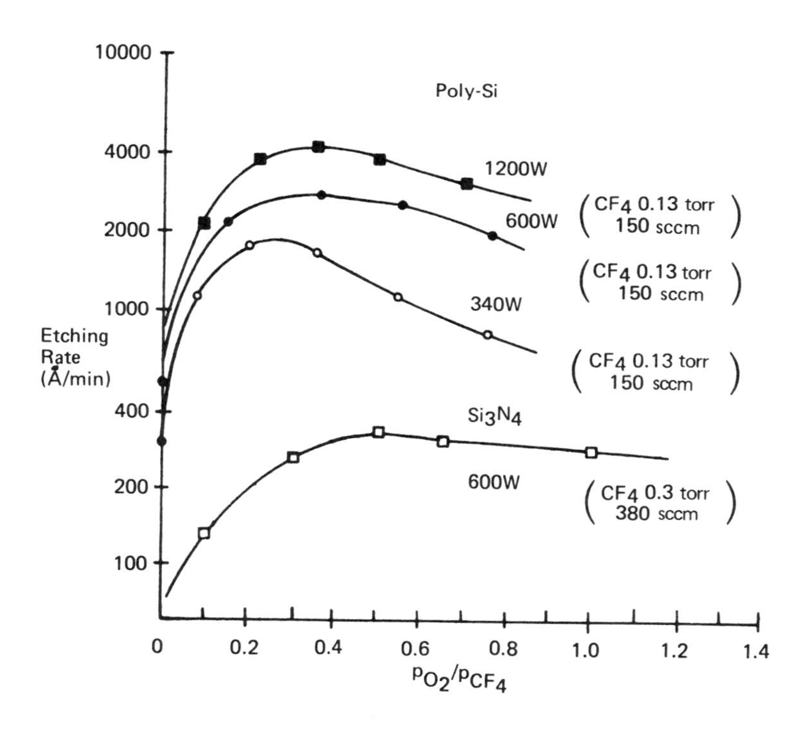

Figure 7-26. Etch rates of polysilicon and silicon nitride in the chemical dry etching system (Tokuda Seisakusho)

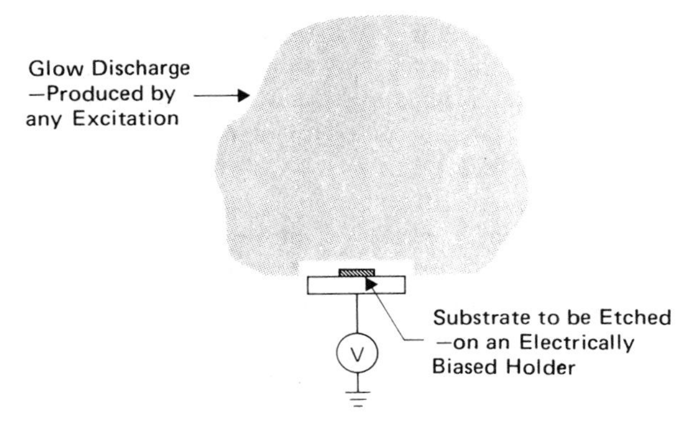

Figure 7-27. The principle of triode plasma etching

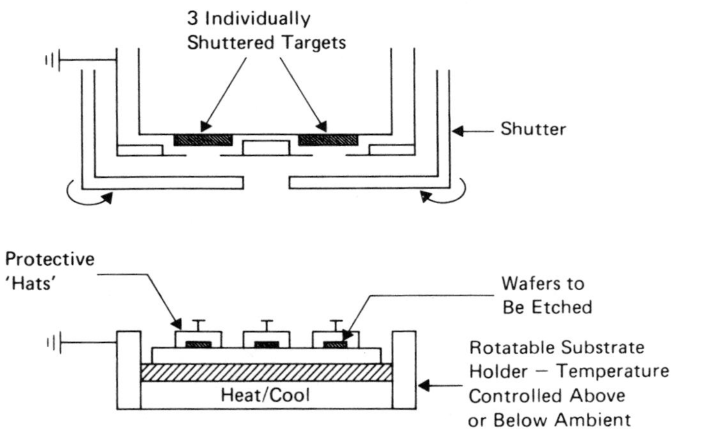

Figure 7-28. Schematic of a cold cathode triode etching system (Chapman and Minkiewicz 1979)

Similar etch rates and etch behaviour are obtained (Figure 7-29) in the triode systems as in diode systems, although at lower ion energies. But as an additional feature, the profile of the etch pit can be controlled by the ion energy (Figure 7-30), lending credence to the etching model involving a chemical component and an ion-assisted component. (A microwave system with the wafer in contact with the glow has been reported by Suzuki et al. 1977, and gives consistent results).

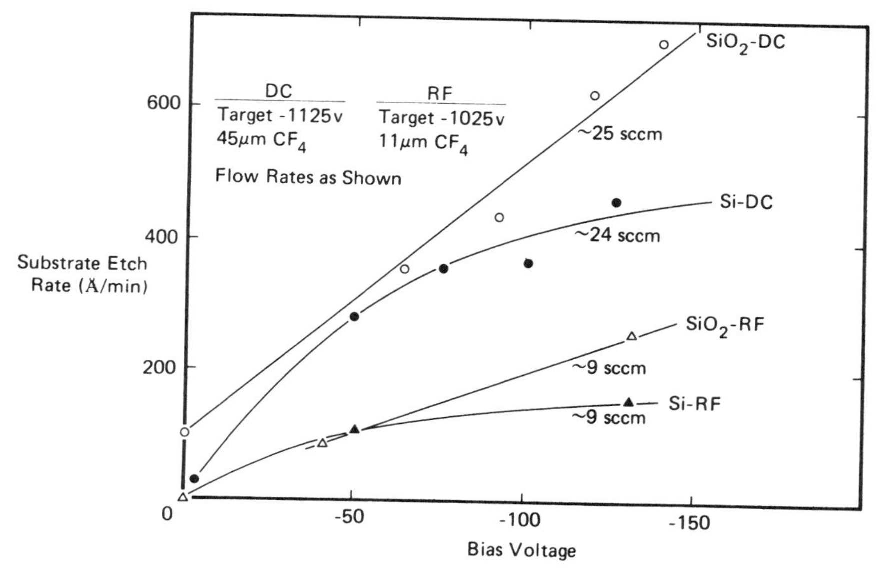

Figure 7-29. Etch rates of Si and SiO_2 versus substrate bias, using a carbon target. The target is rf or dc excited, as shown; the wafers are rf biased in each case to the dc offset indicated (Chapman and Minkiewicz 1979)

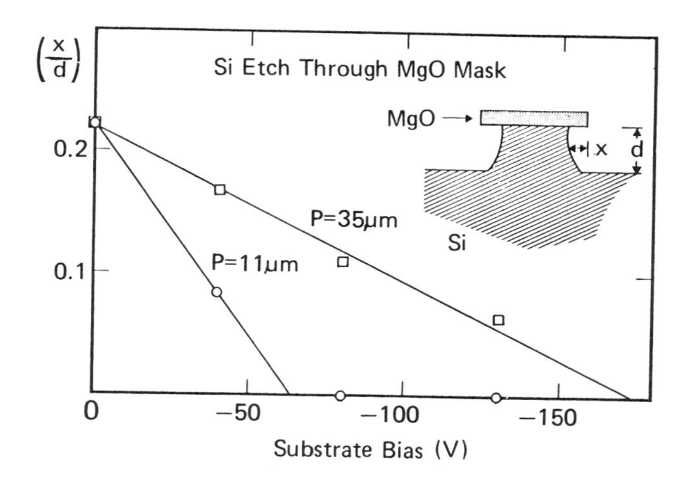

Figure 7-30. Etch isotropy in the cold cathode triode system. The figure shows the scaled dimension (x/d) as a function of substrate bias for two pressures. The substrate bias needed to generate vertical side walls is a strong function of the system pressure (Chapman and Minkiewicz 1979)

GAS FLOW RATE EFFECTS

The dependence of plasma etch rates on the flow rate of the reaction gas seems to be a well-known phenomenon in all the various types of plasma etching systems, although there are rather few examples in the published literature. The general observation, examples of which are shown in Figures 7-31, 7-32, 7-33, and 7-34, is that the etch rate rises rapidly with increasing flow rate, reaches a maximum, and then decreases at higher flow rates. The precise shape of the dependence seen, and the absolute flow rates and etch rates involved, will vary from system to system and from material to material; in some cases, only portions of the dependence will be observed. This flow rate dependence occurs because of a superposition of two different effects:

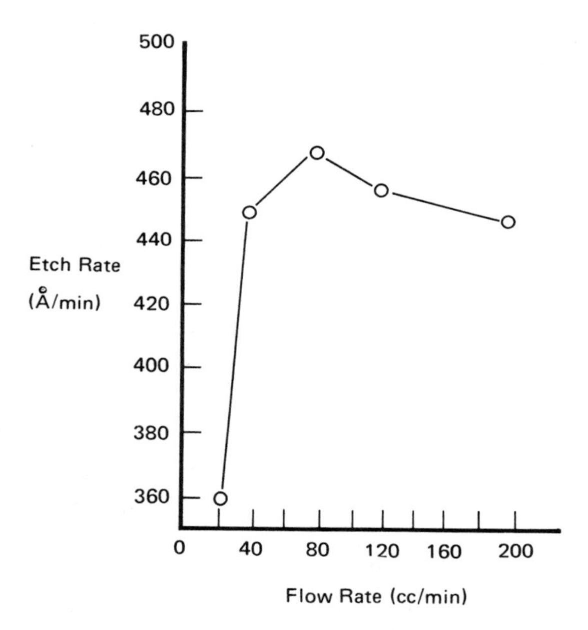

Figure 7-31. Variation of quartz etch rate with flow rate (Pickar et al. 1975)

Low Flow Rate Region

The reason for the behaviour at low flow rates is actually rather simple. As we have discussed earlier in the chapter, one deduces from mass spectrometer data that the most significant silicon product of the etching process is SiF_4 and this is common to the etching of silicon itself, and of the silicon compounds commonly used in microelectronics (oxide, nitride etc.) The etching process for all these

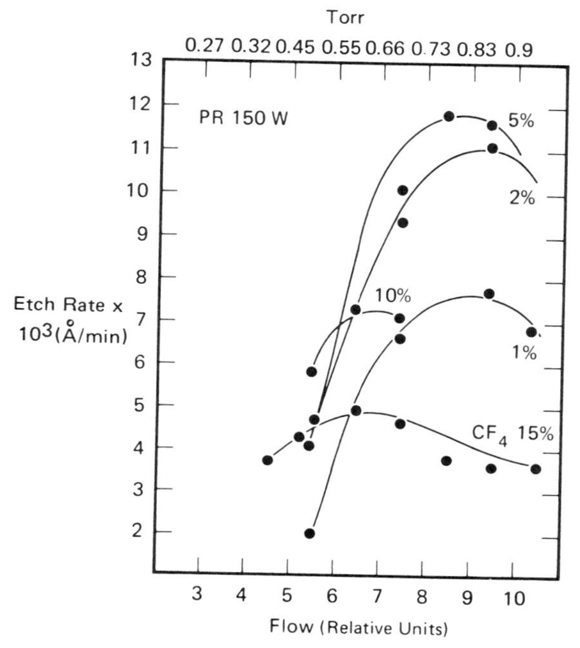

Figure 7-32. Variation of silicon etch rate with flow rate in mixtures of CF_4 and oxygen in the percentage shown (Nakane)

materials therefore consists essentially of converting CF_4 to SiF_4, although there are of course other products containing fluorine. If one knows the area being etched and its etch rate, then one can calculate the number of molecules of Si etched per unit time. In silicon there are $5.00\ 10^{14}$ silicon atoms per square centimetre per Å depth; for silica the average equivalent is $2.66\ 10^{14}$. Similarly, one can convert the CF_4 flow rate into the system into a flux of CF_4 molecules per unit time; 1 standard cc/min (sccm) is equivalent to $2.69\ 10^{19}$ molecules/min. We can then define a *utilization factor* U as:

$$U = \frac{\text{Etch rate of Si atoms}}{\text{Input flux of } CF_4 \text{ molecules}} \tag{1}$$

which gives the fraction of CF_4 molecules converted to SiF_4, assuming this to be the only silicon product.

If

A_S = the substrate area (cm^2)

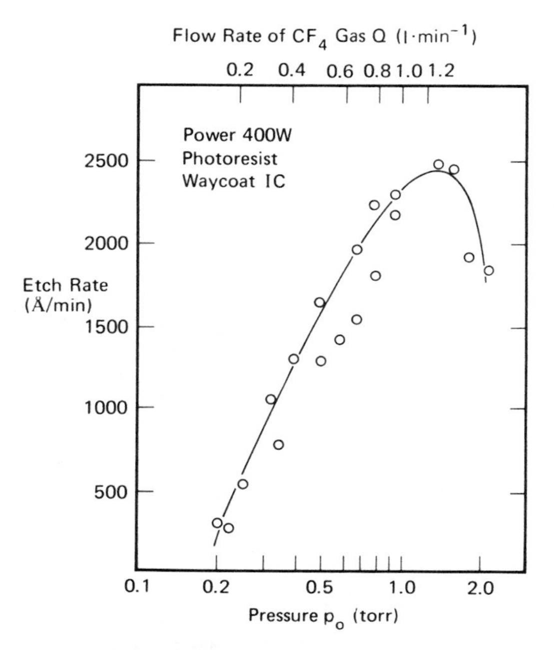

Figure 7-33. Variation of photoresist etch rate with flow (Mimura)

$$Q = \text{flow rate (sccm)}$$
$$E = \text{etch rate (Å/min)}$$

then

$$U = 1.85\ 10^{-5}\ A_S E/Q \text{ for Si} \tag{2}$$
$$= 9.89\ 10^{-6}\ A_S E/Q \text{ for } SiO_2 \tag{3}$$

If we now calculate utilization factors for plasma etching situations, they are sometimes found to be unexpectedly high, even approaching unity in some cases. As an example, if a 10 inch silicon electrode is etching at 1000 Å/min in a CF_4/O_4 mixture with a CF_4 flow rate of 20 sccm, then the utilization factor is 0.47. With high utilization factors, reaction products can dominate the composition of the discharge (Figure 7-35). Utilization factors are found to be high in situations where there is a large flow rate effect for low flows. In other words, this low flow rate effect is simply due to an inadequate supply of reactant gas.

Figure 7-36 illustrates how extensively the etching gas is consumed under high utilization factor conditions. The figure shows a repeated mass spectrometer trace of the mass 69 (CF_3^+) and 85 (SiF_3^+) peaks. At the commencement of etching (t = O), the 69 peak is large but there is no 85 peak. As etching proceeds, the etching gas is consumed so that the 69 peak decreases, and simultaneously the 85 peak of the reaction product rises; eventually the peaks reach equal height, giving a measure of the steady state concentrations.

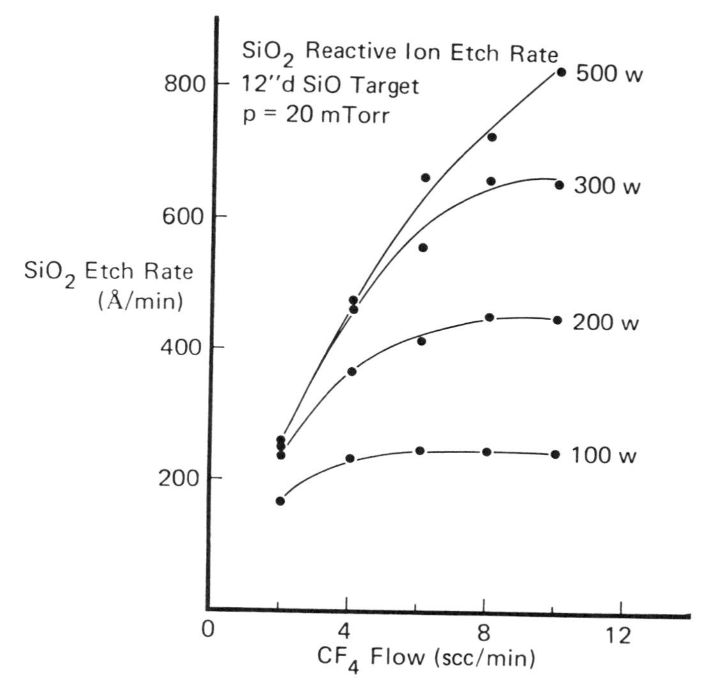

Figure 7-34. Variation of etch rate of sputtered quartz with flow rate of CF_4 for various power inputs, in a reactive ion etching system (Chapman et al. 1979)

It is difficult to analyze relevant results shown in the literature because of inadequate information, particularly concerning substrate area. Even when the substrate area is apparently known, there is often a problem because of the additional effective substrate area presented by quartz chambers, windows, etc. The limiting case, which we would not expect to attain, would be 100% utilization – i.e. all of the input CF_4 converted to SiF_4. For given experimental conditions, one can calculate a 100% utilization line (etch rate being linearly dependent on flow under these circumstances). With the data of Figure 7-34, the various etch rate curves for different conditions asymptotically approach this line at low flows, as expected (Figure 7-37).

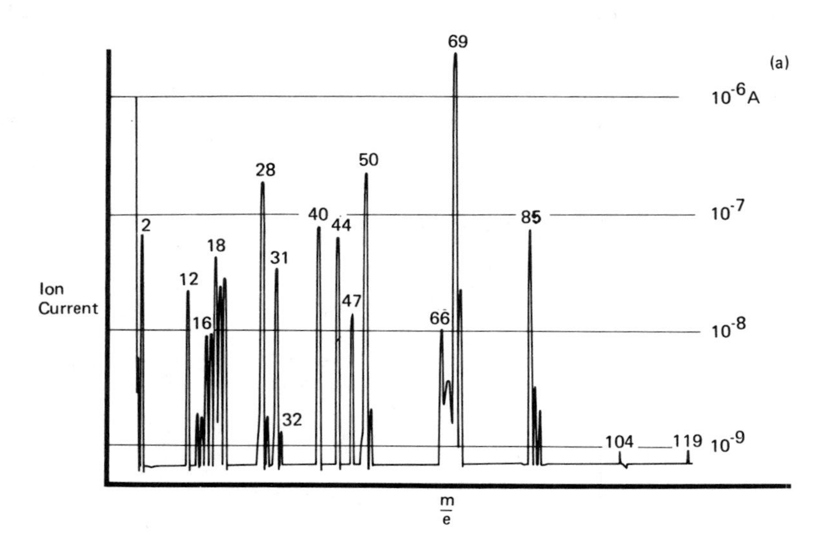

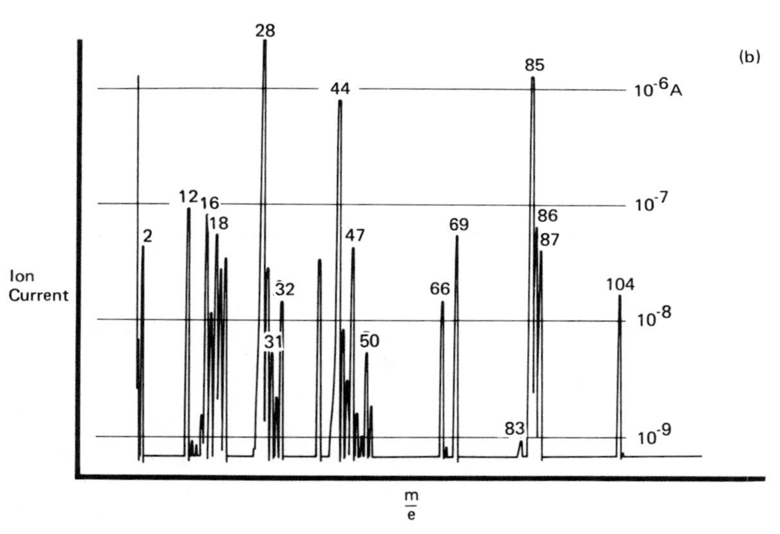

Figure 7-35. Mass spectrometer data (n.b. log scale)

(a) low utilization factor
(b) high utilization factor
(Chapman et al. 1979)

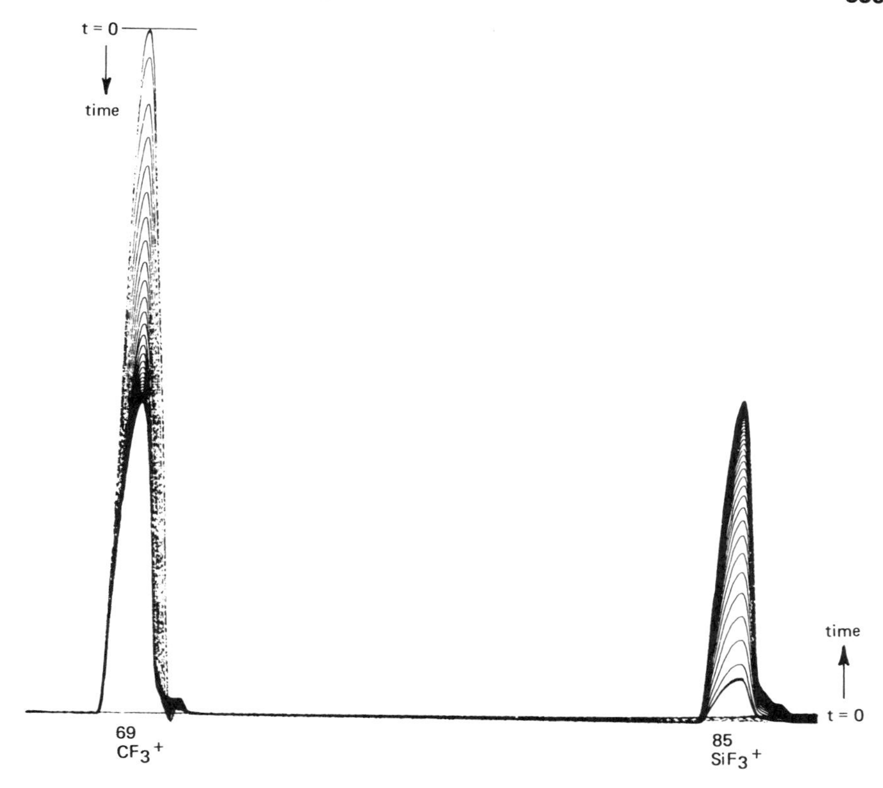

Figure 7-36. Time dependence of the CF_3^+ and SiF_3^+ mass spectrometer peak heights after etching commences

High Flow Rate Region

The high flow rate region does not appear to have any flow rate dependencies as dramatic as those for high utilization rates. We can predict, however, that as flow rates increase, the accompanying increase in pumping speed (for constant pressure operation) will eventually produce a situation where active species can be pumped away before they have the opportunity to react. By analyzing the competitive consumption of active species by etching and by pumping, an etch rate is predicted (Chapman and Minkiewicz 1978) which varies as $(1 + x)^{-1}$, where x is proportional to flow rate. The results suggest a reaction probability $\sim 10^{-2}$, consistent with the results of Coburn and Winters (1979a).

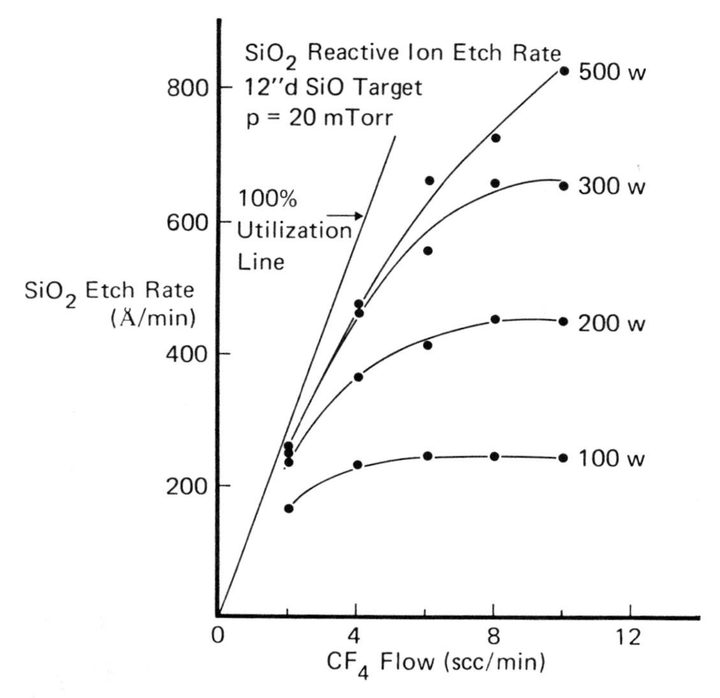

Figure 7-37. Showing the 100% utilization line superimposed on Figure 7-34, for complete conversion of CF_4 to SiF_4 (Chapman et al. 1979)

Overall Flow Rate Dependence

The overall flow rate dependence is the product of the two effects described above. The etch rate would be determined by the generation rate of reactive species in the discharge, but is limited by lack of reactant gas at low flows and by pumping of active species at high flows (Figure 7-38).

It is surprising how frequently flow rates affect the etching process, sometimes in rather unobvious ways. Lack of reproducibility, from system-to-system or from run-to-run, is a common complaint about plasma etching; e.g. Figure 7-39 apparently shows two quite different dependencies of etch rate on input power, one linearly dependent and the other almost independent. These curves were obtained in the *same* system under different conditions; curve (ii) relates to a flow rate limited situation, whilst curve (i) is generation rate limited. So one has to be very careful about interpretation of data.

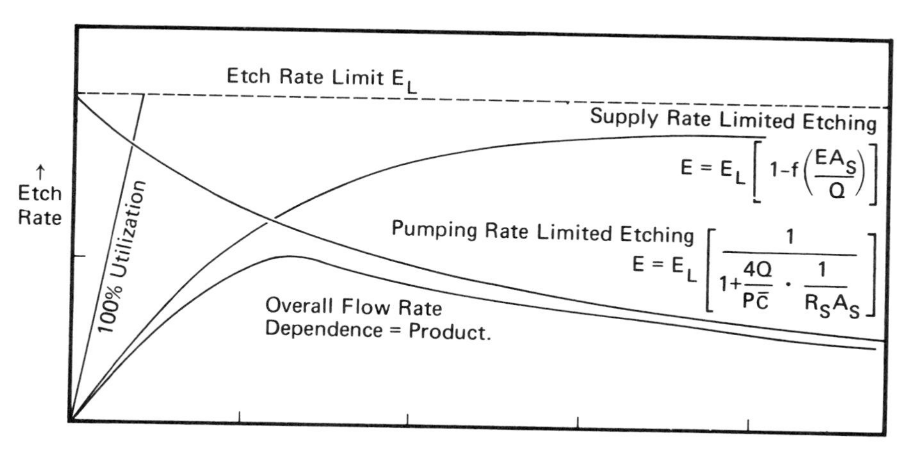

Figure 7-38. Theoretical etch rate versus flow rate for the flow-rate-limited and pumping-speed-limited cases, and their product – the generalized flow rate dependence. The etch rate limit E_L is determined by the generation rate of active species; p is the system pressure, $\bar{c}$ the mean speed of the gas molecules, and R_S the reaction probability at the wafer (Chapman and Minkiewicz 1978)

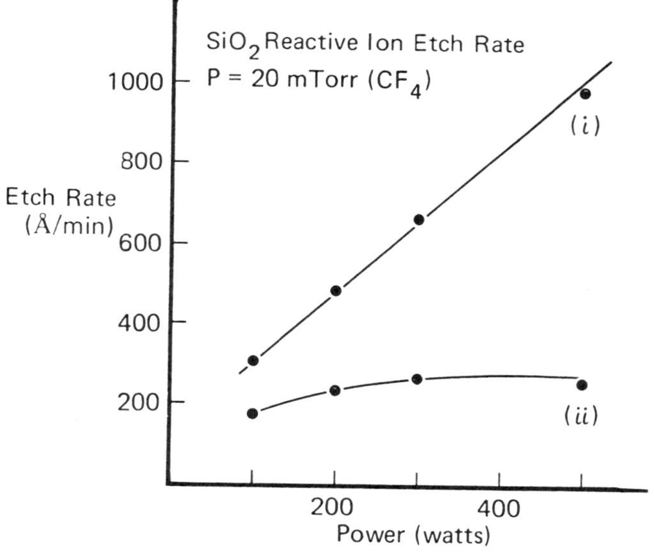

Figure 7-39. Etch rates of SiO_2 as a function of input power: (i) at 20 sccm CF_4, and (ii) at 2 sccm CF_4. From the same series of experiments as Figure 7-34 (Chapman et al. 1979)

ETCHING OF ALUMINIUM AND ALUMINIUM ALLOYS

By comparison with silicon, the etching of aluminium and its alloys is very poorly understood. Aluminium cannot be etched in fluorine-based gases since aluminium fluoride has an extremely low vapour pressure. As a result, chlorine-based gases are almost always used. And this is one reason why the aluminium etching system has been rather little researched; chlorine is extremely corrosive and most people either fear for their analytical equipment or are still repairing it from a previous encounter with chlorine! It is assumed that $AlCl_3$ is the volatile reaction product, although this product is not observed by mass spectrometry unless special precautions are taken to ensure that the aluminium trichloride is not condensed before spectrometric sampling (UTI 1977).

There are several reports of successful aluminium etching in both high pressure diode (Poulsen 1977, Herndon and Burke 1977, Heinecke 1978, Bersin 1978) (Figure 7-40) and reactive ion etching systems (Schaible et al. 1978) (Figure 7-41), but apparently not in barrel reactors. Various gases, including Cl_2, Br_2, HCl, HBr, CCl_4, and BCl_3 have been used. It seems that significant ion bombardment is required to plasma etch the aluminium, although whether this is really to activate the aluminium-chlorine reaction, or simply to remove the surface aluminium oxide coating, is less clear. It has been proposed that the oxide is reduced by the use of BCl_3 (Poulsen 1977), CCl_4 (Heinecke 1978), or etched

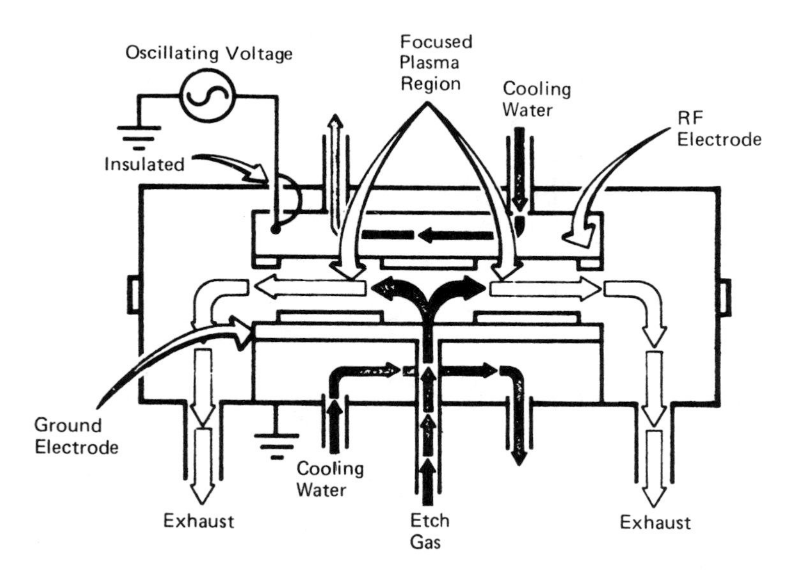

Figure 7-40. Diagram of IPC (Dionex) Planar Reactor for anisotropic etching; operating position – closed (Bersin 1978)

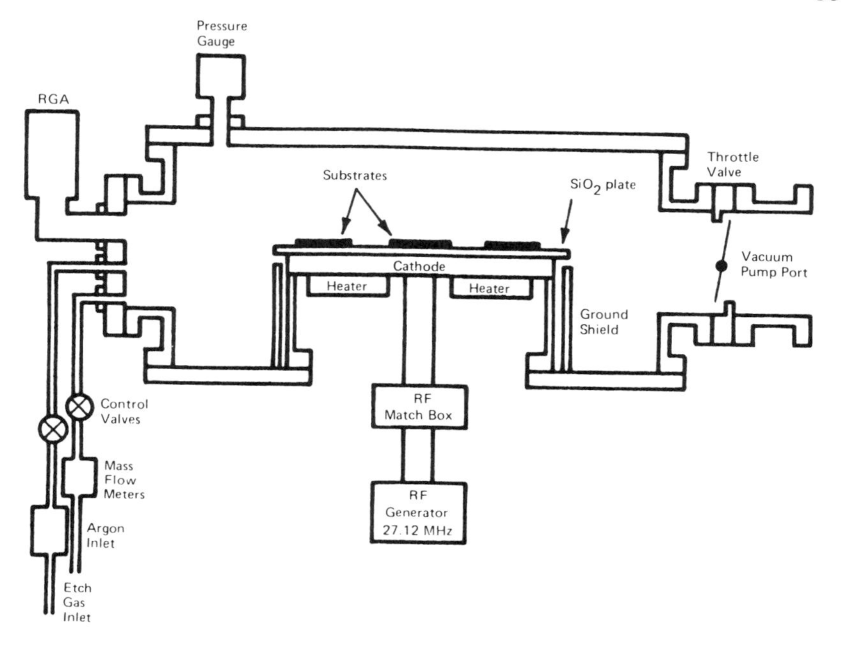

Figure 7-41. Schematic of RIE apparatus for Al-Cu-Si etching (Schaible et al. 1978)

away by sputtering. The use of significant ion bombardment is manifested in the almost exclusive finding of directional etching in aluminium (Figure 7-42).

It seems that there is far more difficulty in etching with chlorine-based rather than fluorine-based gases. These difficulties are manifested primarily as non-reproducibilities and non-uniformities. Various recipes are offered to overcome these problems, for example by a CF_4 plasma preclean of the wafer immediately prior to etching or by an initial high power etch.

Aluminium and aluminium-silicon seem to be easier to etch than the alloy containing copper, though all these alloys have been etched. The low volatility of copper chlorides makes the effectiveness of wafer heating to 200°C (Figure 7-41) during the etching of copper-containing alloys apparently easy to understand, although the etching has also been achieved without deliberate wafer heating (Herndon and Burke 1977) albeit with post-treatment which probably serves to remove copper residues, amongst others.

Metal etching may need such post-treatment. One proposal (Herndon and Burke 1977) is that Al-metallized wafers should be water rinsed immediately after etching, whilst the residue on Al-Cu-Si metallized wafers should be removed

Figure 7-42. Plasma etched microwave transistor metallization 2.0 μm thick with gaps 1.2 μm wide (Heinecke 1978)

by rinses in nitric acid, a wetting agent, and deionized water. A second proposal is to follow a CCl_4 - He plasma etching process with an in situ oxygen plasma clean (Dionex). Pitting in the sidewalls of RIE-etched aluminium has also been recently reported (Schaible and Schwartz, 1979).

SILICON ETCHING IN CHLORINE DISCHARGES

In passing, it is interesting to note that silicon, as well as aluminium, can be etched in chlorine-based discharges with some interesting differences from the fluorine-based systems (Schwartz and Schaible 1979). Silicon appears to be etched by an ion-assisted mechanism so that vertical walls, without undercut, are obtained. Silicon oxide is etched slowly, possibly only by physical sputtering, so that reasonably high Si:SiO_2 etch rate ratios can be obtained. Although silicon and silicon dioxide should both form silicon tetrachloride, there appears to be a large activation barrier in the case of the oxide.

MONITORING OF THE ETCHING PROCESS

It is very useful to monitor the etching process as it proceeds, both to ensure the integrity and reproducibility of the process and to be able to detect an *end point*, so as not to etch too far.

The most popular of the available techniques appears to be optical. One can observe the interference generated in a laser beam reflected from a dielectric layer being etched (Bondur et al. 1978, Clark and Bondur 1978, Busta et al. 1978, 1979) or one can observe the emission intensity of a particular species in the discharge (Degenkolb et al. 1976, Hirobe and Tsuchimoto 1978).

We have already seen that mass spectrometry is capable of analyzing the composition of the plasma system, and it has also been proposed as an in-process monitor and end point detector (Bunyard and Raby 1977). Analyzing the effluent in this way has a disadvantage of slower response than the optical technique.

A more novel method for detecting the end point has been proposed by Ukai and Hanazawa (1979). For the etching of aluminium in CCl_4, they have looked for the change in impedance of the glow discharge (monitored by measuring the target voltage) as the composition changes when an interface is reached, e.g. at an end point. They found good correlation with an optical technique.

The interested reader is referred to the literature for further information.

PLASMA DEPOSITION

We should briefly mention the process of *plasma deposition* or *plasma-enhanced deposition*. Some chemical deposition processes have long been carried out in the gas or vapour phase. There are a whole family of processes differing in detail, and collectively referred to as *chemical vapour deposition* (CVD) techniques. Plasma deposition is essentially the extension and enhancement of some of these processes, by using a plasma to dissociate and activate the reaction gases. The detail of the stoichiometry of the films is somewhat different by the two methods, but more importantly (at least for the semiconductor industry) the plasma process can be accomplished at lower substrate temperatures. For example, silicon nitride can be prepared from $SiCl_4$ and NH_3 by conventional CVD in the temperature range 800°-1400°C. In the corresponding plasma deposition technique, the substrate temperature can be reduced below 400°C.

Many other coatings can be deposited by plasma deposition. For example, amorphous aluminium oxide films have been deposited onto silicon substrates by the decomposition of aluminium trichloride in an rf oxygen plasma Katto and Koga 1971). Rand (1979) has reviewed the salient features of plasma deposition.

CONCLUSION

In this chapter, we have seen some of the chemical properties of glow discharges. Our treatment certainly wasn't comprehensive, and several important topics such as plasma anodisation and plasma oxidation have been omitted, unfortunately. Similarly, there were several omissions from the discussion about the uses of the physical phenomena in glow discharges, in Chapter 6. Of course, such divisions are rarely clear-cut, and processes such as reactive sputtering rely on the synergism between the physical and chemical aspects of the discharge.

In this book I have attempted to describe some of the basic properties of glow discharges and some of their applications. My intention was to give the reader some familiarity with the technology, rather than to write a definitive text book. I believe that what I wrote is largely correct, although there are probably exceptions to many of the statements; I'm not a plasma physicist, and I don't have enormous amounts of experience with all of the topics discussed herein. The book is essentially a collection of my learning, thoughts, confusions and experiences during several years in the practical implementation of vacuum and plasma processing. So don't take my comments too seriously, but use them as one perspective; I hope they are of some help to you.

REFERENCES AND BIBLIOGRAPHY

A. K. Agajanian, *Semiconducting Devices; A Bibliography of Fabrication Technology, Properties, and Applications*, IFI/Plenum, New York and London (1976)

R. L. Bersin, "Chemically Selective, Anisotropic Plasma Etching", Solid State Technology **21**, 4, 117 (1978)

J. A. Bondur, "Dry Process Technology (reactive ion etching)", J. Vac. Sci. Tech. **13**, 1023 (1976)

J. A. Bondur, W. R. Case, and H. A. Clark, "Interferometric Method for in situ Plasma Etch Rate Monitoring", ECS Abstract 303, **78-1** (1978)

G. B. Bunyard and B. A. Raby, "Plasma Process Development and Monitoring via Mass Spectrometry", Solid State Tech. **20**, 53 (1977)

H. H. Busta, R. E. Lajos, and D. A. Kiewit, "Control Plasma Etching", Ind. Res./Develop., p. 133, June (1978); Solid State Tech., p. 61, February (1979)

B. N. Chapman and V. J. Minkiewicz, previously unpublished results (1976)

B. N. Chapman, J. R. Lyerla, L. A. Pedersen, and V. J. Minkiewicz, previously unpublished results (1977)

B. N. Chapman and V. J. Minkiewicz, "Flow Rate Effects in Plasma Etching", J. Vac. Sci. Tech. **15**, 239 (1978)

B. N. Chapman, T. A. Hansen, and V. J. Minkiewicz, "The Implications of Flow Rate Dependencies in Plasma Etching", Electronic Materials Conference, Santa Barbara, California (1978); J. Appl. Phys., to be published (1979)

B. N. Chapman and H. F. Winters, "Etching Method Employing Radiation", Pending U.S. Patent Application (1978)

B. N. Chapman and V. J. Minkiewicz, "Triode Plasma Etching", Appl. Phys. Letters **34**, 192 (1979)

B. N. Chapman, "Triode Systems for Plasma Etching", IBM Tech. Discl. Bull. **21**, 5006 (1979)

T. J. Chuang, "Infrared Chemiluminescence form Surface Reactions", J. Vac. Sci. Tech. **16**, 2 (1979)

H. A. Clark, "Plasma Processing at Moderate Vacuum", Solid State Tech. **19**, 6, 51 (1976)

H. A. Clark and J. A. Bondur, "Applications of Interferometric Plasma Etch Rate Monitoring", ECS Abstract 304, **78-1** (1978)

J. W. Coburn and E. Kay, "A Mass Spectrometric Study of a CF_4 Plasma Etching Glow Discharge", Proc. 7th Intern. Vac. Congr. & 3rd Intern. Conf. Solid Surfaces, Vienna (1977)

J. W. Coburn, H. F. Winters, and T. J. Chuang, "Ion-Surface Interactions in Plasma Etching", J. Appl. Phys. **48**, 3532 (1977)

J. W. Coburn and H. F. Winters, "Mechanisms in Plasma Etching", J. Vac. Sci. Tech. **15**, 327 (1978)

J. W. Coburn and H. F. Winters, "Ion and Electron-Assisted Gas-Surface Chemistry – an Important Effect in Plasma Etching", J. Appl. Phys., to be published (1979a)

J. W. Coburn and H. F. Winters, "Plasma Etching – a Discussion of Mechanisms", J. Vac. Sci. Tech. **16**, 391 (1979b)

E. O. Degenkolb, C. J. Mogab, M. R. Goldrick, and J. E. Griffiths, "Spectroscopic Study of Radiofrequency Oxygen Plasma Stripping of Negative Photoresists. I. Ultraviolet Spectrum", Applied Spectroscopy **30**, 520 (1976)

D. J. DiMaria, L. M. Ephrath, and D. R. Young, J. Appl. Phys., to be published (1979)

Dionex (Formerly International Plasma Corp.) Operating Bulletin No. 78 – "Aluminum Etching Process"

G. Dittmer, "Plasma Polymerization of Coatings in Vacuum", Ind. Res./Develop., p. 169, September (1978)

L. M. Ephrath, "Selective Etching of Silicon Dioxide Using Reactive Ion Etching with CF_4-H_2", J. Electrochem. Soc., to be published (1979)

D. L. Flamm, "Measurement and Mechanisms of Atomic Fluorine Production in the Plasma Oxidation of CF_4 and C_2F_6", AVS Meeting, Danvers, Massachusetts (1978)

D. L. Flamm, C. J. Mogab, and E. R. Sklaver, "Reaction of Fluorine Atoms with CF_4", J. Appl. Phys., to be published (1979)

W. R. Harshbarger, R. A. Porter, T. A. Miller, and P. Norton, "A Study of the Optical Emission from an R. F. Plasma During Semiconductor Etching", Applied Spectroscopy **31**, 3, 201 (1977)

R. A. H. Heinecke, "Plasma Etching of Films at High Rates", Solid State Technology **21**, 4, 104 (1978)

T. O. Herndon and R. L. Burke, "Plasma Etching of Aluminum", Interface '77 Kodak Microelectronics Symposium, Monterey, California (1977)

U. Hetzler and E. Kay, "Conduction Mechanisms in Plasma Polymerized Tetrafluoroethylene Films", J. Appl. Phys. **49**, 5617 (1978)

K. Hirobe and T. Tsuchimoto, "End Point Detectability in Plasma Etching", ECS Abstract 302, **78-1** (1978)

Y. Horiike and M. Shibagaki, "A New Chemical Dry Etching", Proc. 7th Conf. on Solid State Devices, Tokyo, 1975; Supp. to Japan J. Appl. Phys. **15**, 13 (1976)

N. Hosokawa, R. Matsuzaki, and T. Asamaki, "RF Sputter-Etching by Fluorochloro-hydrocarbon Gases", Japan J. Appl. Phys., Suppl. 2, Pt. 1, 435 (1974)

H. G. Hughes, W. L. Hunter, and K. Ritchie, J. Electrochem. Soc. **120**, 99 (1973)

S. M. Irving, K. E. Lemons, and G. E. Bobos, "Gas Plasma Vapor Etching Process", U.S. Patent 3 615 956 (1971)

A. Jacob, "Process for use in the Manufacture of Semiconductive Devices", U.S. Patent 3 806 365 (1974)

A. Jacob, "Fluorocarbon Composition for use in Plasma Removal of Photoresist Material from Semiconductor Devices", U.S. Patent 3 951 843 (1976)

W. E. Jones and E. G. Skolnik, "Reactions of Fluorine Atoms", Chemical Reviews **76**, 563 (1976)

H. Katto and Y. Koga, "Preparation and Properties of Aluminum Oxide Films Obtained by Glow Discharge Technique", J. Electrochem. Soc. **118**, 1619 (1971)

H. W. Lehmann and R. Widmer, "Profile Control by Reactive Sputter Etching", J. Vac. Sci. Tech. **15**, 319 (1978)

H. M. Manasevit and F. L. Morritz, "Gas Phase Etching with Sulfur Fluorides", J. Electrochem. Soc. **114**, 204 (1967)

J. L. Mauer, J. S. Logan, L. B. Zielinski, and G. C. Schwartz, "Mechanism of Silicon Etching by a CF_4 Plasma", J. Vac. Sci. Tech. **15**, 1734 (1978)

J. L. Mauer and R. A. Carruthers, "Selective Etching of SiO_2 with a CF_4/H_2 Plasma", Proc. 21st Electron. MaHs. Conf., Boulder, Colorado (1979)

C. M. Meliar-Smith and C. J. Mogab, "Plasma-Assisted Etching Techniques for Pattern Delineation", in *Thin Film Processes*, ed. J.L. Vossen and W. Kern, Academic Press, New York and London (1978)

Y. Mimura, Musashino Electrical Communication Laboratory, Nippon Telegraph and Telephone Public Corporation, 3-9-11 Midori-cho, Musashino-shi, Tokyo 180, Japan (Private Communication)

C. J. Mogab, "The Loading Effect in Plasma Etching", J. Electrochem. Soc. **124**, 1262 (1977)

C. J. Mogab, A. C. Adams, and D. L. Flamm, "Plasma Etching of Si and SiO_2 – the Effect of Oxygen Additions to CF_4 Plasmas", J. Appl. Phys. **49**, 3796 (1978)

C. J. Mogab, to be published (1979)

N. Morosoff, W. Newton, and H. Yasuda, J. Vac. Sci. Tech. **15**, 1815 (1978)

H. Nakane, Tokyo Ohka Kogyo Co. Ltd., 150, Nakamaruko, Nakahara-ku, Kawasaki, 211, Japan.

K. A. Pickar, R. G. Poulsen, W. D. Westwood, A. Aitken, and J. J. White, U.S. Army Ecom. Interim and Final Reports, Ecom-75-1352-1 (1975)

R. G. Poulsen, "Plasma Etching in Integrated Circuit Manufacture – a Review", J. Vac. Sci. Tech. **14**, 266 (1977)

M. Rand, "Plasma-Promoted Deposition of Thin Inorganic Films", J. Vac. Sci. Tech. **16**, 420 (1979)

R. F. Reichelderfer, J. M. Welty, and J. F. Battey, "The Ultimate By-Products of Stripping Photoresist in an Oxygen Plasma", J. Electrochem. Soc. **124**, 1926 (1977)

A. R. Reinberg, "Radial Flow Reactor", U.S. Patent 3 757 733 (1973)

A. R. Reinberg, "RF Plasma Deposition of Inorganic Films for Semiconductor Applications", Electrochem. Soc., San Francisco Mtg., May (1974)

A. R. Reinberg, "Plasma Processing with a Planar Reactor", Circuits Mfg., p 25, April (1979)

P. M. Schaible, W. C. Metzger, and J. P. Anderson, "Reactive Ion Etching of Aluminum and Aluminum Alloys in an RF Plasma Containing Halogen Species", J. Vac. Sci. Tech. **15**, 334 (1978)

P. M. Schaible and G. C. Schwartz, "Preferential Lateral Chemical Etching in Reactive Ion Etching of Aluminum and Aluminum Alloys", J. Vac. Sci. Tech. **16**, 377 (1979)

G. C. Schwartz and P. M. Schaible, "Reactive Ion Etching of Silicon", J. Vac. Sci. Tech. **16**, 410 (1979)

G. Smolinsky and D. L. Flamm, "The Plasma Oxidation of CF_4 in a Tubular Alumina, Fast-Flow Reactor", J. Appl. Phys., to be published (1979)

K. Suzuki, S. Okudaira, N. Sakudo, and I. Kanomata, "Microwave Plasma Etching", Japan. J. Appl. Phys. **16**, 11, 1979 (1977)

Tokuda Seisakusho Ltd., commercial literature

K. Ukai and K. Hanazawa, "End Point Determination of Aluminum Reactive Ion Etching by Discharge Impedance Monitoring", J. Vac. Sci. Tech. **16**, 385 (1979)

Uthe Technology Inc., private communication (1977)

N. S. Viswanathan, private communication (1977)

J. L. Vossen, "Glow Discharge Phenomena in Plasma Etching and Plasma Deposition", J. Electrochemical Soc. **126**, 319 (1979)

A. Zafiropoulo, "Dry vs. Wet; Plasma Etching/Stripping" Circuits Manufacturing, p. 42, April (1976)

H. F. Winters and J. W. Coburn, "The Etching of Silicon with XeF_2 Vapor" Appl. Phys. Letters, to be published (1979)

Appendices

Appendix 1 refers to Chapter 1, etc.

APPENDIX 1

Useful Expressions in Gas Behaviour

One gram molecule of any gas contains 6.02 10^{23} molecules and occupies 22.4 litres at STP (Avogadro), and hence:

(a) Hydrogen atom mass is 1.66 10^{-24} g.

(b) There are 3.54 10^{16} molecules/cc of any gas at 1 torr and 0°C.

$$\tfrac{1}{2} m\overline{c^2} = \frac{3kT}{2}$$

$$k = 1.38\ 10^{-16}\ \text{erg/deg K (Boltzmann)}$$

$$\bar{c} = \left(\frac{8kT}{\pi m}\right)^{1/2} \quad \text{cf.}\ \overline{c^2} = \left(\frac{3kT}{m}\right)$$

$$\frac{dn}{dc} = 4n\left(\frac{1}{\pi}\right)^{1/2}\left(\frac{m}{2kT}\right)^{3/2} c^2 \exp\left(-\frac{mc^2}{2kT}\right) \quad \text{(Maxwell-Boltzmann)}$$

$$p = \frac{nm\overline{c^2}}{3} = nkT$$

1 atmosphere = 760 torr = 760 mm Hg.

1 mtorr = 1 10^{-3} torr = 1 10^{-3} mmHg = 1 μm Hg = 1μ

cgs pressure unit = 1 dyne/cm^2

1 dyne/cm^2 = 0.75 mtorr = 0.1 Pa

[1 Pa = 1 pascal = 1 N/m^2]

Impingement flux = $\frac{n\bar{c}}{4}$/unit area

$\sim$ 1 monolayer/sec @ 10^{-6} torr

Mean free path λ

$\sim$ 5 mm for argon @ 10 mtorr

Energy transfer function (elastic collisions) $\frac{4m_1 m_2}{(m_1+m_2)^2}$

Gas flow:		
	viscous	$\lambda < 0.01$ d
	transition	0.01 d $< \lambda <$ d
	molecular	d $< \lambda$

$$Q = pS$$

gas flow = pressure x pumping speed

1 sccm = 2.69 10^{19} molecules/minute.

79.05 sccm = 1 torr litre/second.

Residence time $\tau = \frac{V}{S} = \frac{pV}{pS} = \frac{pV}{Q}$

Conductance $F = \frac{Q}{\Delta p}$

APPENDIX 2

The following data is reproduced, with permission, from *Compilation of Low Energy Electron Collision Cross Section Data* by Dr. L.J. Kieffer.

Effective Line Excitation Cross Sections of Argon I

Wavelength (Angstroms)	Transition[1]	E (eV)	σ^2
4190/4191[a]	4s[1 1/2]° - 5p[2 1/2]	19.6	4.86 -19
	4s'[1/2]° - 5p'[1 1/2]		-19
4200[a]	4s[1 1/2]° - 5p[2 1/2]	22.6	6.74 -19
4259[a]	4s'[1/2]° - 5p'[1/2]	22.8	5.42 -19
4272[a]	4s[1 1/2]° - 5p[1 1/2]	24.6	8.89 -19
4300[a]	4s[1 1/2]° - 5p[2 1/2]	24.2	7.19 -19

Wavelength (Angstroms)	Transition[1]	E (eV)	σ[2]
4510[a]	4s'[1/2]° - 5p[1/2]	24.9	8.52 -19
4522[a]	4s'[1/2]° - 5p[1/2]	19.4	2.30 -19
4596[a]	4s'[1/2]° - 5p[1 1/2]	19.0	7.35 -19
4628[a]	4s'[1/2]° - 5p[2 1/2]	18.9	4.92 -19
4702[a]	4s'[1/2]° - 5p[1/2]	19.3	2.61 -19
5496[b]	4p[2 1/2] - 6d[3 1/2]°	22.3	5.91 -20
5506[b]	4p[2 1/2] - 6d[3 1/2]°	60.	1.22[3] -20
5559[b]	4p[1/2] - 5d[1 1/2]°	21.9	5.06 -20
5572[b]	4p[2 1/2] - 5d'[2 1/2]°	18.2	3.00 -20
5607[b]	4p[1/2] - 5d[1/2]°	21.9	2.26 -20
5621[b]	4p[1 1/2] - 7s'[1/2]°	22.7	4.41 -20
5659[b]	4p[1 1/2] - 8s[1 1/2]°	22.4	3.42 -20
5682[b]	4p[1 1/2] - 6d[2 1/2]°	21.6	1.76 -20
5689[b]	4p[1 1/2] - 5d'[1 1/2]°	20.6	2.37 -20
5740[b]	4p[1 1/2] - 5d'[2 1/2]°	21.8	5.06 -20
5774[b]	4p'[1 1/2] - 7d[1 1/2]°	21.5	2.03 -20
5802[b]	4p[1 1/2] - 6d[1/2]°	25.2	1.31 -20
5834[b]	4p[1 1/2] - 5d'[1 1/2]°	22.6	2.80 -20
5860[b]	4p[1/2] - 6s'[1/2]°	23.1	1.30 -20
5889[b]	4p[2 1/2] - 7s[1 1/2]°	22.2	3.42 -20
5912[b]	4p[1/2] - 4d'[1 1/2]°	18.9	5.47 -20
5929[b]	4p[2 1/2] - 7s[1 1/2]°	22.5	2.68 -20
5943[b]	4p[2 1/2] - 7s[1 1/2]°	22.8	1.37 -20
5972[b]	4p'[1 1/2] - 7s'[1/2]°	23.1	1.49 -20
5987[b]	4p[2 1/2] - 5d[3 1/2]°	21.9	1.72 -20
5999[b]	4p[2 1/2] - 5d[2 1/2]°	23.0	1.61 -20
6013[b]	4p[2 1/2] - 5d[1 1/2]°	18.9	1.18 -20
6025[b]	4p'[1 1/2] - 7s'[1/2]°	22.9	1.72 -20

Wavelength (Angstroms)	Transition[1]	E (eV)	σ^2
6032[b]	4p[2 1/2] - 7s'[1/2]°	18.8	1.99 -19
6043[b]	4p[2 1/2] - 5d[3 1/2]°	22.0	1.55 -19
	4p'[1 1/2] - 6d[2 1/2]°		
6053[b]	4p[1/2] - 4d'[1 1/2]°	21.8	1.49 -20
6059[b]	4p[1/2] - 4d'[2 1/2]°	22.0	3.87 -20
6091[b]	4p[1/2] - 6d[1/2]°	23.5	1.72 -20
	4p'[1/2] - 8d[1/2]°		
6099[b]	4p[1 1/2] - 7s[1 1/2]°	22.4	1.61 -20
6106[b]	4p[1 1/2] - 5d'[2 1/2]°	22.5	5.99 -20
6127[b]	4p[2 1/2] - 5d[1/2]°	22.0	1.72 -20
6145[b]	4p'[1 1/2] - 5d'[2 1/2]°	24.4	5.60 -20
6155[b]	4p[1 1/2] - 7s[1 1/2]°	22.4	1.07 -21
	4p'[1 1/2] - 5d'[1 1/2]°		
6173[b]	4p[1 1/2] - 5d[2 1/2]°	23.7	6.10 -20
6212[b]	4p[1 1/2] - 5d[2 1/2]°	22.6	5.03 -20
6297[b]	4p'[1/2] - 5d'[1 1/2]°	23.8	6.56 -20
6308[b]	4p[1 1/2] - 5d[1 1/2]°	22.7	5.79 -20
6365[b]	4p[1 1/2] - 5d[1/2]°	24.0	3.11 -20
6385[b]	4p[1/2] - 6s[1 1/2]°	23.1	3.76 -20
6416[b]	4p[1/2] - 6s[1 1/2]°	23.0	7.06 -20
6466[b]	4p[1/2] - 5d[1 1/2]°	60.0	6.45[3] -20
6538[b]	4p[2 1/2] - 4d'[1 1/2]°	21.8	3.95 -20
6605[b]	4p[2 1/2] - 4d'[2 1/2]°	22.0	5.06 -20
6753[b]	4p[1/2] - 4d[1 1/2]°	22.0	5.42 -19
6871[b]	4p[1/2] - 4d[1/2]°	23.5	5.65 -19
6965[b]	4s[1 1/2] - 4p'[1/2]°	21.3	1.65 -18
7067[b]	4s[1 1/2]° - 4p'[1 1/2]	19.2	1.37 -18

Wavelength (Angstroms)	Transition[1]	E (eV)	σ^2
7273[b]	4p[2 1/2] - 4d[3 1/2]°	19.[illegible]	3.51 -19
7353[b]	4p[2 1/2] - 4d[3 1/2]°	24.9	5.36 -19
	4p[1 1/2] - 6s[1 1/2]°		
7372[b]	4s[1 1/2]° - 4p'[1 1/2]°	23.9	1.95 -18
7383[b]	4s[1 1/2]° - 4p'[1 1/2]	19.5	4.07 -18
7504[b]	4s'[1/2]° - 4p'[1/2]	22.6	6.61 -18
7515[b,c]	4s[1 1/2]° - 4p[1/2]	19.4	6.22 -18
		28.0	5.40 -19
7635[b]	4s[1 1/2]° - 4p[1 1/2]	21.4	1.752 -17
7723/7724[b,c]	4s[1 1/2]° -4p[1 1/2]	19.4	1.55 -18
	4s'[1/2]° - 4p'[1/2]	28.0	9.22[3,4] -19
7948[c]	4s'[1/2]° - 4p'[1 1/2]	28.0	5.50[3] -19
8006/8015[c]	4s[1 1/2]° - 4p[1 1/2]	20.2	7.25 -19
	4s[1 1/2]° - 4p[2 1/2]		
8115[c]	4s[1 1/2]° - 4p[2 1/2]	28.0	1.22[3] -18
8264.5[c]	4s'[1/2]° - 4p'[1/2]	20.0	8.44 -19
8408[c]	4s'[1/2]° - 4p'[1 1/2]	20.9	2.06 -18
8425[c]	4s[1 1/2]° - 4p[2 1/2]	20.0	2.48 -18

Effective Line Excitation Cross Sections of Argon II

Wavelength (Angstroms)	Transition[1]	E (eV)	σ^2
4332[a]	$3d\,^4D - 4p\,^4P°$	50.1	5.57 -19
4371[a]	$3d\,^4D - 4p\,^4P°$	54.5	3.79 -19
4426[a]	$4s\,^4P - 4p\,^4D°$	60.3	2.12 -19
4430[a]	$4s\,^4P - 4p\,^4D°$	60.2	1.30 -19
4579[a]	$4s\,^2P - 4p\,^2S°$	59.8	1.79 -19
4609[a]	$4s'\,^2D - 4p'\,^2F°$	60.2	1.54 -19
4657[a]	$4s\,^2P - 4p\,^2P°$	60.0	7.30 -20
4726[a]	$4s\,^2P - 4p\,^2D°$	59.9	1.91 -19

Notes

a. Data taken from Fischer (1933). The method used to obtain the normalized factor was through Fischer's comparison of the excitation function for the 4358 Å line with that obtained by Hanle and Schaffernicht (1930) at 60 eV.

b. Data taken from Hermann (1936). The maxima of the cross sections were read and converted to cm^2 by dividing by the number of atoms/cc at 1 torr pressure and 0° C (3.539 10^{16} atoms/cc).

c. Data taken from Volkova (1959) and Volkova and Devyatov (1959)

1. Transitions according to Striganov and Sventitskii (1968)

2. σ is in units of cm^2 with the order of magnitude indicated in the column. The cross section was measured at the maximum unless otherwise noted.

3. The cross section was not measured at the maximum.

4. The data were normalized by multiplying by the gas kinetic cross section (6.15 10^{-16} cm^2).

Basic Data

The following data is reproduced, with permission, from *Basic Data of Plasma Physics, 1966* by Prof. S.C. Brown.

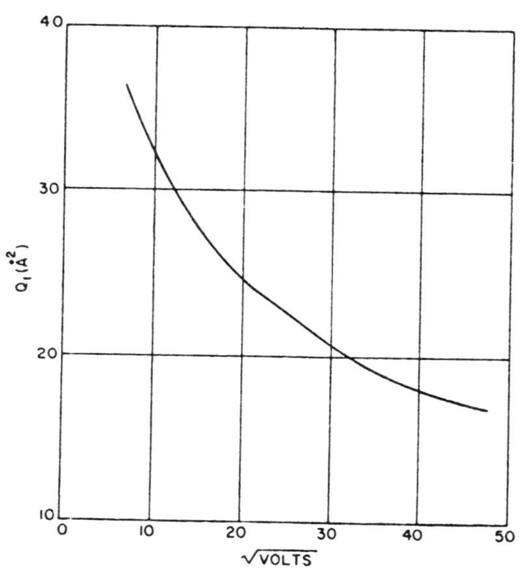

Charge-transfer cross sections for A^+ in argon.
J. B. Hasted, M. Hussain (1964)

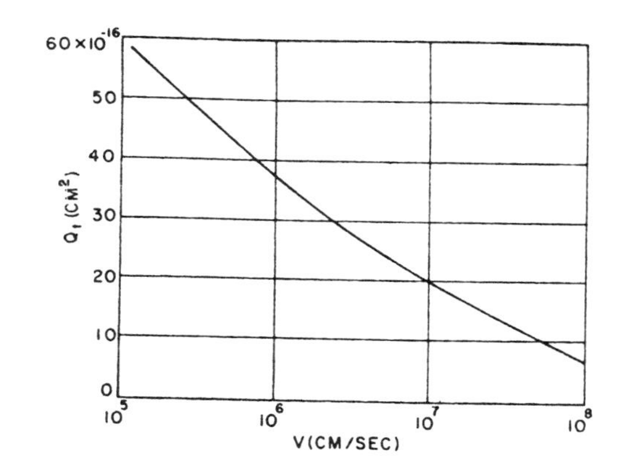

Charge-transfer collision cross sections for A^+ in A.
B. M. Smirnov, M. I. Chibisov (1965)

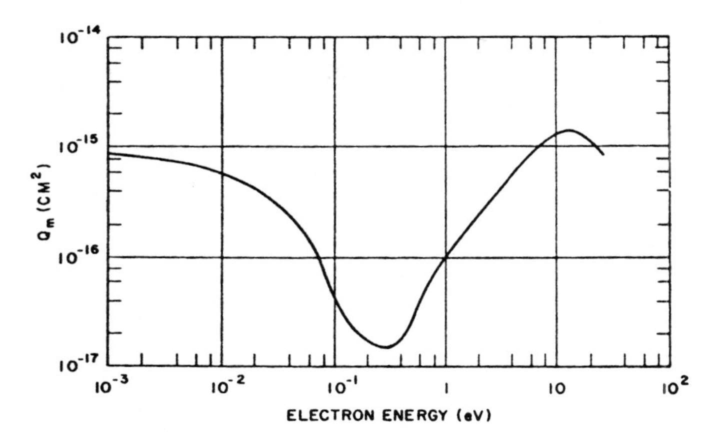

Momentum-transfer cross sections for electrons in argon.
L. S. Frost, A. V. Phelps (1964)

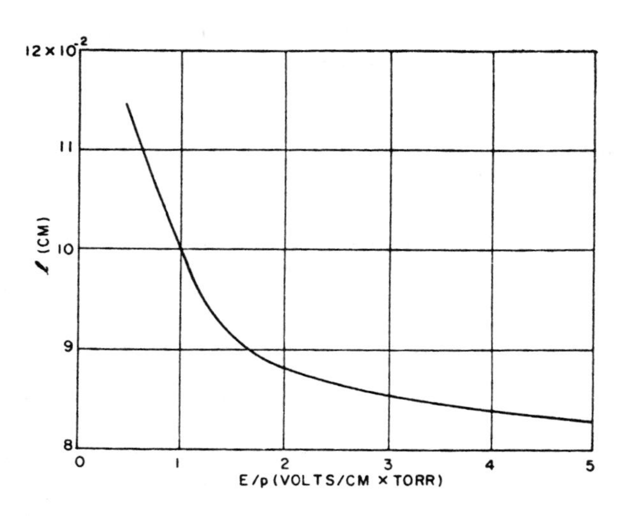

Mean free path of electrons in argon as a function of E/p.
O. J. Orient (1965)

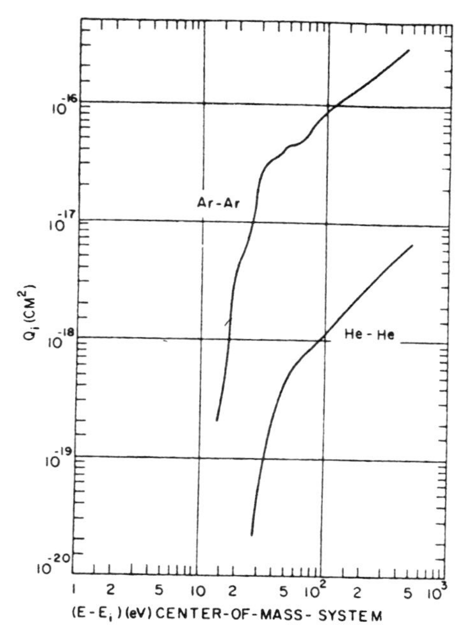

Total ionization cross section for A-A collisions and He-He collisions as a function of excess center of mass energy.
H. C. Hayden, R. C. Amme (1966)

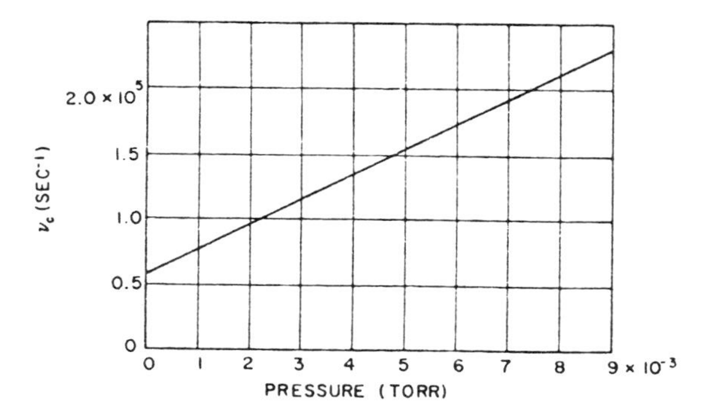

Collision frequency of A^+ in A as a function of pressure.
D. Wobschall, J. R. Graham, Jr., D. P. Malone (1963)

APPENDIX 4

Ionization Rate of Metastables

If the density of metastables is n^*, and the ionization cross-section is q above the threshold E_t, then the ion production rate I_m per unit volume per unit time is

$$I_m = \int_{E_t}^{\infty} n^* q \left(\frac{2E}{m_e}\right)^{1/2} dn_e(E)$$

by comparison with the ionization of ground state atoms. If the cross-section q is constant above threshold, then substituting for the Maxwell-Boltzmann distribution,

$$I_m = n_e n^* q \left(\frac{8}{\pi m_e}\right)^{1/2} (kT_e)^{1/2} \int_{E_t/kT_e}^{\infty} \frac{E}{kT_e} \exp - \frac{E}{kT_e} \, d\left(\frac{E}{kT_e}\right)$$

The integration can be made by parts to become

$$I_m = n_e n^* q \left(\frac{8}{\pi m_e}\right)^{1/2} (kT_e)^{1/2} \left(\frac{E_t}{kT_e} + 1\right) \exp - \frac{E_t}{kT_e}$$

Some values obtained for the ionization of argon metastables are shown in Table 4-6 of Chapter 4.

Energy Dependence of the Electron Scattering Cross-Section

Consider the Coulomb interaction between an electron of velocity v and a comparatively stationary charge Q of mass M (Figure A4-1). Let the time origin $t = O$ be at the point of closest approach b. Provided the interaction is weak enough, the electron will not be deflected and the charge Q will not move sensibly during the interaction; by symmetry, the net effect on Q will then be to give it motion in the y axis perpendicular to the electron velocity.

The force f_y in the y direction at time t due to the Coulomb interaction is given by

$$f_y = \frac{Qe^2}{4\pi \epsilon_0 r^2} \sin \phi$$

But force is equal to the rate of change of momentum P_y in the y direction, therefore

$$P_y = \int_{-\infty}^{+\infty} f_y \, dt = \int_{0}^{\pi} f_y \frac{dt}{d\phi} \, d\phi$$

Since $\dfrac{vt}{b} = \dfrac{1}{\tan \phi}$

then $\dfrac{v}{b} \dfrac{dt}{d\phi} = - \dfrac{1}{\sin^2 \phi}$

Also $\dfrac{1}{r} = \dfrac{\sin \phi}{b}$

$$\therefore P_y = \int_{0}^{\pi} \frac{Qe^2}{4\pi \epsilon_0} \frac{\sin^2 \phi}{b^2} \sin \phi \frac{-b}{v \sin^2 \phi} \, d\phi$$

$$= \frac{Qe^2}{4\pi \epsilon_0 bv} \int_{0}^{\pi} - \sin \phi \, d\phi$$

$$= \frac{Qe^2}{2\pi \epsilon_0 bv}$$

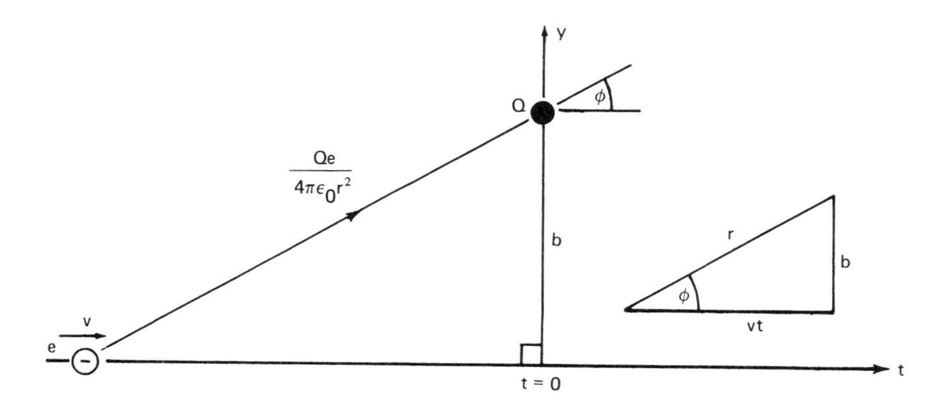

Figure A4-1. Coulomb interaction between mobile charge e with velocity v, and charge Q.

The energy E transferred is $P_y^2/2M$, which can be expressed in terms of the initial electron energy $E_o = \frac{1}{2} m_e v^2$ as

$$E = \frac{Q^2 e^4 m_e}{16\pi^2 \epsilon_o^2 b^2 M E_o}$$

so that the energy transfer, and hence the collision cross-section, *decreases* with increasing electron energy E_o. At the same time, the incident electron will be deflected less; i.e. it will be *forward scattered*.

Electron-Electron (and Electron-Ion) Collisions

The normal radial dependence of the Coulomb potential V(r) due to a charge q,

$$V(r) = \frac{q}{4\pi \epsilon_o r}$$

becomes the *screened* or *shielded Coulomb potential*:

$$V(r) = \frac{q}{4\pi \epsilon_o r} \exp\left(-\frac{r}{\lambda_D}\right)$$

when the screening effect of a plasma is considered (see Chapter 3, "Debye Shielding"). In principle, the screened Coulomb potential still has infinite range, although there is an effective range where the Coulomb potential becomes comparable with the potential kT_e/e due to fluctuations in the plasma caused by electron thermal motion. It is usual to put the maximum range equal to one Debye length.

It is common to quantify the interaction by introducing an *impact parameter* b_o (Figure A4-2) which leads to 90° scattering. This is then used to define a cross-section q_e given, not simply by πb_o^2 but, allowing for the long range shielded interaction, by:

$$q_e = 6\pi \overline{b_o^2} \ \ell n \Lambda$$

where

$$\Lambda = 1.24 \ 10^7 \left(\frac{T_e^{\ 3}}{n_e}\right)^{1/2}$$

and

$$\overline{b_o} = \frac{5.56 \ 10^{-6}}{T_e}$$

all in MKS units (Mitchner and Kruger 1973). Because of the long-range of the Coulomb interaction, there are many more small-angle than large-angle collisions, and their cumulative effect is included in the multiplicative factor of $\ell n \Lambda$, where

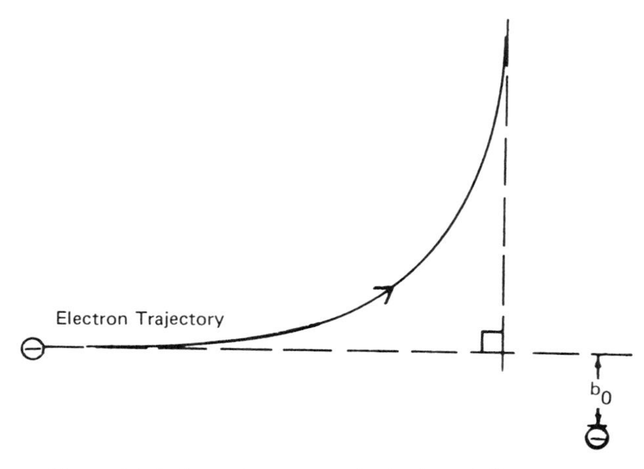

Figure A4-2. Impact parameter b_o leads to 90° scattering

Λ is $\lambda_D/\overline{b_o}$. For a plasma of 2 eV and 10^{10} cm^{-3}, these expressions yield $\overline{b_o}$ = 2.4 Å and $q_e = 1.4\ 10^{-13}$ cm^2, still much larger than any other of the cross-sections we have been looking at. Note that q_e refers to both electron-electron and electron-ion interactions, and that it is primarily dependent on the electron temperature; the density is involved only in Λ which appears in the weak variation of $\ell n\ \Lambda$.

The collision cross-section q_e can further be used to predict a *collision frequency* $\overline{\nu_e}$ given by

$$\overline{\nu_e} = 3.64\ 10^{-6}\ n_e \frac{\ell n\ \Lambda}{T_e^{3/2}} \qquad \text{(MKS)}$$

which amounts to $1.3\ 10^5$ per second for our example with T_e = 23 200K, and compares directly with the value of $4\ 10^5$ per second derived in Chapter 1, "Collision Frequency", for atom-atom collisions in an ideal gas at 50 millitorr. That result might lead one to expect quite efficient energy transfer amongst the electrons, but this is not so. The electron-electron interactions are primarily weak small angle scatterings with little energy interchange.

We can use a classical derivation of the energy interchange (Birkhoff 1958). We have already shown that in a single collision between two electrons, the energy is given by:

$$E = \frac{e^4}{16\pi^2\ \epsilon_o^2\ b^2\ E_o}$$

The energy exchange between the incident fast electron of energy E_o and the $n_e\ 2\pi\ b\ db$ electrons per unit length with impact parameters between b and b + db,will be (still in MKS units)

$$E = \frac{e^4}{16\pi^2\ \epsilon_o^2\ b^2\ E_o}\ n_e\ 2\pi\ b\ db$$

per unit length of travel of the fast electron. For all electrons in the glow, this amounts to

$$E = \frac{n_e\ e^4}{8\pi\ \epsilon_o^2\ E_o} \int_{b_{min}}^{b_{max}} \frac{db}{b}$$

$$= \frac{n_e\ e^4}{8\pi\ \epsilon_o^2\ E_o}\ \ell n\ \frac{b_{max}}{b_{min}}$$

This expression appears to have unlimited possibilities since it goes to infinity as b_{min} goes to zero. However, this is just an error introduced by the oversimplifying assumption that the interactions are so weak that the slow electron hardly moves during the collision. This is mostly true, and the assumption is justified. Even when b is small, the slow electron is repelled away so that the closest approach is rarely small and hence the interaction is weak; b_{min} is taken care of by using the obvious criterion introduced by Bohr (1948) that the maximum energy transfer possible in a single collision is 100%. Hence b_{min} is given by

$$E_o = \frac{e^4}{16\pi^2\ \epsilon_o^2\ b_{min}^2\ E_o}$$

$$\text{i.e.} \quad b_{min} = \frac{e^2}{4\pi\ \epsilon_o\ E_o}$$

For a 100 eV electron, b_{min} will be equal to 0.14Å. We have already decided that b_{max} will be equal to λ_D, which has a value of 100 μm for our example. Now b_{max}/b_{min} is equivalent to λ_D/b_{min} and so is very similar to the parameter $\Lambda = \lambda_D/\overline{b_o}$ introduced earlier. As with Λ, b_{max}/b_{min} appears only as a natural logarithm, and the final result for the energy transfer is only weakly dependent on the ratio b_{max}/b_{min}, explaining the rather cavalier estimations of b_{max} and b_{min}. For our example, $\ell n\ (b_{max}/b_{min})$ will be 15.8.

To return to the expression for the energy loss of the fast electron, the pre-exponential simplifies to $3.34\ 10^{-55}\ n_e/E_o$ joules per metre when we substitute the values of the constants. If we insert a value of $n_e = 10^{16}/m^3$ and $\ell n\ (b_{max}/b_{min}) = 15.8$, then the energy loss of a 100 eV electron should be $2.05\ 10^{-4}$ eV/cm – almost insignificant and clearly inadequate to explain the energy loss of the primary electrons.

Ion and Electron Velocities

Values calculated from $E = \frac{1}{2} mv^2$ are given in cm/sec.

	Energy (eV)				
	0.1	1	10	100	1000
Electron	$1.9\ 10^7$	$5.9\ 10^7$	$1.9\ 10^8$	$5.9\ 10^8$	$1.9\ 10^9$
Argon ion or neutral	$6.9\ 10^4$	$2.1\ 10^5$	$6.9\ 10^5$	$2.1\ 10^6$	$6.9\ 10^6$

Basic Data

The following data is reproduced, with permission, from *Basic Data of Plasma Physics, 1966* by Prof. S.C. Brown.

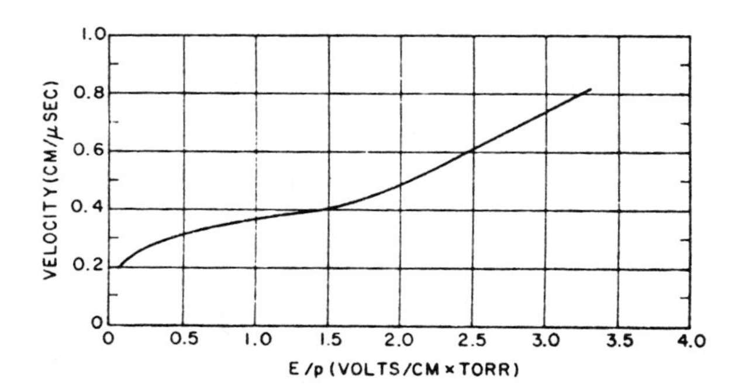

Drift velocity of electrons in argon.
J. C. Bowe (1960)

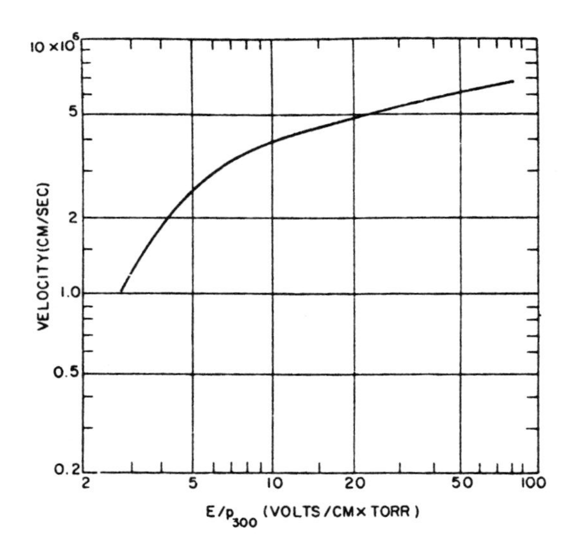

Electron drift velocity in argon.
J. M. Anderson (1964)

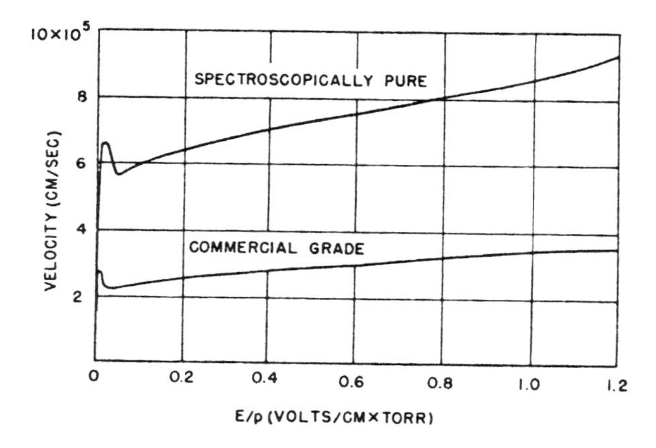

Electron drift velocity as a function of E/p in argon.
A. A. Vorob'ev, B. A. Ivanov, A. P. Komar, V. A. Korolev (1960)

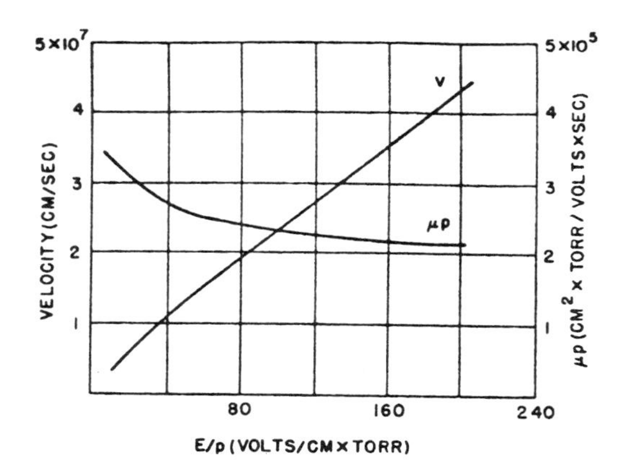

Drift velocity and mobility of electrons in argon.
V. E. Golant (1959)

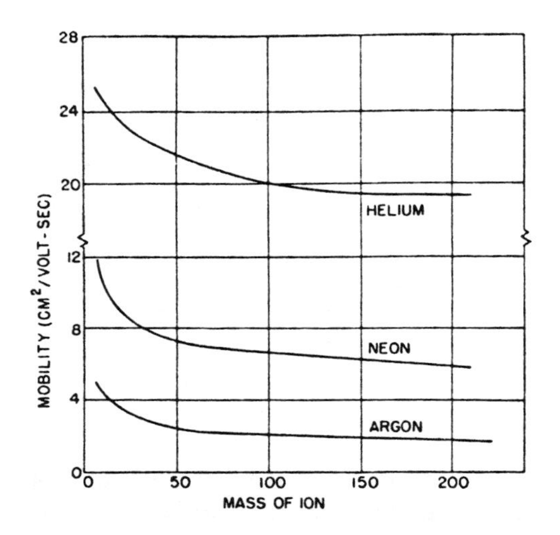

Mobility as a function of ion mass in He, Ne, and A.*
L. M. Chanin, M. A. Biondi (1957)

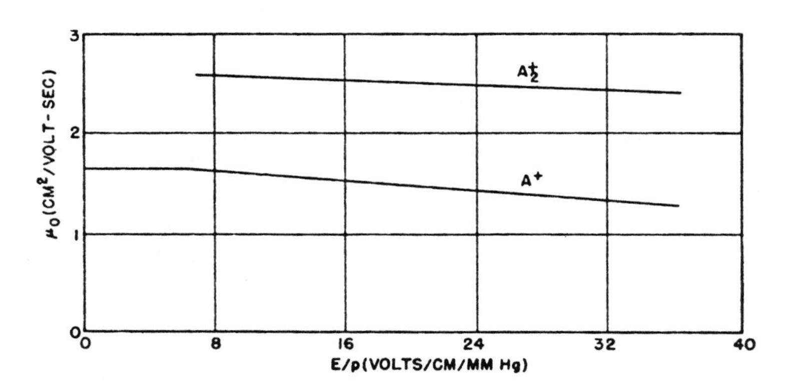

Mobility of A^+ and A_2^+ in argon.*
M. A. Biondi, L. M. Chanin (1954)

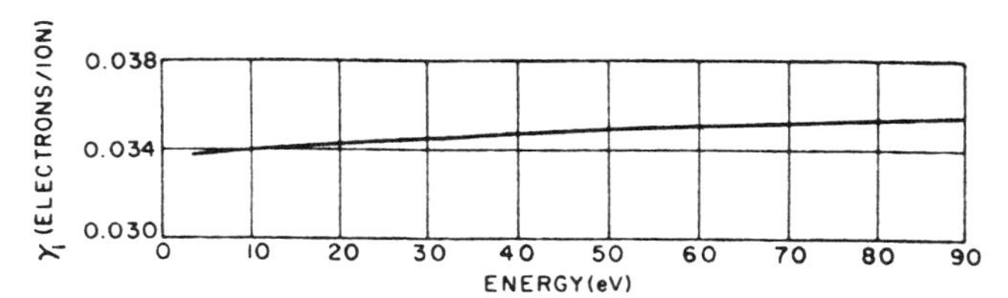

Secondary electron emission coefficient as a function of kinetic energy of A^+ ions incident on clean (111) face of nickel.
Y. Takeishi, H. D. Hagstrum (1965)

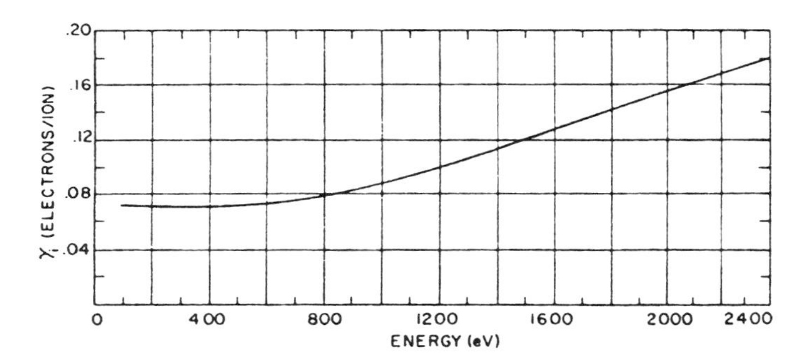

Secondary electron emission coefficients for A^+ on clean molybdenum.
P. Mahadevan, G. D. Magnuson, J. K. Layton, C. E. Carlston (1965)

APPENDIX 6

Ion Mass Spectrometry Data

The following data, previously unpublished, is presented by permission of John Coburn (1979). The results are of the mass spectrum of ions bombarding the substrate in a sputtering system, with order of magnitude increases in amplifier gain. The sputtering target was half copper and half tantalum (see Chapter 6).

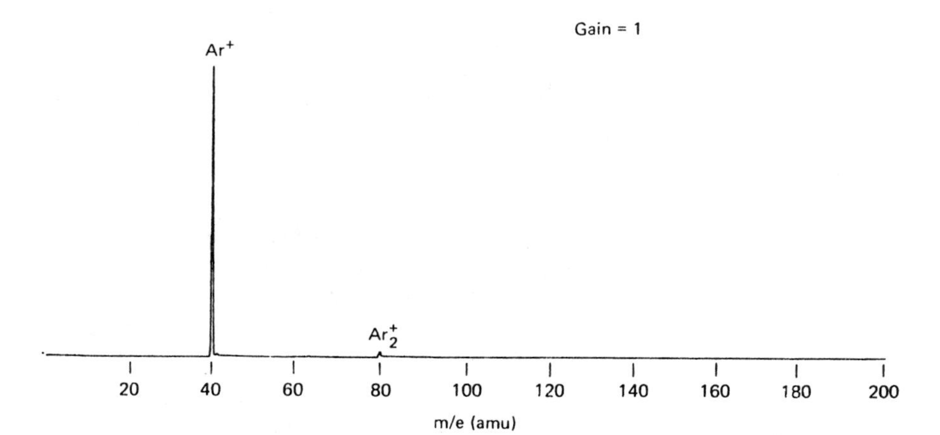
Gain = 1
Ar^+
Ar_2^+
20
40
60
80
100
120
140
160
180
200
m/e (amu)

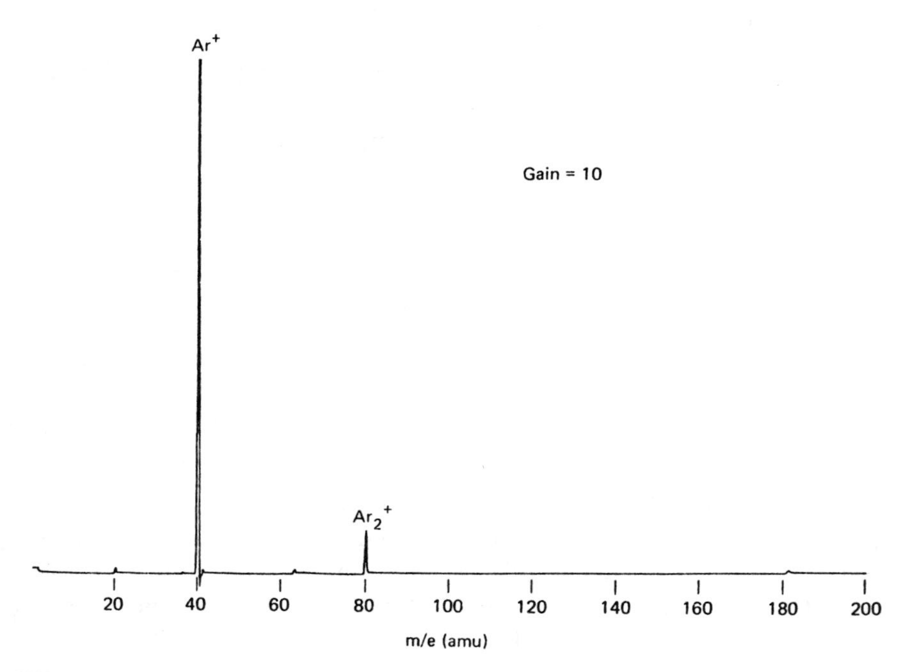
Ar^+
Gain = 10
Ar_2^+
20
40
60
80
100
120
140
160
180
200
m/e (amu)

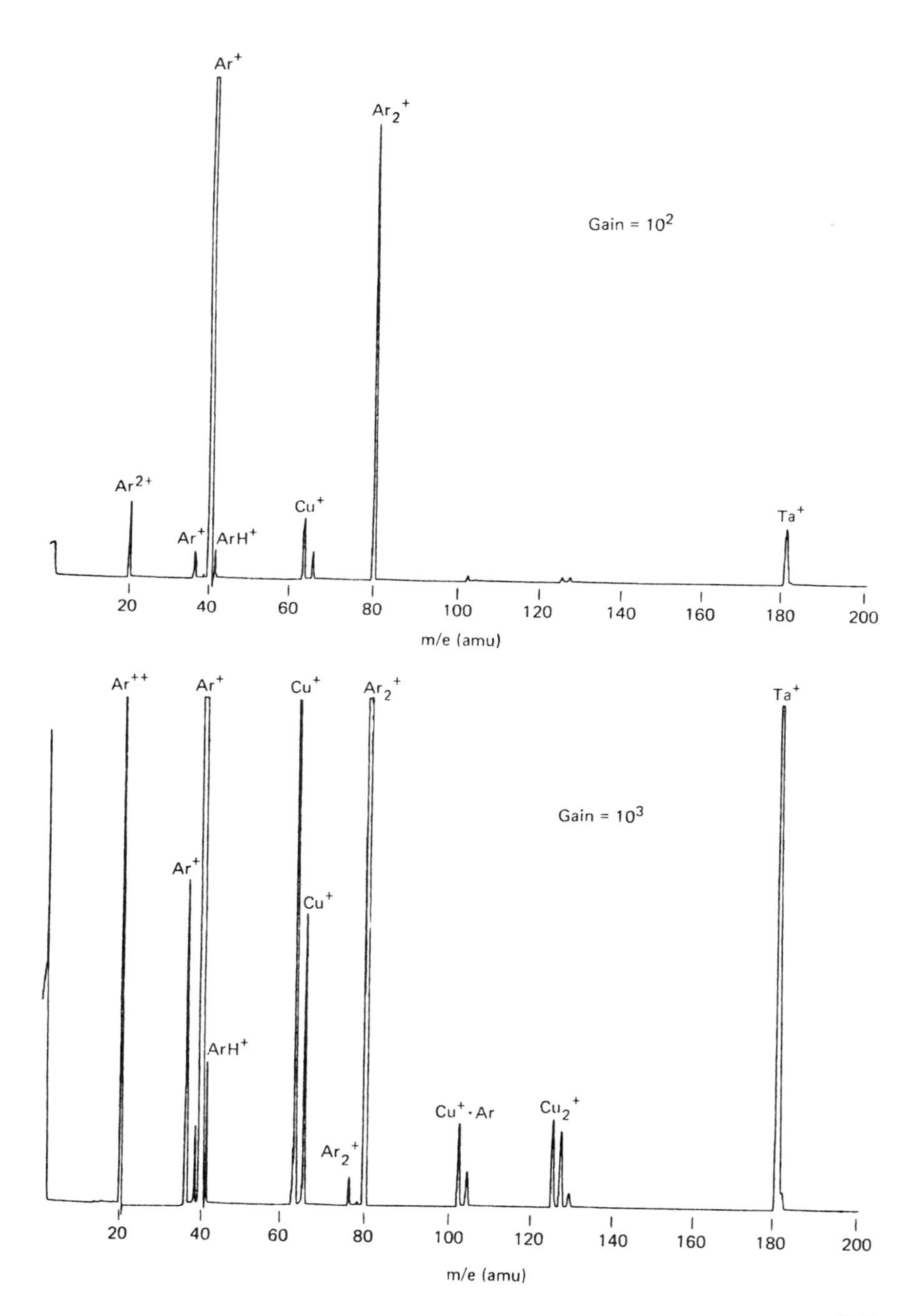
Ar$^+$
Ar$_2$$^+$
Gain = 10^2
Ar^{2+}
Ar$^+$
ArH$^+$
Cu$^+$
Ta$^+$
20
40
60
80
100
120
140
160
180
200
m/e (amu)
Ar^{++}
Ar$^+$
Cu$^+$
Ar$_2$$^+$
Ta$^+$
Gain = 10^3
Ar$^+$
Cu$^+$
ArH$^+$
Cu$^+$·Ar
Cu$_2$$^+$
Ar$_2$$^+$
20
40
60
80
100
120
140
160
180
200
m/e (amu)

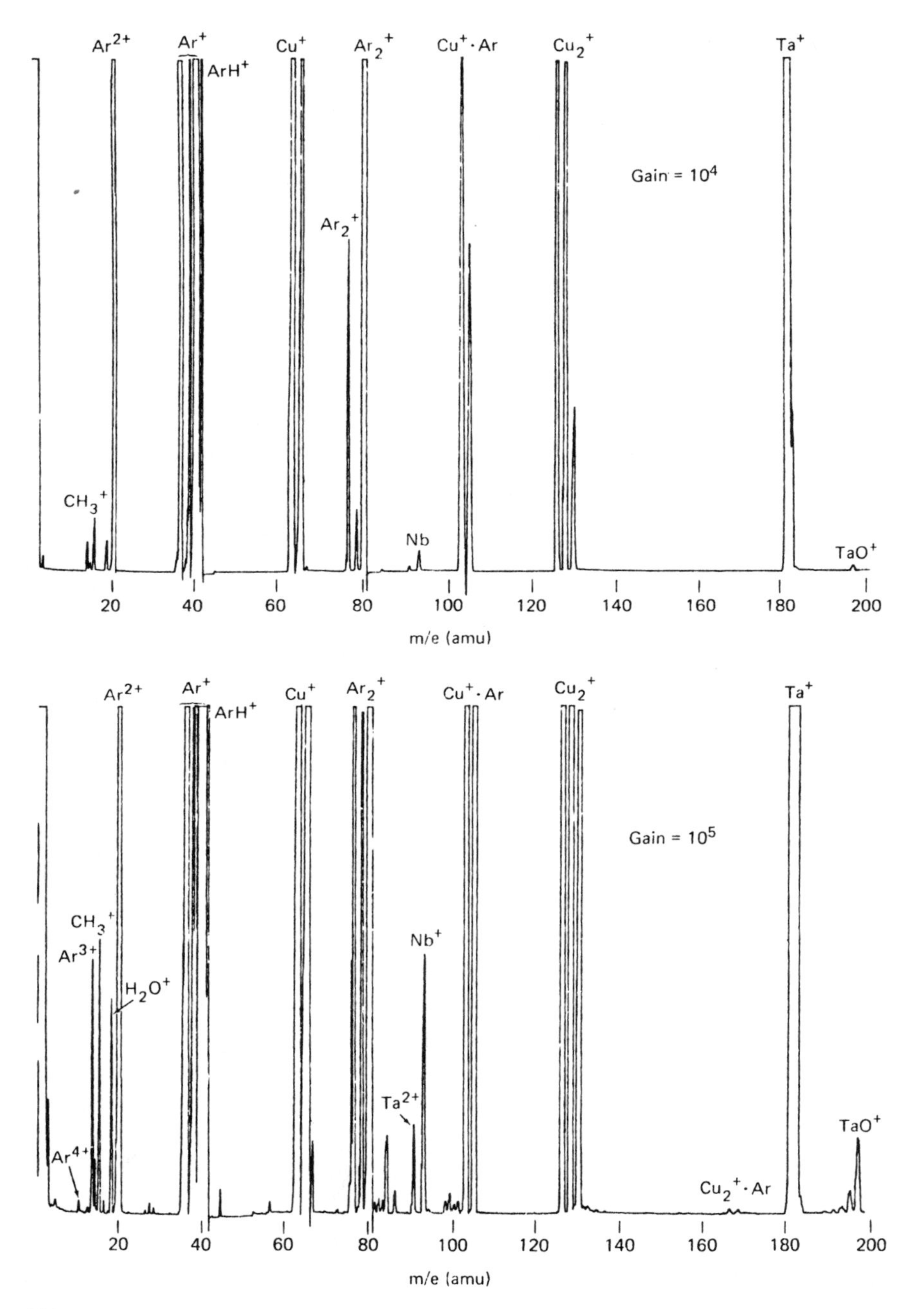

Ar2+
Ar+
ArH+
Cu+
Ar2+
Cu+·Ar
Cu2+
Ta+
Gain = 10^4
Ar2+
CH3+
Nb
TaO+
20
40
60
80
100
120
140
160
180
200
m/e (amu)
Ar2+
Ar+
ArH+
Cu+
Ar2+
Cu+·Ar
Cu2+
Ta+
Gain = 10^5
CH3+
Ar3+
H2O+
Nb+
Ta2+
Ar4+
TaO+
Cu2+·Ar
20
40
60
80
100
120
140
160
180
200
m/e (amu)

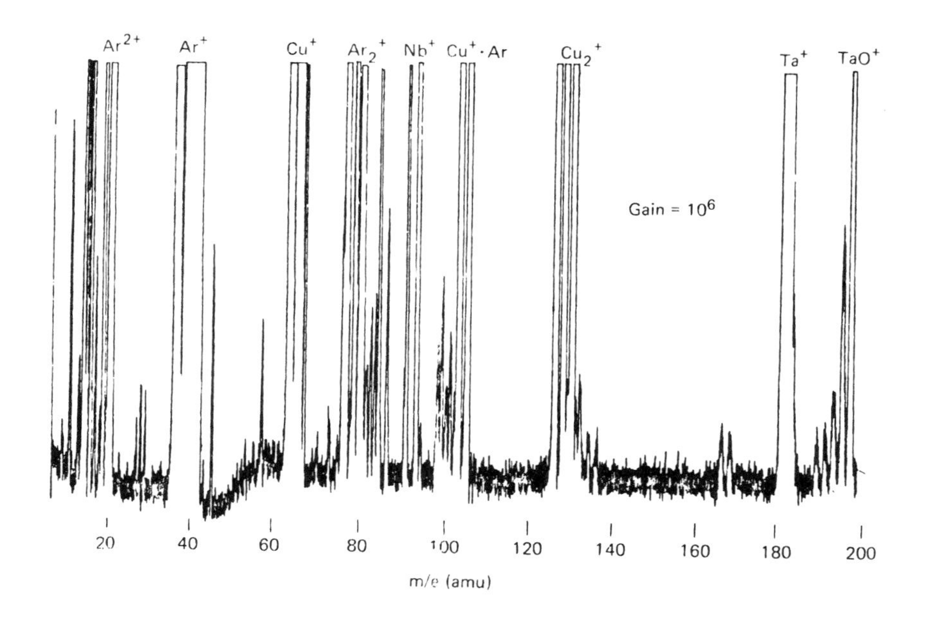
Ar^{2+}
Ar^{+}
Cu^{+}
Ar$_2^{+}$
Nb^{+}
Cu^{+}·Ar
Cu$_2^{+}$
Ta^{+}
TaO^{+}
Gain = 10^6
20
40
60
80
100
120
140
160
180
200
m/e (amu)

Sputtering Yield Data

The following sputtering yield data for argon is reproduced from General Mills Report 2309 (1962), published by permission of Prof. G.K. Wehner and Litton Industries, Beverly Hills. The Report is a compilation from Wehner (1957), Wehner (1958), Laegried and Wehner (1961), Rosenberg and Wehner (1962). Note that the yield S is given as $S/(1 + \gamma)$, as is much of the other yield data shown, where γ is the secondary electron emission coefficient.

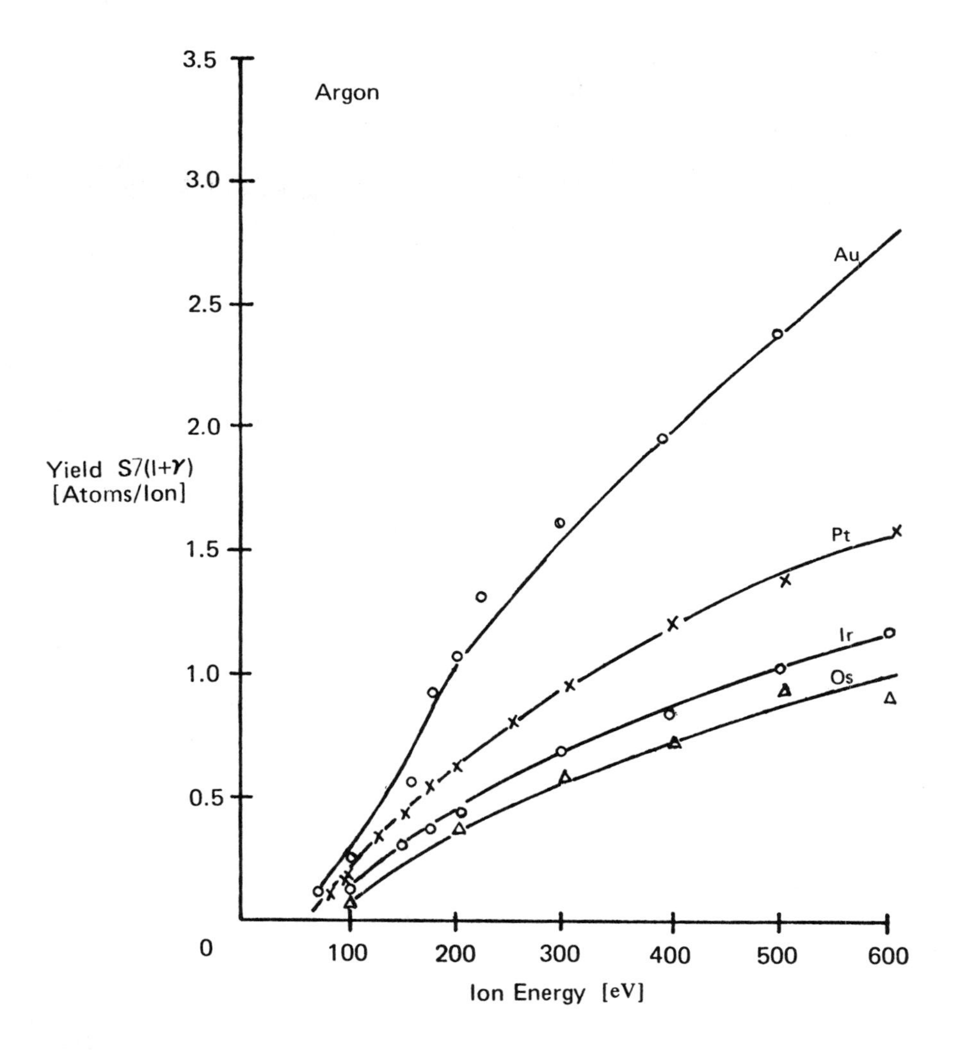

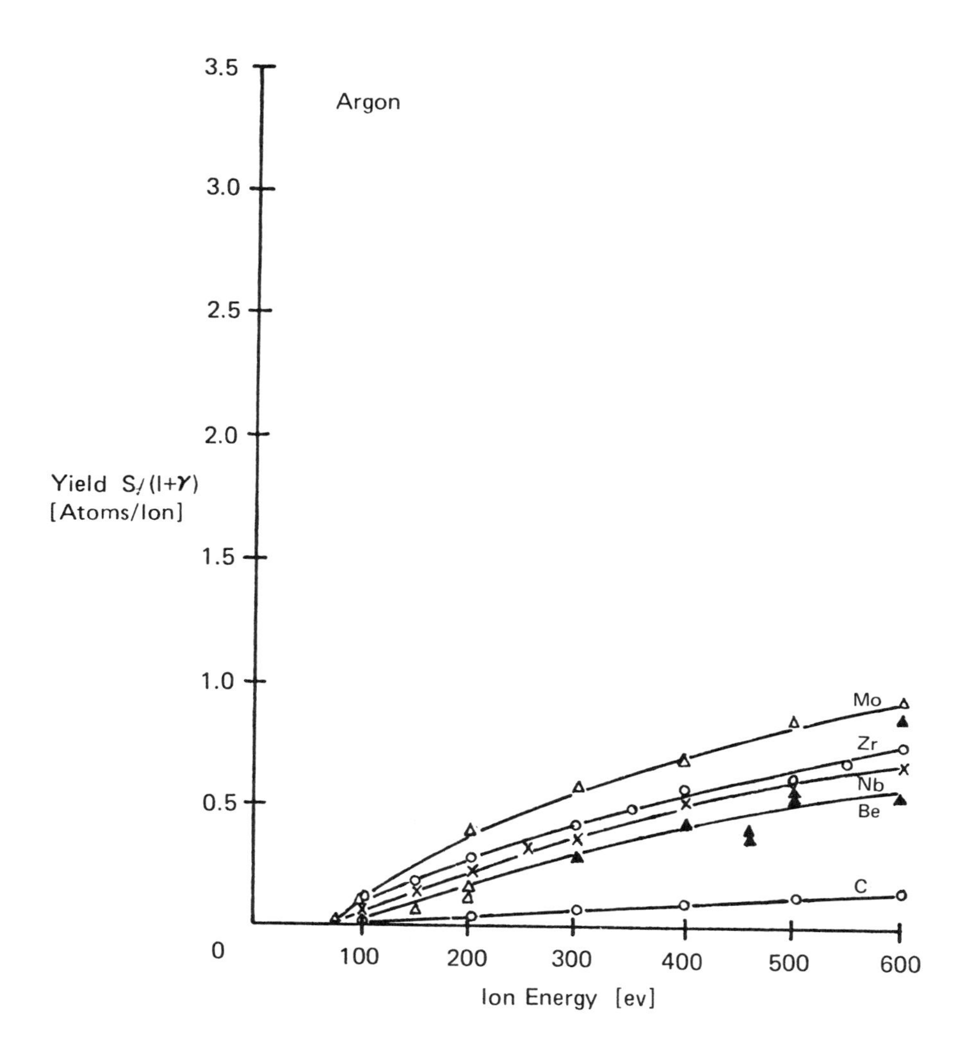
Argon
3.5
3.0
2.5
2.0
1.5
1.0
0.5
0
Yield S/(l+γ)
[Atoms/Ion]
Mo
Zr
Nb
Be
C
100
200
300
400
500
600
Ion Energy [ev]

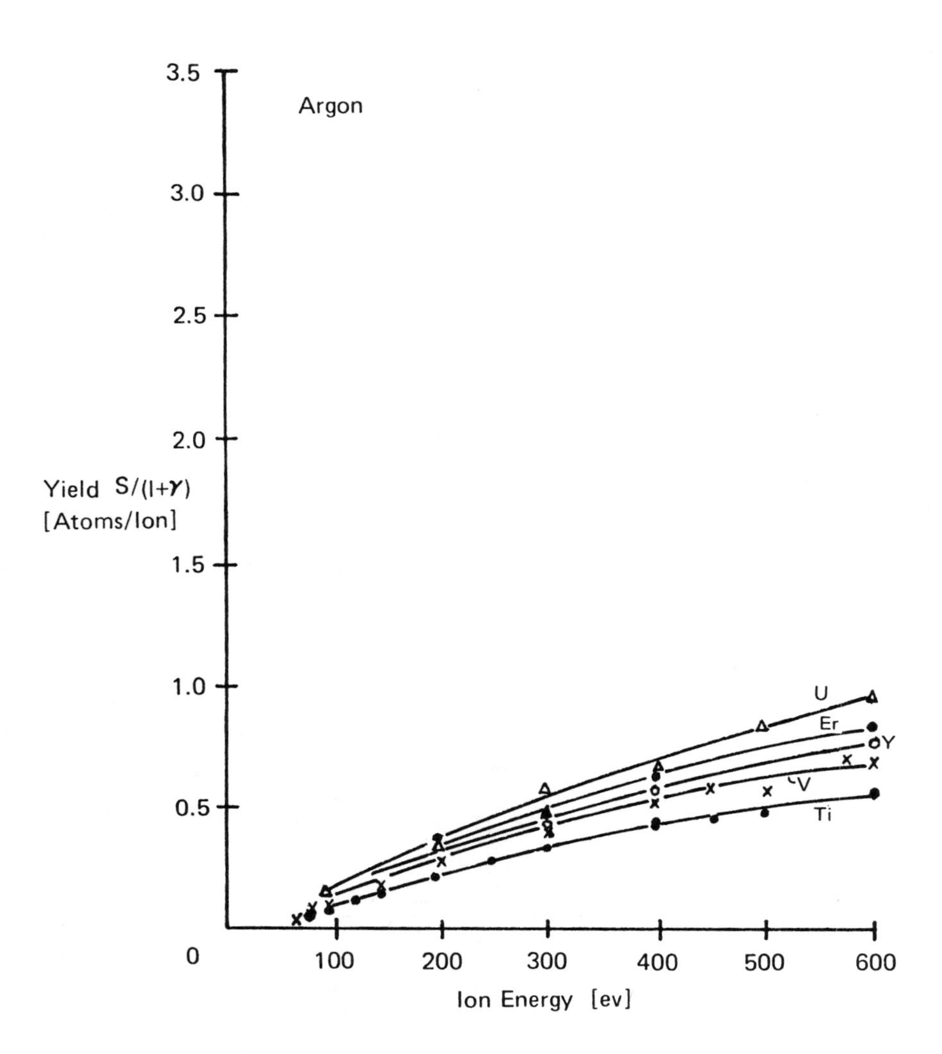

Argon
Yield S/(1+γ)
[Atoms/Ion]
3.5
3.0
2.5
2.0
1.5
1.0
0.5
0
100
200
300
400
500
600
Ion Energy [ev]
U
Er
Y
V
Ti

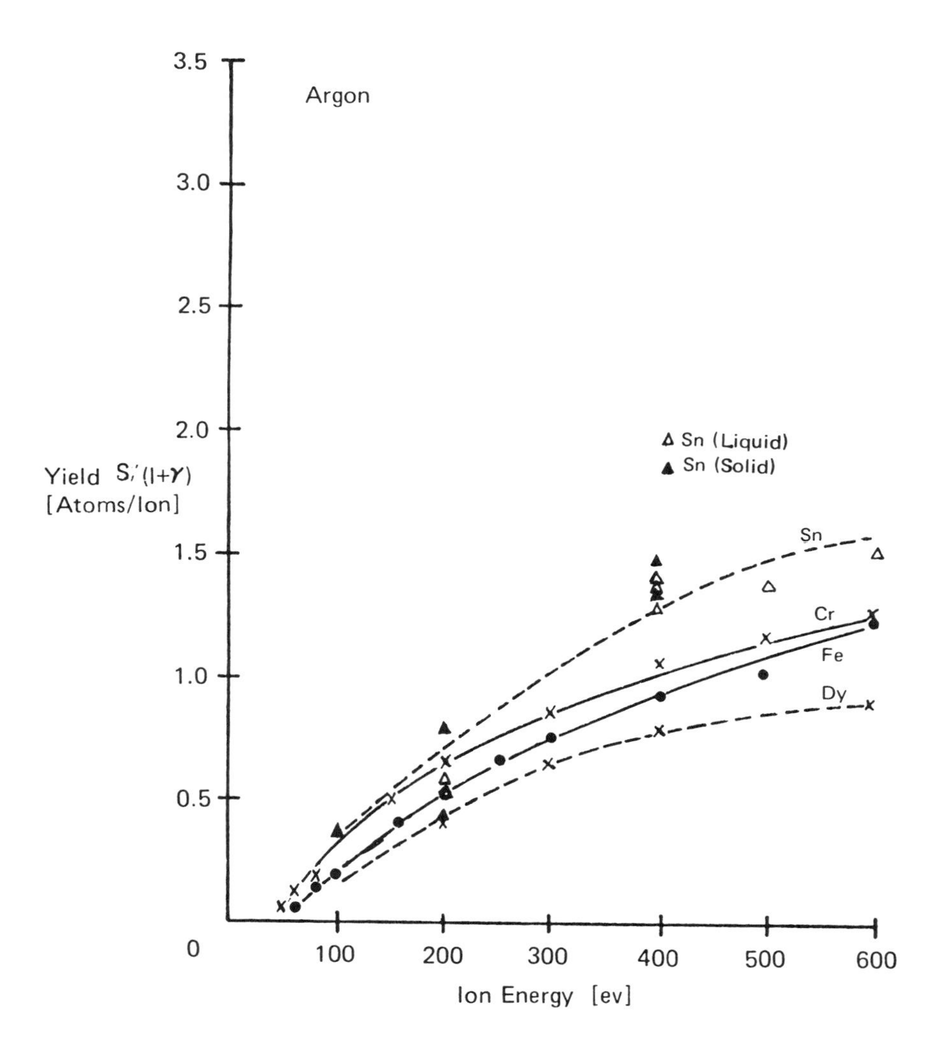

Argon
3.5
3.0
2.5
2.0
1.5
1.0
0.5
0
Yield S/(1+γ)
[Atoms/Ion]
Sn (Liquid)
Sn (Solid)
Sn
Cr
Fe
Dy
100
200
300
400
500
600
Ion Energy [ev]

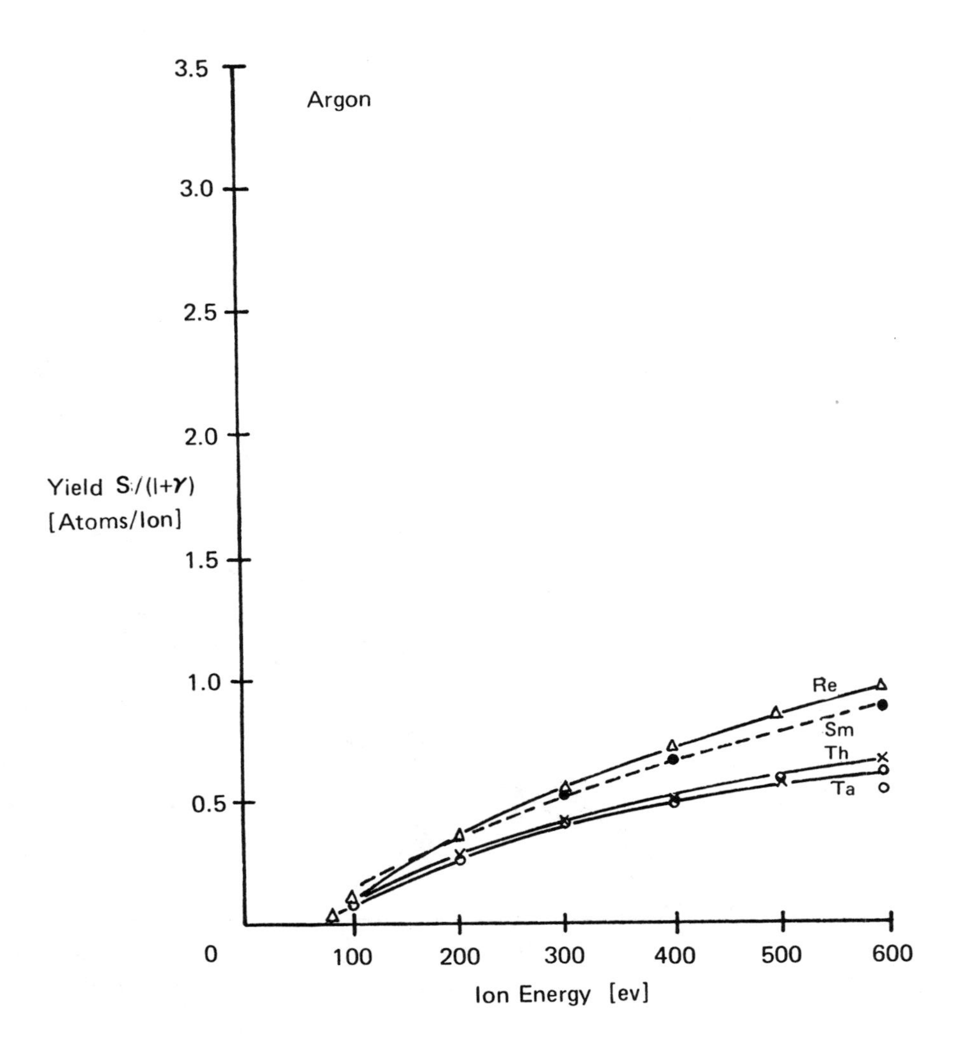

Argon
Yield S/(1+γ)
[Atoms/Ion]
3.5
3.0
2.5
2.0
1.5
1.0
0.5
0
100
200
300
400
500
600
Ion Energy [ev]
Re
Sm
Th
Ta

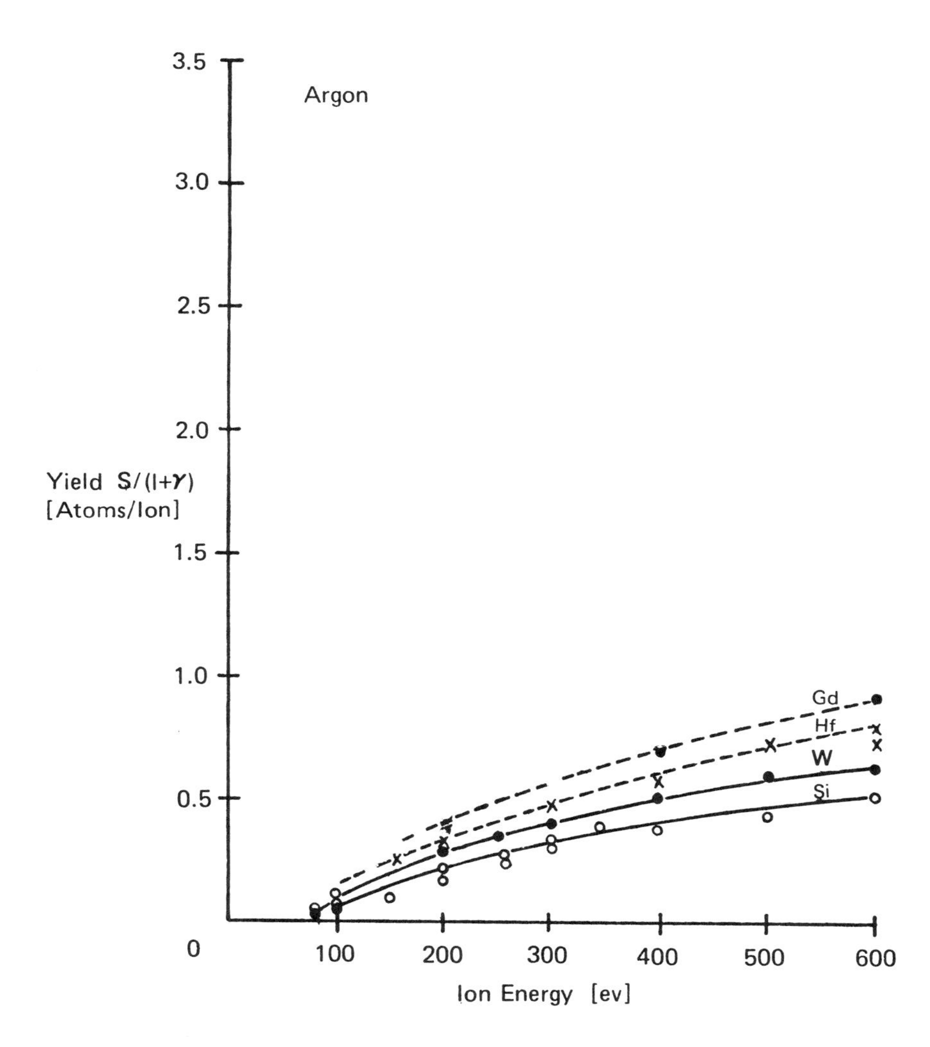
3.5
3.0
2.5
2.0
1.5
1.0
0.5
0
Argon
Yield S/(l+γ)
[Atoms/Ion]
Gd
Hf
W
Si
100
200
300
400
500
600
Ion Energy [ev]

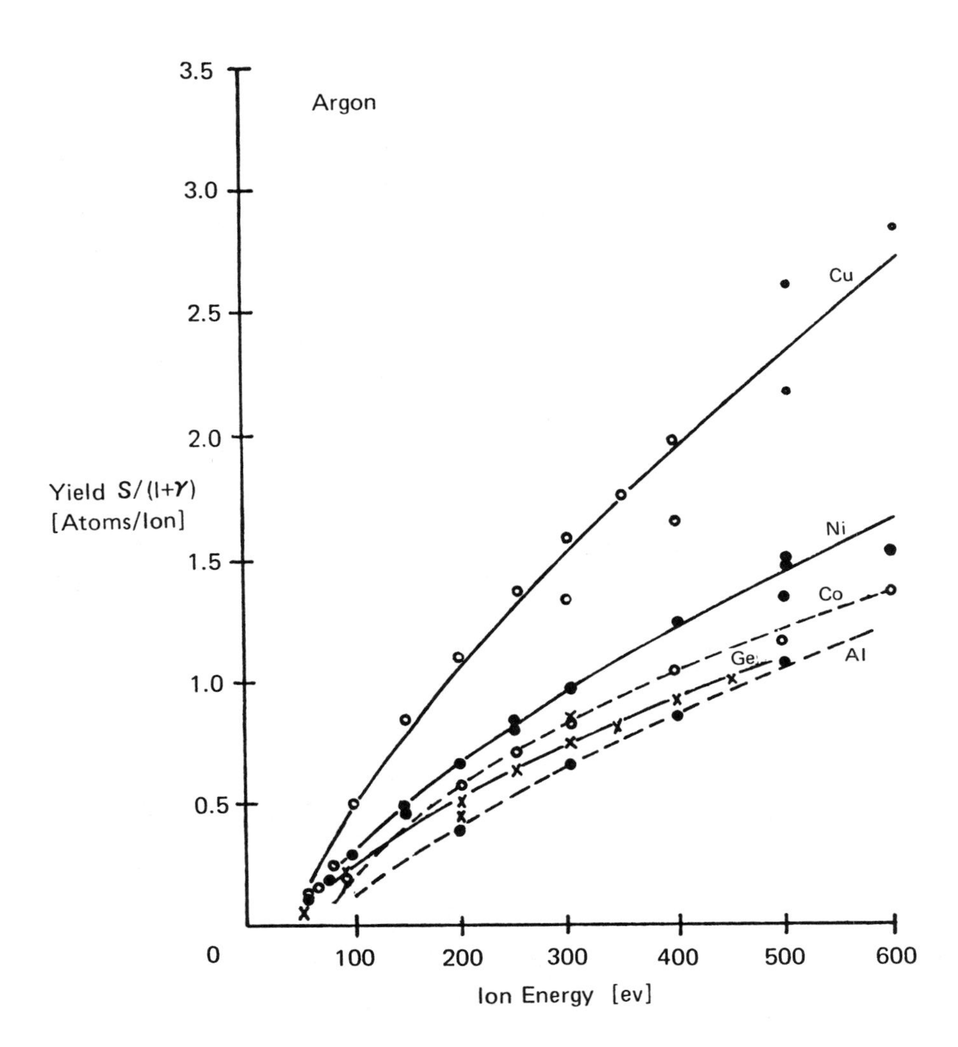
3.5
3.0
2.5
2.0
1.5
1.0
0.5
0
Argon
Yield S/(I+γ)
[Atoms/Ion]
Cu
Ni
Co
Ge
Al
100
200
300
400
500
600
Ion Energy [ev]

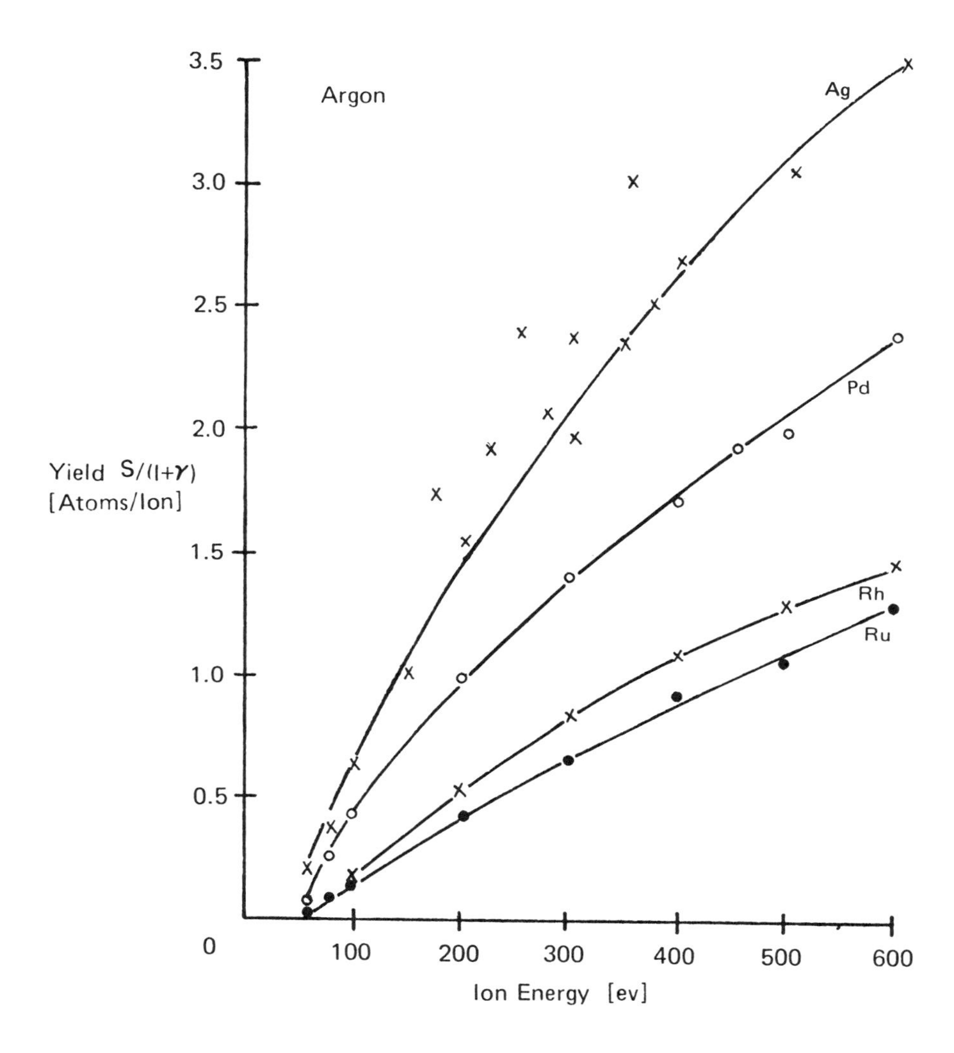

Argon
Ag
Pd
Rh
Ru
Yield S/(1+γ)
[Atoms/Ion]
3.5
3.0
2.5
2.0
1.5
1.0
0.5
0
100
200
300
400
500
600
Ion Energy [ev]

Sputtering Yields and Thresholds

(Stuart & Wehner 1962)

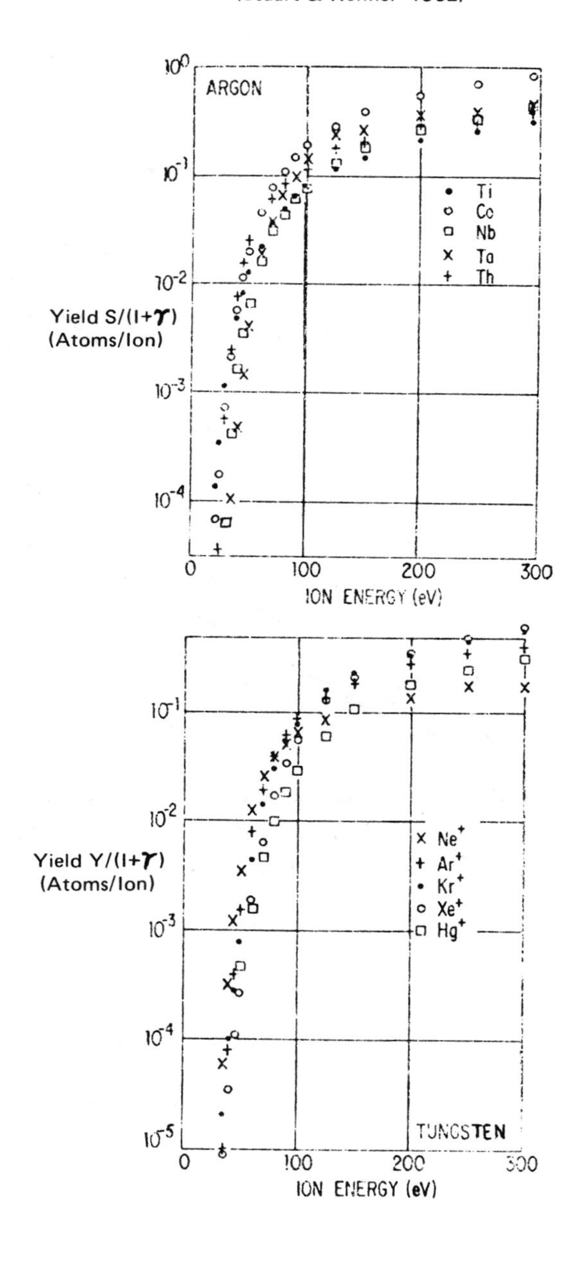

More Sputtering Yield Data

The following data is reproduced, with permission, from *Basic Data of Plasma Physics, 1966* by Prof. S.C. Brown.

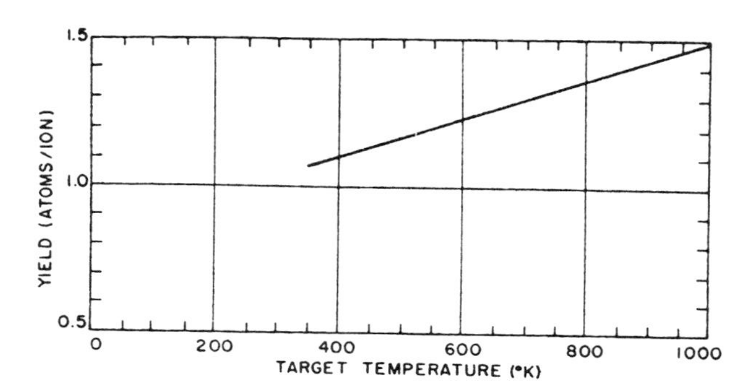

Sputtering yield of polycrystalline tantalum bombarded by 5-keV A^+ ions.
C. E. Carlston, G. D. Magnuson, A. Comeaux, P. Mahadevan (1965)

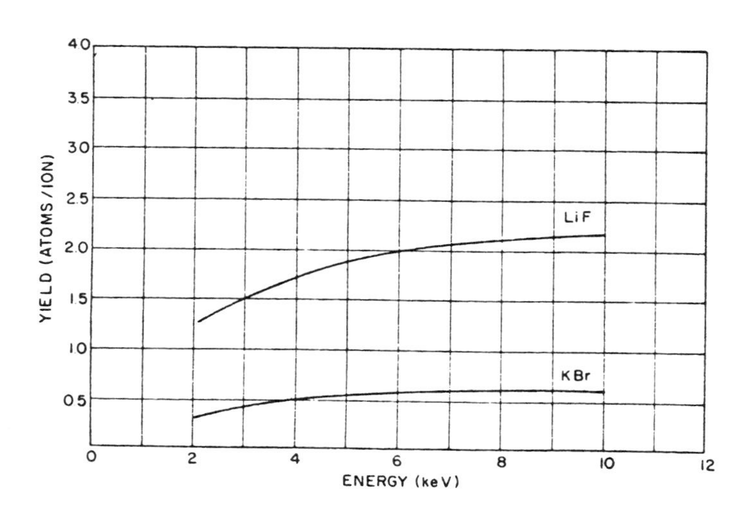

Sputtering yields for the (100) surface of LiF and KBr under normally incident A^+ ions.
B. Navinšek (1965)

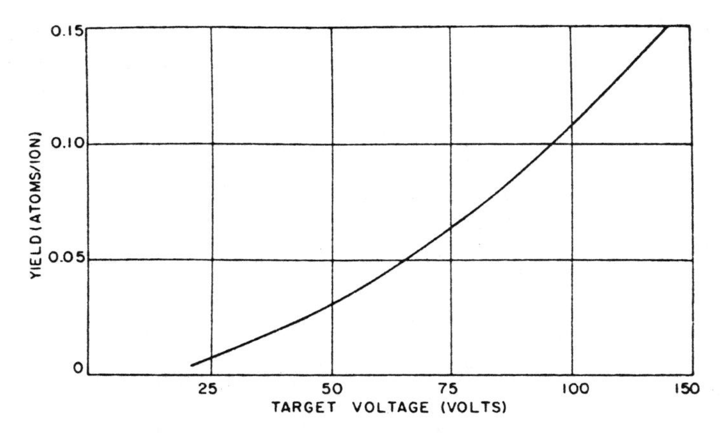

Sputtering yield of (111) aluminum bombarded by A^+ ions.
E. J. Zdanuk, S. P. Wolsky (1965)

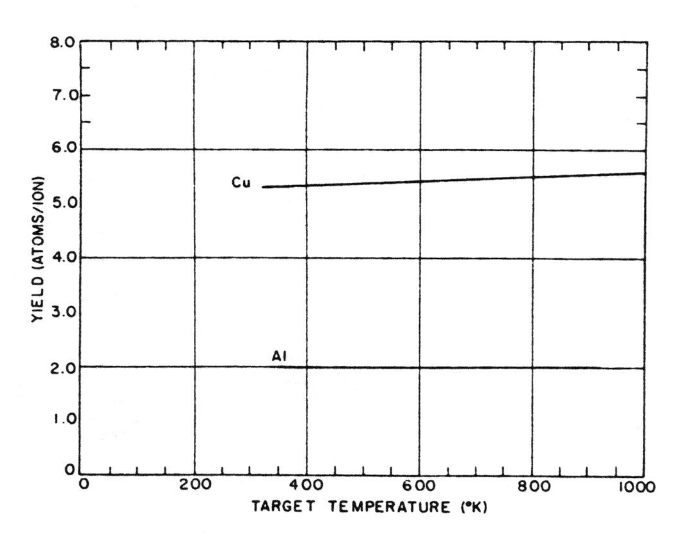

Sputtering yields of polycrystalline Cu and Al bombarded by 5-keV A^+ ions.
C. E. Carlston, G. D. Magnuson, A. Comeaux, P. Mahadevan (1965)

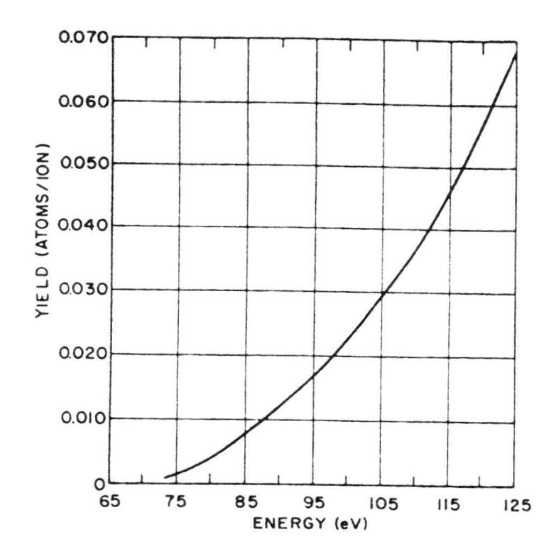

Sputtering yield for A^+ ions on silicon.
E. B. Henschke, S. E. Derby (1963)

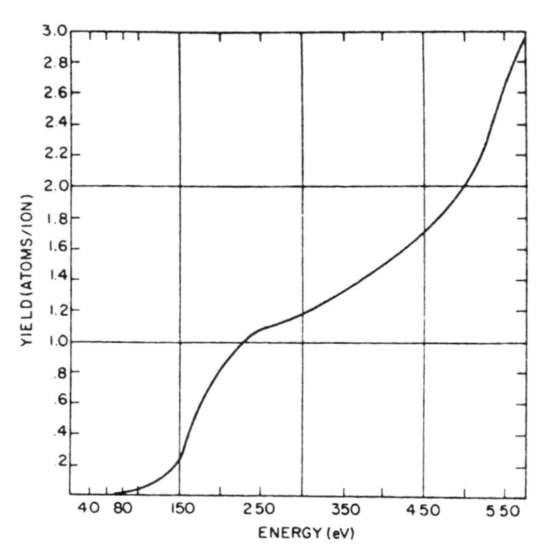

Sputtering yield curves for A^+ ions on chromium.
E. B. Henschke, S. E. Derby (1963)

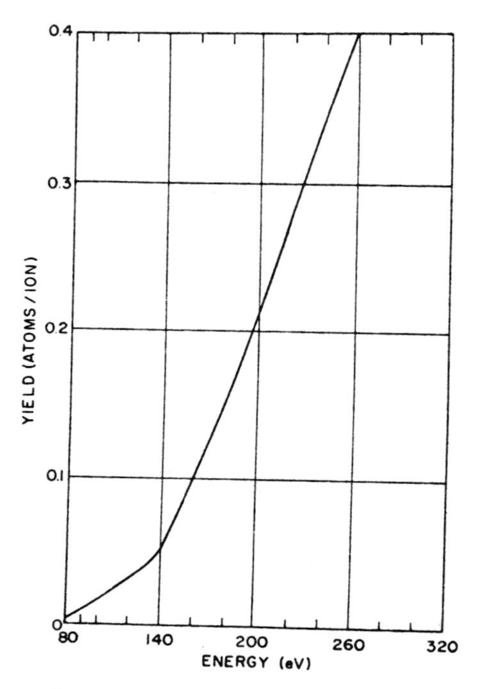

Sputtering yield for A^+ ions on titanium.
E. B. Henschke, S. E. Derby (1963)

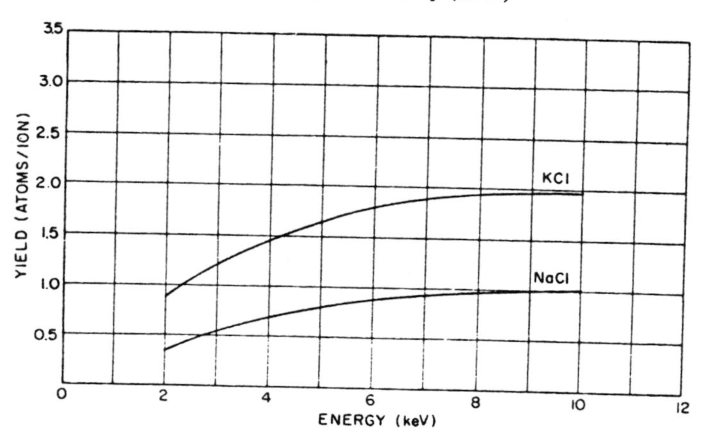

Sputtering yields for the (100) surface of KCl and NaCl under normally incident A^+ ions.
B. Navinšek (1965)

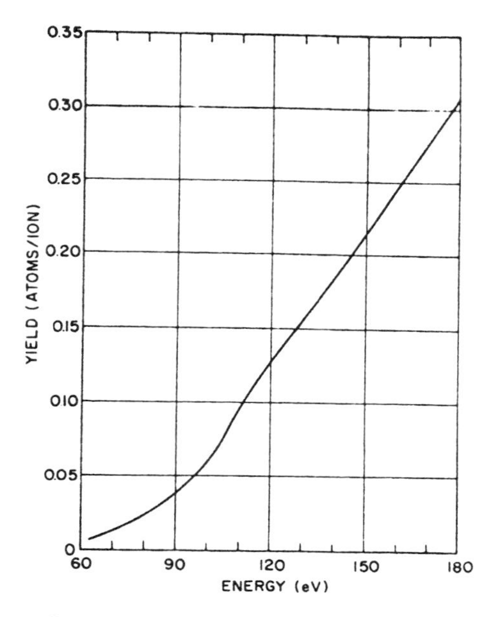

Sputtering yield for A^+ ions on cobalt.
E. B. Henschke, S. E. Derby (1963)

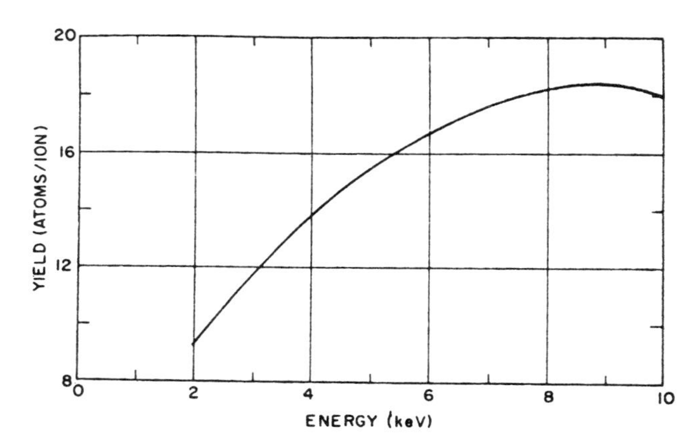

Sputtering yield of a Cu (100) plane bombarded by A^+ ions incident in a [111] direction.
G. D. Magnuson, C. E. Carlston (1963)

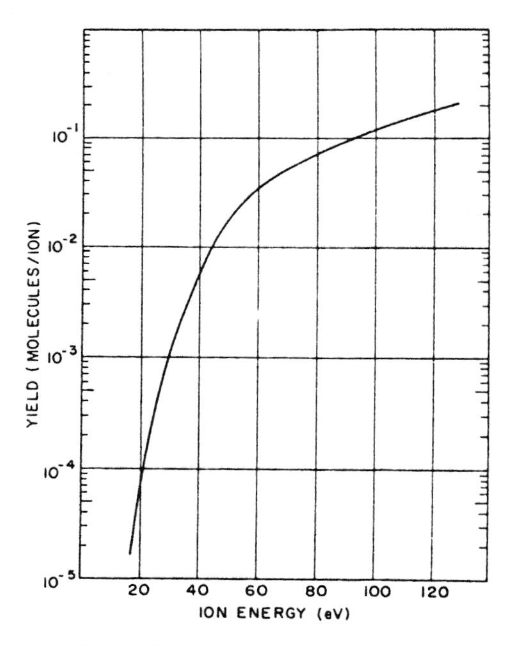

Sputtering yields of A^+ on quartz.
G. V. Jorgenson, G. K. Wehner (1965)

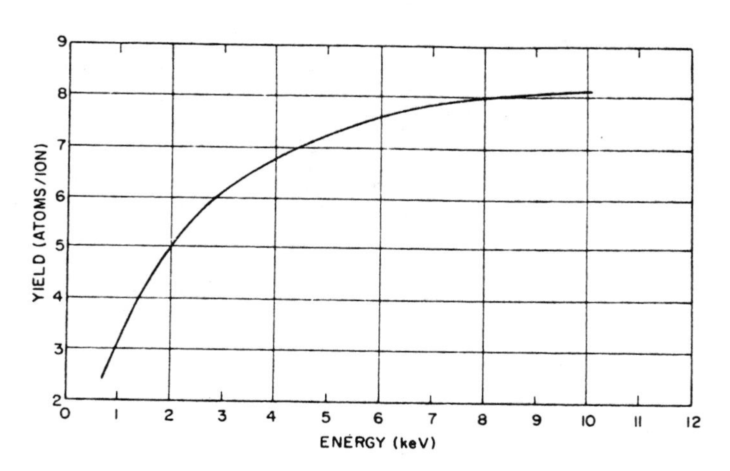

Sputtering yield of a Cu (311) plane bombarded at normal incidence by A^+ ions.
G. D. Magnuson, C. E. Carlston (1963)

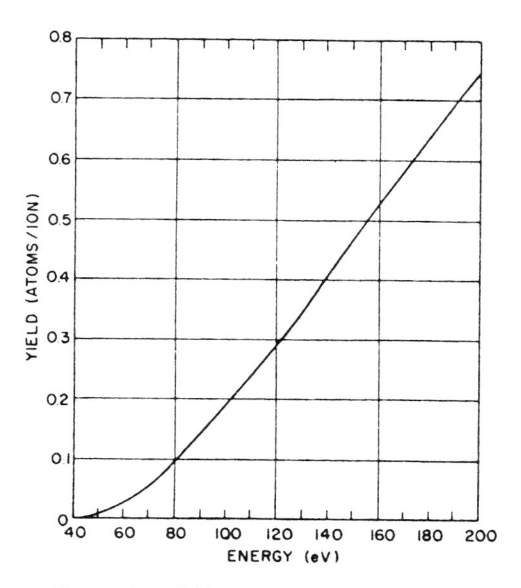

Sputtering yield for A^+ ions on copper.
E. B. Henschke, S. E. Derby (1963)

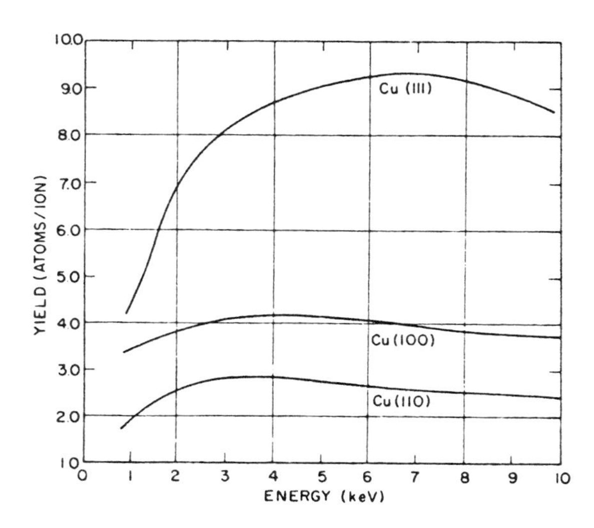

Sputtering yield of the three low-index planes of Cu bombarded by normal-incidence A^+ ions.
C. D. Magnuson, C. E. Carlston (1963)

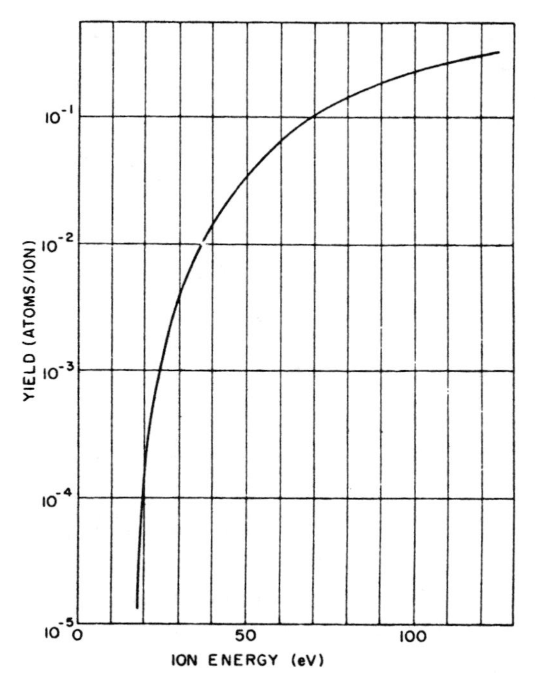

Sputtering yield of A^+ on copper.
J. R. Woodyard, C. B. Cooper (1964)

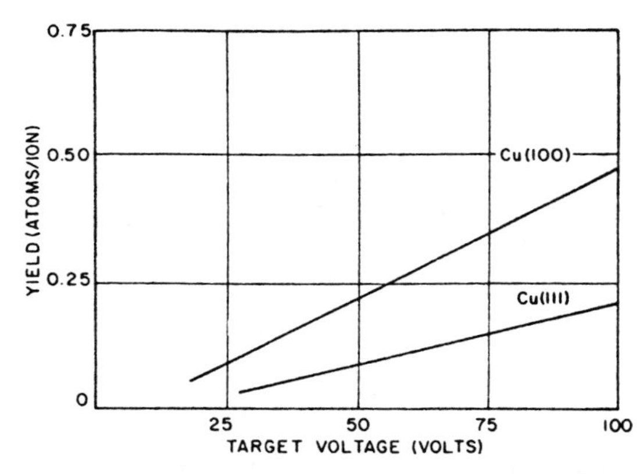

Sputtering yield of (100) and (111) copper bombarded by A^+ ions.
E. J. Zdanuk, S. P. Wolsky (1965)

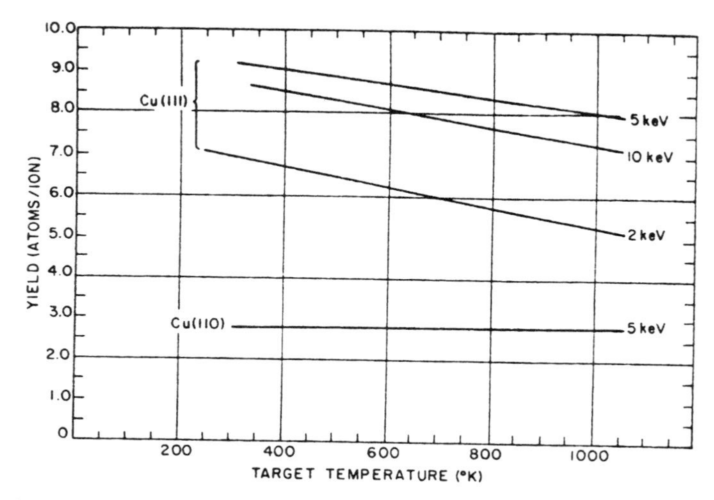

Sputtering yields of two copper single crystals bombarded by A^+ ions.
C. E. Carlston, G. D. Magnuson, A. Comeaux, P. Mahadevan (1965)

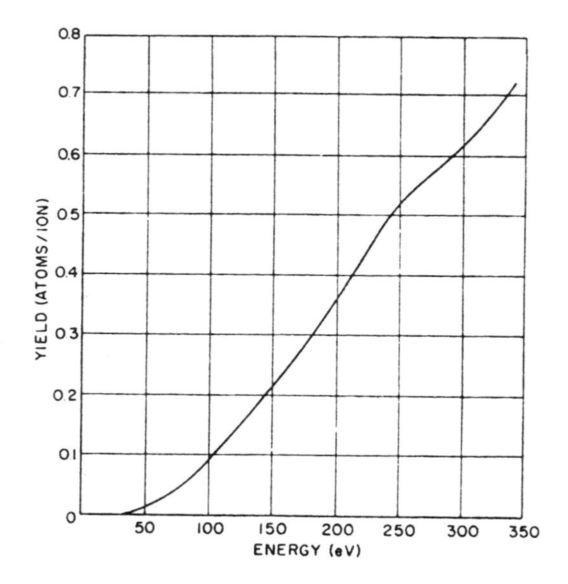

Sputtering yield for A^+ ions on germanium.
E. B. Henschke, S. E. Derby (1963)

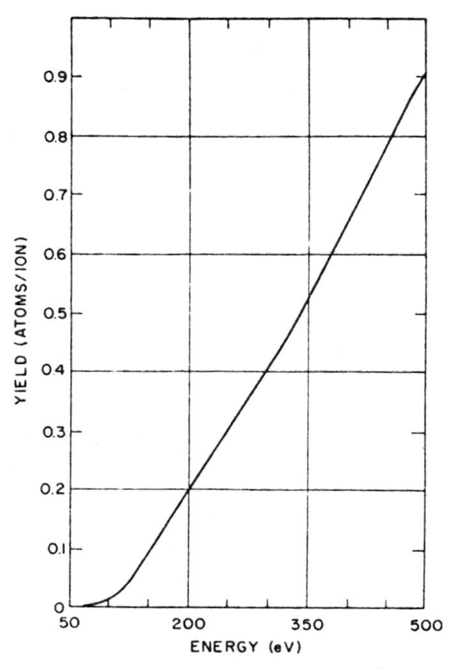

Sputtering yield curve for A^+ ions on molybdenum.
E. B. Henschke, S. E. Derby (1963)

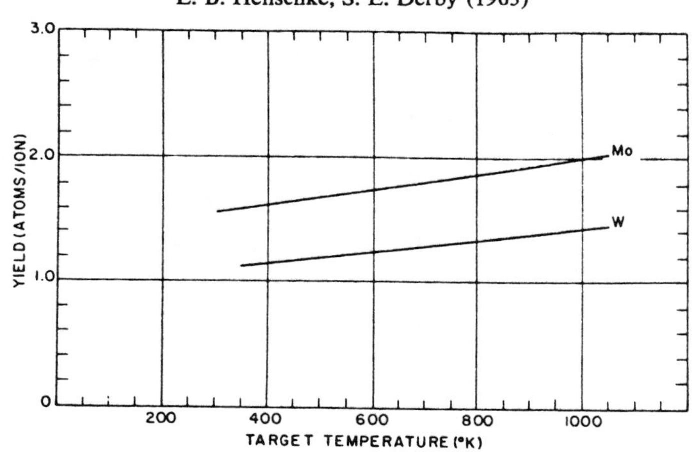

Sputtering yields of polycrystalline Mo and W bombarded by 5-keV A^+ ions.
C. E. Carlston, G. D. Magnuson, A. Comeaux, P. Mahadevan (1965)

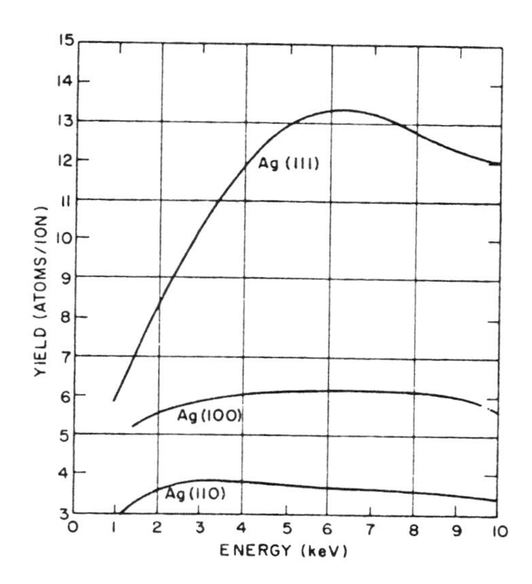

Sputtering yield of the three low-index planes of Ag bombarded by normal-incidence A^+ ions.
G. D. Magnuson, C. E. Carlston (1963)

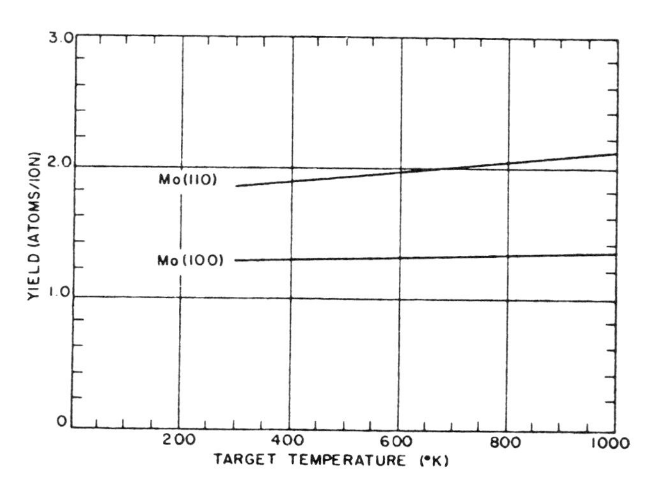

Sputtering yields of two Mo single crystals bombarded by 5-keV A^+ ions.
C. E. Carlston, G. D. Magnuson, A. Comeaux, P. Mahadevan (1965)

Sputtering Yields for Argon Ion Bombardment at Various Energies

Most of the values quoted are actually $S/(1 + \gamma)$ and, because of various uncertainties, should be regarded as guidance only. Superscripts are codes for references at the foot of the table; 30-500 eV are reference (b)

	a	Bombarding argon ion energy in eV														
	Threshold (eV)	← b →									b,c		g			f	
		30	40	50	60	80	100	200	300	500	600	1000	1100	2900	5000	5600	10000
Ag	15				0.22		0.63	1.58	2.20		3.40				7[e]		8.8[k]
Al	13						0.11	0.35	0.65		1.24	2[d]			2.0[l]		
Al_2O_3											0.18	0.04[n]	0.05	0.17		0.18	
Au	20			0.035			0.32	1.07	1.65	2.43		4[d]			7.9[m]		
Be	15					0.05	0.074	0.18	0.29		0.80						
Bi_2O_3											1.32						
CdS (1010)								0.5[i]			1.2[i]						
Co	25				0.048		0.15	0.57	0.81		1.36						
Cr	22		0.026				0.30	0.67	0.87		1.30						
Cr_2O_3											0.18						
Cu	17				0.10		0.48	1.10	1.59		2.30	2.75[d]					
Fe	20				0.064		0.20	0.53	0.76		1.26	1.25[d]					
Fe_2O_3											0.71						
GaAs (110)								0.4[i]			0.9[i]						
GaP								0.4[i]			0.95[i]						
GaSb (111)								0.4[j]			0.9[j]	1.2[j]					
Ge	25	0.017					0.22	0.50	0.74		1.22					2.47	
Hf			0.004				0.16	0.35	0.48		0.83						
In_2O_3											0.57						
InSb								0.25[i]			0.55[i]						
Ir					0.019		0.12	0.43	0.70		1.17						
K Br (100)															0.55[o]		0.6[o]

	a Threshold (eV)	Bombarding argon ion energy in eV															
		b 30	40	50	60	80	100	200	300	500	b,c 600	1000	g 1100	2900	5000	f 5600	10000
K Cl (100)															1.6[o]		1.95[o]
La_2O_3											0.37						
LiF															1.9[o]	1.38	2.2[o]
MgO											0.36					0.40	
Mo	24				0.027		0.13	0.40	0.58		0.93	1.2[d]					
NaCl (110)															0.75[o]		1.0[o]
Nb	25				0.017		0.068	0.25	0.40		0.65						
Nb_2O_5											0.24						
Ni	21				0.067		0.28	0.66	0.95		1.52	2.25[d]					
Os							0.057	0.36	0.56		0.95						
Pb											2.7[h]						
PbS																1.61	
PbTe								0.6[i]			1.4[i]						
Pd	20			0.033			0.42	1.00	1.41		2.39						
Pt	25				0.032		0.20	0.63	0.95		1.56						
Pyrex 7740													0.15	0.43			
Re	35					0.034	0.10	0.37	0.56		0.91						
Rh							0.19	0.55	0.86		1.46						
Ru					0.012		0.14	0.41	0.68		1.30						
Sb_2O_3											1.37						
Si						0.06	0.07	0.18	0.31		0.53						
SiC (0001)											0.41[l]						
SiC (poly)											1.8						
SiO_2											1.34	0.13[n]	0.16	0.50		0.36	
SnO_2											0.96						
Ta	26				0.01		0.10	0.28	0.41		0.62						
Ta_2O_5											0.15						

	a Threshold (eV)	Bombarding argon ion energy in eV															
		←				b			→	b,c		g			f	
		30	40	50	60	80	100	200	300	500	600	1000	1100	2900	5000	5600	10000
Th	24				0.017		0.097	0.27	0.42		0.66						
Ti	20						0.081	0.22	0.33		0.58				1.7[p]		2.1[p]
TiO_2											0.96						
U	23						0.14	0.35	0.59		0.97						
V	23				0.03		0.11	0.31	0.41		0.70						
V_2O_5											0.45						
W	33				0.008		0.068	0.29	0.40		0.62				1.1[l]		
ZnO									1.18								
ZnS																1.30	
Zr	22				0.027		0.12	0.28	0.41		0.75						
ZrO_2											0.32						

(a) Stuart and Wehner 1962
(b) Laegreid and Wehner 1961
(c) Vossen 1979
(d) Weijsenfeld 1966
(e) Pitkin et al. 1960
(f) Bach 1968
(g) Maissel 1966
(h) Keywell 1955
(i) Comas and Cooper 1966
(j) Wolsky et al. 1962
(k) Almen and Bruce 1961
(l) Carlston et al. 1965
(m) Robinson and Southern 1967
(n) Davidse and Maissel 1967
(o) Navinsek 1965
(p) Kurbatov 1968

REFERENCES

O. Almen and G. Bruce, Nucl. Instr. Methods **2**, 257 (1961)

J.M. Anderson, Phys. Fluids **7**, 1517 (1964)

H. Bach, Naturwissenschaften **55**, 439 (1968)

M.A. Biondi and L.M. Chanin, Phys. Rev. **94**, 910 (1954)

R.D. Birkhoff, in *Handbuch der Physik,* Vol. XXXIV, ed. S. Flügge, Springer-Verlag, Berlin (1958)

N. Bohr, Kgl. danske Vid. Sels., mat-fys. Medd. **18**, 8 (1948)

J.C. Bowe, Phys. Rev. **117**, 1411 (1960)

S.C. Brown, *Basic Data of Plasma Physics, 1966,* M.I.T. Press, Cambridge, Mass. (1966)

C.E. Carlston, G.D. Magnuson, A. Comeaux, and P. Mahadevan, Phys. Rev. **138**, 759 (1965)

L.M. Chanin and M.A. Biondi, Phys. Rev. **107**, 1219 (1957)

J.W. Coburn, unpublished results (1979)

J. Comas and C.B. Cooper, J. Appl. Phys. **37**, 2820 (1966)

P.D. Davidse and L.I. Maissel, J. Vac. Sci. Tech. **4**, 33 (1967)

O. Fischer, Z. Physik **86**, 646 (1933)

L.S. Frost and A.V. Phelps, Phys. Rev. **136**, 1538 (1964)

V.E. Golant, Soviet Phys. Tech. Phys. **4**, 680 (1959)

W. Hanle and W. Schaffernicht, Ann. Physik **6**, 905 (1930)

J.B. Hasted and M. Hussain, Proc. Phys. Soc. **83**, 911 (1964)

H.C. Hayden and R.C. Amme, Phys. Rev. **141**, 30 (1966)

E.B. Henschke and S.E. Derby, J. Appl. Phys. **34**, 2458 (1963)

O. Hermann, Ann. Physik **25**, 143 (1936)

G.V. Jorgenson and G.K. Wehner, J. Appl. Phys. **36**, 2672 (1965)

F. Keywell, Phys. Rev. **97**, 1611 (1955)

L.J. Kieffer, JILA Information Center Report No. 7, Boulder, Colorado (1969)

O.K. Kurbatov, Soviet Phys. Tech. Phys. English Transl. **12**, 1328 (1968)

N. Laegreid and G. K. Wehner, J. Appl. Phys. **32**, 365 (1961)

G.D. Magnuson and C.E. Carlston, J. Appl. Phys. **34**, 3267 (1963)

P. Mahadevan, G.D. Magnuson, J.K. Layton, and C.E. Carlston, Phys. Rev. **140**, 1407 (1965)

L.I. Maissel, in *Physics of Thin Films*, Vol. 3, Academic, New York (1966)

B. Navinsek, J. Appl. Phys. **36**, 1678 (1965)

O.J. Orient, Can. J. Phys. **43**, 422 (1965)

E.T. Pitkin, M.A. MacGregor, V. Salemme, and R. Bierge, ARL Tr-60-299 OTS, Dept. Commerce, Washington DC (1960)

M.T. Robinson and A.L. Southern, J. Appl. Phys. **38**, 2969 (1967)

D. Rosenberg and G.K. Wehner, J. Appl. Phys. **33**, 1842 (1962)

B.M. Smirnov and M.I. Chibisov, Soviet Phys. Tech. Phys. **10**, 88 (1965)

A.R. Striganov and N.S. Sventitski, *Tables of Spectral Lines of Neutral and Ionized Atoms*, IFI/Plenum, New York (1968)

R.V. Stuart and G.K. Wehner, J. Appl. Phys. **33**, 2345 (1962)

Y. Takeishi and H.D. Hagstrum, Phys. Rev. **137**, 641 (1965)

L.M. Volkova, Bull. Acad. Sci. USSR Phys. Ser. English Transl. **23**, 957 (1959)

L.M. Volkova and D.M. Devyatov, Opt. Spectroscopy USSR English Transl. **7**, 480 (1959)

A.A. Vorobev, B.A. Ivanov, A.P. Komar and V.A. Korolev, Soviet Phys. Tech. Phys. **4**, 1148 (1960)

J.L. Vossen, private communication

G.K. Wehner, Phys. Rev. **108**, 35 (1957)

G.K. Wehner, Phys. Rev. **112**, 1120 (1958)

C.H. Weijsenfeld, Thesis, Rijksuniversiteit Te Utrecht (1966)

D. Wobschall, J.R. Graham Jr., and D.P. Malone, Phys. Rev. **131**, 1565 (1963)

S.P. Wolsky, D. Shooter, and E.J. Zdanuk, p. 164, Trans. 9th National Symp. on Vac. Technology, Pergamon, New York (1962)

J.R. Woodyard and C.B. Cooper, J. Appl. Phys. **35**, 1107 (1964)

E.J. Zdanuk and S.P. Wolsky, J. Appl. Phys. **36**, 1683 (1965)

Index